Human-Robot Interaction: Emerging Technologies

Human-Robot Interaction: Emerging Technologies

Edited by Martha Dixon

CLANRYE INTERNATIONAL
www.clanryeinternational.com

Clanrye International,
750 Third Avenue, 9th Floor,
New York, NY 10017, USA

ISBN: 978-1-63240-930-0

Cataloging-in-Publication Data

Human-robot interaction : emerging technologies / edited by Martha Dixon.
p. cm.
Includes bibliographical references and index.
ISBN 978-1-63240-930-0
1. Human-robot interaction. 2. Robots. 3. Robotics--Human factors. 4. Human-computer interaction.
5. Human-machine systems. I. Dixon, Martha.
TJ211.49 .H86 2020
629.892 401 9--dc23

For information on all Clanrye International publications
visit our website at www.clanryeinternational.com

CLANRYE INTERNATIONAL

Printed in China.

Contents

Preface

Every book is initially just a concept; it takes months of research and hard work to give it the final shape in which the readers receive it. In its early stages, this book also went through rigorous reviewing. The notable contributions made by experts from across the globe were first molded into patterned chapters and then arranged in a sensibly sequential manner to bring out the best results.

Human-robot interaction (HRI) is a field which studies the interactions between a human and a robot. It is a multidisciplinary field that is based on the principles of artificial intelligence, robotics, human-computer interaction, design, natural language understanding and social sciences. Robots and humans are increasingly sharing the workspace due to the ever widening applicability of robots in different domains. The study of HRI becomes pertinent in this changing landscape, to improve utility and evaluate the benefits or risks of such close interaction. The goal of HRI is to create an easy and intuitive communication with the medium of gestures, speech and facial expressions. Research in HRI is exploring the boundaries of motion planning, human perception, human-robot coordination, etc. An emerging direction of robot application has been the use of assistive robots in schools, homes and hospitals to aid in physical activities, improve education, or act as companions. Military utility, space exploration, rehabilitation and elderly care, search and rescue, etc. are the other domains where HRI can be vital. There has been rapid progress in human-robot interaction and its applications are finding their way across multiple industries. This book traces the progress of this field and highlights some of its key concepts and applications. Researchers and students in this field will be assisted by this book.

It has been my immense pleasure to be a part of this project and to contribute my years of learning in such a meaningful form. I would like to take this opportunity to thank all the people who have been associated with the completion of this book at any step.

Editor

1

Hobbit: Providing Fall Detection and Prevention for the Elderly in the Real World

Markus Bajones ,[1] David Fischinger,[1] Astrid Weiss,[1] Daniel Wolf,[1] Markus Vincze,[1] Paloma de la Puente,[2] Tobias Körtner,[3] Markus Weninger,[3] Konstantinos Papoutsakis,[4] Damien Michel,[4] Ammar Qammaz,[4] Paschalis Panteleris,[4] Michalis Foukarakis,[4] Ilia Adami,[4] Danai Ioannidi,[4] Asterios Leonidis,[4] Margherita Antona,[4] Antonis Argyros,[4] Peter Mayer,[5] Paul Panek,[5] Håkan Eftring,[6] and Susanne Frennert[6]

[1]*Automation and Control Institute (ACIN), TU Wien, Vienna, Austria*
[2]*Universidad Politécnica de Madrid, Madrid, Spain*
[3]*Akademie für Altersforschung am Haus der Barmherzigkeit, Vienna, Austria*
[4]*Institute of Computer Science, FORTH, Heraklion, Greece*
[5]*Institute for Design and Assessment of Technology, TU Wien, Vienna, Austria*
[6]*Department of Design Sciences, Lund University, Lund, Sweden*

Correspondence should be addressed to Markus Bajones; markus.bajones@tuwien.ac.at

Academic Editor: Brady King

We present the robot developed within the Hobbit project, a socially assistive service robot aiming at the challenge of enabling prolonged independent living of elderly people in their own homes. We present the second prototype (Hobbit PT2) in terms of hardware and functionality improvements following first user studies. Our main contribution lies within the description of all components developed within the Hobbit project, leading to autonomous operation of 371 days during field trials in Austria, Greece, and Sweden. In these field trials, we studied how 18 elderly users (aged 75 years and older) lived with the autonomously interacting service robot over multiple weeks. To the best of our knowledge, this is the first time a multifunctional, low-cost service robot equipped with a manipulator was studied and evaluated for several weeks under real-world conditions. We show that Hobbit's adaptive approach towards the user increasingly eased the interaction between the users and Hobbit. We provide lessons learned regarding the need for adaptive behavior coordination, support during emergency situations, and clear communication of robotic actions and their consequences for fellow researchers who are developing an autonomous, low-cost service robot designed to interact with their users in domestic contexts. Our trials show the necessity to move out into actual user homes, as only there can we encounter issues such as misinterpretation of actions during unscripted human-robot interaction.

1. Introduction

While socially assistive robots are considered to be potentially useful for society, they can provide the highest value to older adults and homebound people. As reported in [1], future robot companions are expected to be

(1) strong machines that can take over burdensome tasks for the user,

(2) graceful and soft machines that will move smoothly and express immediate responses to their users,

(3) sentient machines that offer multimodal communication channels and are context-aware and trustable.

More and more companies and research teams present service robots with the aim of assisting older adults (e.g., Giraff (http://www.giraff.org), Care-O-Bot (http://www.care-o-bot.de), and Kompai (https://kompai.com)) with services such as entertainment, medicine reminders, and video telephony. Requirement studies on needs and expectations of older adults towards socially assistive robots [2] indicate that they expect them to help with household chores (e.g., cleaning the kitchen, bath, and toilet), lifting heavy objects, reaching for and picking up objects, delivering objects, and so forth. However, most of these tasks cannot satisfyingly be performed by state-of-the-art robotic platforms; hardly any companion robot fulfills the requirements mentioned above and only very few robots entered private homes of older adults so far. One of the biggest challenges is offering sufficient useful and social functionalities in an autonomous and safe manner to achieve the ultimate goal of prolonging independent living at home. The ability of a robot to interact autonomously with a human requires sophisticated cognitive abilities including perception, navigation, decision-making, and learning. However, research on planners and cognitive architectures still faces the challenge of enabling flexibility and adaptation towards different users, situations, and environments while simultaneously being safe and robust. To our conviction, for successful long-term human-robot interaction with people in their private homes, robotic behavior needs to be above all safe, stable, and predictable. During our field trials, this became increasingly evident, as the users failed to understand the robot's behavior during some interaction scenarios.

In this article, we present the Hobbit PT2 platform, referred to in the remainder of this article as Hobbit. A former version of Hobbit has been presented in detail in [3]. Hobbit is a socially assistive robot that offers useful personal and social functionalities to enable independent living at home for seniors. To the best of our knowledge, the Hobbit trials mark the first time a social service robot offering multifunctional services was placed in users' homes, operated autonomously and whose usage was not restricted by a schedule or any other means. The main contribution of this paper is twofold. First, we give a description of the hardware that is based on improvements derived from the first user trials on the previous version of Hobbit. Second, we describe the implemented functionality and its integration into the behavior coordination system. The building blocks of the behavior coordination system are based on a set of hierarchical state-machines implemented using the SMACH framework [4]. Each behavior was built upon simpler building blocks, each responsible for one specific task (e.g., speech and text output, arm movements, and navigation) to add up to the complex functionalities presented in Sections 3.3 and 4. Finally, we present the lessons learned from the field trials in order to support fellow researchers in their developments of autonomous service robots for the domestic environment. We evaluated Hobbit during 371 days of field trials with five platforms with older adults in their private homes in Austria, Greece, and Sweden. However, details on the field trials will be published elsewhere.

The paper proceeds as follows. Section 2 reflects on relevant related work on behavior coordination for social service robots and on studies of such robots outside of the laboratory environment. In Section 3, we give an overview on the project vision for Hobbit and its historical development up to the Hobbit PT2 platform, followed by a detailed description of its hardware and interaction modalities. Section 4 presents the behavior coordination system. We outline how we developed the interaction scenarios and transferred them into an implementable behavior concept. Section 5 presents an overview on the field trials. Lessons learned from the development and testing of Hobbit and a summary and conclusions are provided in Sections 6 and 7.

2. Related Work

Moving towards autonomous service robots, behavior coordination systems constitute an important building block to fulfill the requirements of action planning, safe task execution, and integration of human-robot interaction. HAMMER from Demiris and Khadhouri [5] is built upon the concept of using multiple forward/backward control loops, which can be used to predict the outcome of an action and compare this against the actual result of the action. Through this design, it is possible to choose the action with the highest probability of reaching the desired outcome, which has successfully been used in a collaboratively controlled wheelchair system [6], in order to correct the user's input to avoid an erroneous situation. Cashmore et al. [7] introduced ROSPlan, a framework that uses a temporal planning strategy for planning and dispatching robotic actions. Depending on the needs, a cost function can be optimized for planning in a certain manner (e.g., time- or energy-optimized). However, the constructed plan is up until now only available as a sequence of executed actions and observed events, but no direct focus is put on the human, besides modeling the user as means to acquire some event (e.g., moving an object from one location to another). Mansouri and Pecora [8] incorporate temporal and spatial reasoning in a robot tasked with pick and place in environments suited for users. In the context of ALIAS, Goetze et al. [9] designed their *dialogue manager* for the tasks of emergency call, a game, e-ticket event booking, and the navigation as state-machines. However, there are still significant research challenges regarding how to incorporate humans into the planning stages and decide when the robot needs to adapt to the user instead of staying with the planned task.

Most of those behavior coordination and planning systems treat the human as an essential part of the system [6] (e.g., for command input) and rely on the user to execute actions planned by the coordination system [10]. Such systems only work under the precondition that the robot will execute a given task for the user independently of the user input [8]. A crucial aspect, however, to successfully integrate a multifunctional service robot into a domestic environment is that it needs not only to react to user commands but also to proactively offer interaction and adapt to user needs (e.g., the user wanting a break from the robot or a proactive suggestion for an activity they could perform together). Our proposed solution is based on state-machines, which reflect turn-taking in the interaction, providing adaptations within certain states (e.g., voice dialogues) or situations (e.g., user approach). We integrated the possibility not only to handle robot-driven actions on a purely scheduled basis but also to adapt this scheduling and actions based on the user's commands.

2.1. State of the Art: Robotic Platforms. According to a study conducted by Georgia Tech's Healthcare Robotics Lab, people with motor impairment drop items on average 5.5 times a day. Their small tele-operated Dusty (http://pwp.gatech.edu/hrl/project_dusty/) robots are developed for that purpose: picking up objects from the floor, which they achieve with a scoop-like manipulator. Cody, a robotic nurse assistant, can autonomously perform bed (sponge) baths. Current work focuses on GATSBII (http://www.robotics.gatech.edu), a willow Garage PR2, as a generic aid for older adults at home. The Care-O-Bot research platforms developed at the Fraunhofer Institute (IPA) are designed as general purpose robotic butlers, with a repertoire from fetching items to detecting emergency situations, such as a fallen person. Also from Fraunhofer is Mobina (https://www.ipa.fraunhofer.de/de/referenzprojekte/MobiNa.html), a small (vacuum-sized) robot specifically performing fallen person detection and video calls in emergency. Carnegie Mellon University's HERB (https://personalrobotics.ri.cmu.edu/) is another general-purpose robotic butler. It serves as the main research platform at the Personal Robotics Lab, which is part of the Quality of Life Technology (QoLT) Center. KAIST in Korea has been developing their Intelligent Sweet Home (ISH) smart home technology including intelligent wheelchairs, intelligent beds, and robotic hoists [11]. Their system also employs the bimanual mobile robot Joy to act as an intermediary between these systems and the end user. Robotdalen (http://www.robotdalen.se), a Swedish public-private consortium, has funded the development of needed robotic products such as Bestic (http://www.camanio.com/en/products/bestic/), an eating device for those who cannot feed themselves; Giraff, a remote-controlled mobile robot with a camera and monitor providing remote assistance and security; or TrainiTest, a rehabilitation robot that measures and evaluates the capacity of muscles and then sets the resistance in the robot to adapt to the users' individual training needs. Remote presence robots have recently turned up in a variety of forms, from simple Skype video chats on a mobility platform (Double Robotics (https://www.doublerobotics.com/)) to serious medical assistance remote presence robots such as those provided by the partnership between iRobot and InTouch Health (https://www.intouch-health.com/about/press-room/2012/InTouch-Health-and-iRobot-to-Unveil-the-RP-VITA-Telemedicine-Robot.html), Giraff, and VGo Communications' postop pediatric at-home robots (http://www.vgocom.com/) for communication with parents, nurses, doctors, and patients.

Another class of robots aims more specifically at well-being of older adults. The recently completed FP7 project M obiserv (https://cordis.europa.eu/project/rcn/93537_en.html) aimed to develop solutions to support independent living of older adults as long as possible, in their home or in various degrees of institutionalization, with a focus on health, nutrition, well-being, and safety. These solutions encompass smart clothes for monitoring vital signs, a smart home environment to monitor behavioral patterns (e.g., eating) and detect dangerous events, and a companion robot. The robot's main role is to generally activate, stimulate, and offer structure during the day. It also reminds its user of meals, medication, and appointments and encourages social contacts via video calls. The US NSF is currently running the Socially Assistive Robotics project (https://www.nsf.gov/awardsearch/showAward?AWD_ID=1139078) with partners Yale, University of Southern California, MIT, Stanford, Tufts, and Willow Garage. Their focus is on robots that encourage social, emotional, and cognitive growth in children, including those with social or cognitive deficits. The elder care robot Sil-Bot (http://www.roboticstoday.com/robots/sil-bot) developed at the Center for Intelligent Robotics (CIR) in Korea is devised mainly as an entertainment robot to offer interactive games that have been codeveloped with Seoul National University Medical Center specifically to help prevent Alzheimer's disease and dementia. Sil-Bot is meant to be a companion that helps encourage an active, healthy body and mind. Its short flipper-like arms do not allow for actual manipulation. Another public-private partnership is the EC-funded CompanionAble project (http://www.companionable.net/), which created a robotic assistant for the elderly called Hector. The project integrates Hector to work collaboratively with a smart home and remote control center to provide the most comprehensive and cost-efficient support for older people living at home.

Hoaloha Robotics (http://www.hoaloharobotics.com/) in the United States are planning to bring their elder care robot to market soon. Based on a fairly standard mobile platform offering safety and entertainment, they focus on an application framework that will provide integration of discrete technological solutions like biometric devices, remote doctor visits, monitoring and emergency call services, medication dispensers, online services, and the increasing number of other products and applications already emerging for the assistive care market. Japan started a national initiative in 2013 to foster development of nursing care robots and to spread their use. The program supports 24 companies in developing and marketing their elderly care technologies, such as the 40 cm tall PALRO conversation robot (https://palro.jp/) that offers recreation services by playing games, singing, and dancing together with residents of a care facility. Another example is the helper robot by Toyota, which is mostly remotely controlled from a tablet PC. Going specifically beyond entertainment capabilities, Waseda University's Twendy One (http://www.twendyone.com) is a sophisticated bimanual robot that provides human safety assistance, dexterous manipulation, and human-friendly communication. It can also support a human to lift themselves from a bed or chair. Going even further, the RIBA-II robot (http://rtc.nagoya.riken.jp/RIBA/index-e.html) by RIKEN-TRI Collaboration Center for Human-Interactive Robot Research (RTC) can lift patients of up to 80 kg from a bed to a wheelchair and back. The Pepper robot (https://www.ald.softbankrobotics.com/en/robots/pepper) from Softbank Robotics (Aldebaran) is used in a growing number of projects focusing on human-robot interaction scenarios. Some ADL (activities of daily living) tasks are directly addressed by walking aids, for example [12], and cognitive manipulation training, for example, using exoskeletons [13, 14].

The short overview indicates that individually many ADL tasks are approached. However, they all require different types of robots. The goal of grasping objects from the floor, while at the same time keeping the robot affordable, has led us to design and build the custom Hobbit platform. Moreover, the robot should offer everyday life suitable tasks in a socially interactive manner to be sustainably used by the older adults.

3. The Hobbit Robot

Hobbit is able to provide a number of safety and entertainment functions with low-cost components. The ability to provide many functions with sometimes contradictory requirements for the hardware design creates a demanding challenge on its own. To the best of our knowledge, we are the first to present a robot that operates in users' homes in a fully autonomous fashion for a duration of 21 days per user, while providing an extensive set of functionalities like manipulation of objects with an included arm.

3.1. General Vision. The motivation for Hobbit's development was to create a low-cost, social robot to enable older adults to independently live longer in their own homes. One reason for the elderly to move into care facilities is the risk of falling and eventually inflicted injuries. To reduce this risk, the "must-haves" for the Hobbit robot are *emergency detection* (the robot patrolling autonomously through the flat after three hours without any user activity and checking if the user is well and did not suffer a fall), *emergency handling* (automatic calls to relatives or emergency services), and *fall prevention* (searching and bringing known objects to the user and picking up objects from the floor pointed to by the user and a basic fitness program to enhance the user's overall fitness). Hobbit also provides a *safety check* feature that informs the user about possible risks in specific rooms (e.g., wet floor in the bathroom and slippery carpets on wooden floors) and explains how to reduce such risks.

In science fiction, social robots are often depicted as a butler, a fact that guides the expectations towards such robots. However, as state-of-the-art technology is not yet able to fulfill these expectations, Hobbit was designed to incorporate the Mutual Care interaction paradigm [15] to overcome the robot's downfalls by creating an emotional bond between the users and the robot. The *Mutual Care* concept envisioned that the user and the robot provide help in a reciprocal manner to each other, therefore creating an emotional bond between them, so that the robot not only provides useful assistance but also acts as a companion. The resulting system complexity based on the multifunctionality was considered as acceptable to fulfill the main criteria (*emergency detection and handling*, *fall prevention*, and *providing a feeling of safety*).

3.2. Mutual Care as Underlying Interaction Paradigm. The *Mutual Care* concept was implemented through two different social roles, one that enforces this concept and one that does not. Hobbit started in the Mutual Care-disabled mode during the field trials and changed after 11 days to the Mutual Care mode. The differences between these two modes or *social roles* of the robot were mainly in its dialogues, proactivity, and the proximity in which the robot would remain when the user stops interacting with the robot. In more detail, the main characteristics of the *Mutual Care* mode were the following: (1) return of favor: Hobbit asked if it could return the favor after situations where the user had helped Hobbit to carry out a task, (2) communication style: Hobbit used the user's name in the dialogue and was more human-like such as responding to a reward from the user by saying *You are welcome* instead of *Reward has been received*, (3) proactivity: Hobbit was more proactive and initiated interactions with the user, and (4) presence: Hobbit stayed in the room where the last interaction has taken place for at least 30 minutes instead of heading directly back to the charging station. In order to avoid potential biases, users were not told about the behavioral change of the robot beforehand.

3.3. Development Steps Leading to Hobbit. To gain insight into the needs of elderly living alone, we invited primary users (PU), aged 75 years and older and living alone, and secondary users (SU), who are in regular contact with the primary users, to workshops in Austria (8 PU and 10 SU) and Sweden (25 PU). A questionnaire survey with 113 PU in Austria, Greece, and Sweden and qualitative interviews with 38 PU and 18 SU were conducted. This iterative process [16] not only resulted in the user requirements but also influenced the design and material decisions, which were incorporated into the development of the Hobbit robots as seen in Figure 1. Based on these requirements and laboratory studies with the PT1 platform [17] with 49 users (Austria, Greece, and Sweden), the following main functionalities for Hobbit were selected:

(1) *Call Hobbit*: summon the robot to a position linked to battery-less call buttons

(2) Emergency: call relatives or an ambulance service. This can be triggered by the user from emergency buttons and gesture commands or by the robot during patrolling

(3) *Safety check*: guide the user through a list of common risk sources and provide information on how to reduce them

(4) *Pick up objects*: objects lying on the floor are picked up by the robot with no distinction between known or unknown objects

(5) *Learn and bring objects*: visual learning of user's objects to enable the robot to search and find them within the environment

(6) *Reminders*: deliver reminders for drinking water and appointments directly to the user

(7) *Transport objects*: reduce the physical stress on the user by placing objects on to the robot and letting it transport them to a commanded location

(8) *Go recharging*: autonomously, or by a user command, move to the charging station for recharging

(9) *Break*: put the robot on break when the user leaves the flat or when the user takes a nap

(10) *Fitness*: guided exercises that increase the overall fitness of the user

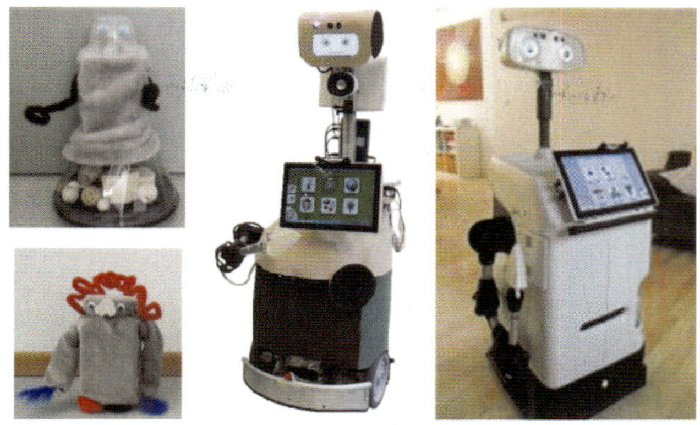

(a) (b) (c)

FIGURE 1: (a–c) First mock-ups designed by secondary users: the first (PT1) and second generation of Hobbit as used during the field trials.

(11) *Entertainment*: brain training games, e-books, and music

3.4. Robot Platform and Sensor Setup. The mobile platform of the Hobbit robot has been developed and built by MetraLabs (http://www.metralabs.com). It moves using a two-wheeled differential drive, mounted close to the front side in driving direction. For stability, an additional castor wheel is located close to the back. To fit all the built-in system components, the robot has a rectangular footprint with a width of 48 cm and a length of 55 cm. For safety reasons, a bumper sensor surrounds the base plate, protecting the hull and blocking the motors when being pressed. This ensures that the robot stops immediately if navigation fails and an obstacle is hit. An additional bumper sensor is mounted below the tablet PC, which provides the graphical user interface. During situations in which the user might not be able to reach the tablet PC (e.g. the person has fallen), a hardware emergency button is located on the bottom front side.

On its right side, the robot is equipped with a 6-DoF arm with a two-finger fin-ray gripper, such that objects lying on the floor can be picked up and placed in a tray on top of the robot's body. Furthermore, the arm can grasp a small turntable stored on the right side of the body, which is used to teach the robot unknown objects.

The robot's head, together with the neck joint with motors for pan and tilt movements, has been developed by Blue Danube Robotics (http://www.bluedanuberobotics.com). It contains two speakers for audio output, two Raspberry Pis with one display each for the eyes of the robot, a temperature sensor, and a RGB-D sensor. This sensor, referred to in the remainder of the paper as *head camera*, is used for obstacle avoidance, for object and gesture recognition, and—in conjunction with the temperature sensor—for user and fall detection. Similar to the previous prototype of the robot [3, 18], the visual sensor setup is completed by a second RGB-D sensor, mounted in the robot's body at a height of 35 cm facing forward. This sensor, referred to in the remainder of the paper as *bottom camera*, is used for localization, mapping, and user following. Figure 2 shows an overview of the Hobbit hardware; a more detailed explanation of the single components is given in the following sections.

3.4.1. Visual Perception System Using RGB-D Cameras. For the visual perception system, Hobbit is equipped with two Asus Xtion Pro RGB-D sensors. The head camera is mounted inside the head and used for obstacle avoidance, object learning and recognition, user detection, and gesture recognition and to detect objects to pick up. Since the head can perform pan and tilt movements, the viewing angle of this camera can be dynamically adapted to a particular task at hand. In contrast, the bottom camera, used for localization, mapping, and user following, is mounted at a fixed position at a height of 35 cm in the front of the robot's body, facing forward. This setup is a trade-off between the cost of the sensor setup (in terms of computational power and money) and the necessary data for safe usage and feature completeness, which we found to be most suitable for the variety of different tasks that require visual perception.

The cameras, which only cost a fraction of laser range sensors commonly used for navigation in robotics, offer a resolution of 640×480 pixels of RGB-D data and deliver useful data in a range of approximately 50 cm to 400 cm. Therefore, our system has to be able to cope with a blind spot in front of the robot. Furthermore, the quality of data acquired with the head camera from an observed object varies depending on the task. For example, in the learning task, an object that is placed on the robot's turntable is very close to the head camera, just above the lower range limit. In the pickup task, on the contrary, the object detection method needs to be able to detect objects at the upper range limit of the camera, where data points are already severely influenced by noise.

Because two of the main goals for the final system were affordability and robustness, we avoided incorporating additional cameras, for example, for visual servoing with the robot's hand. For further details and advantages of our sensor setup for navigation, we refer the reader to [18].

3.4.2. Head and Neck. Besides the head camera, the head contains an infrared camera for distance temperature measurement, two speakers for audio output, and two Raspberry Pis with displays showing the robot's eyes. Through its eyes,

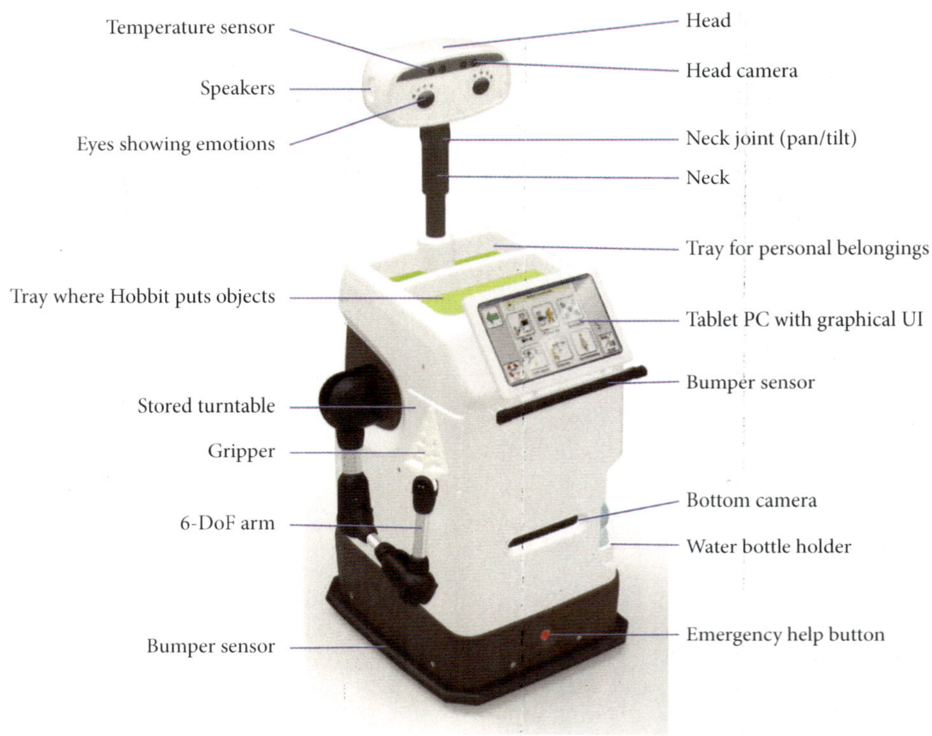

FIGURE 2: Hardware setup of the Hobbit platform.

FIGURE 3: List of emotions shown by Hobbit's eyes.

the robot is able to communicate a set of different emotions to the user, which are shown in Figure 3. The neck joint contains two servo motors, controlling the horizontal and vertical movement of the head.

3.4.3. Arm and Gripper. To be able to pick up objects from the floor or to grab its built-in turntable, Hobbit is equipped with a 6-DoF IGUS arm and a two-finger fin-ray gripper. As a cost-effective solution, the arm joints are moved by stepper motors via Bowden cables; the used fin-ray gripper offers one DoF and is designed to allow form-adaptable grasps. While an additional DoF would increase flexibility and lower the need for accurate self-positioning to successfully grasp objects, for the sake of overall system robustness and low hardware costs, the 6-DoF version was the model of choice for the arm. The arm is not compliant; therefore, cautious behavior implementation with reduced velocities for unsupervised actions was required to minimize the risk of breakage.

4. Behavior Coordination

As Hobbit's goal directly called for an autonomous system running for several weeks, providing interactions on an irregular schedule and on-demand basis, the behavior coordination of the Hobbit robots was designed and implemented in a multistage development process. Based on the workshops with PU and SU and the user study with Hobbit PT1, elderly care specialists designed the specific scenarios. They designed detailed scripts for the 11 scenarios (see Section 3.3) the robot had to perform. Those 11 scenarios were subsequently planned in a flowchart-like fashion, which eased the transition from the design process to the implementation stage.

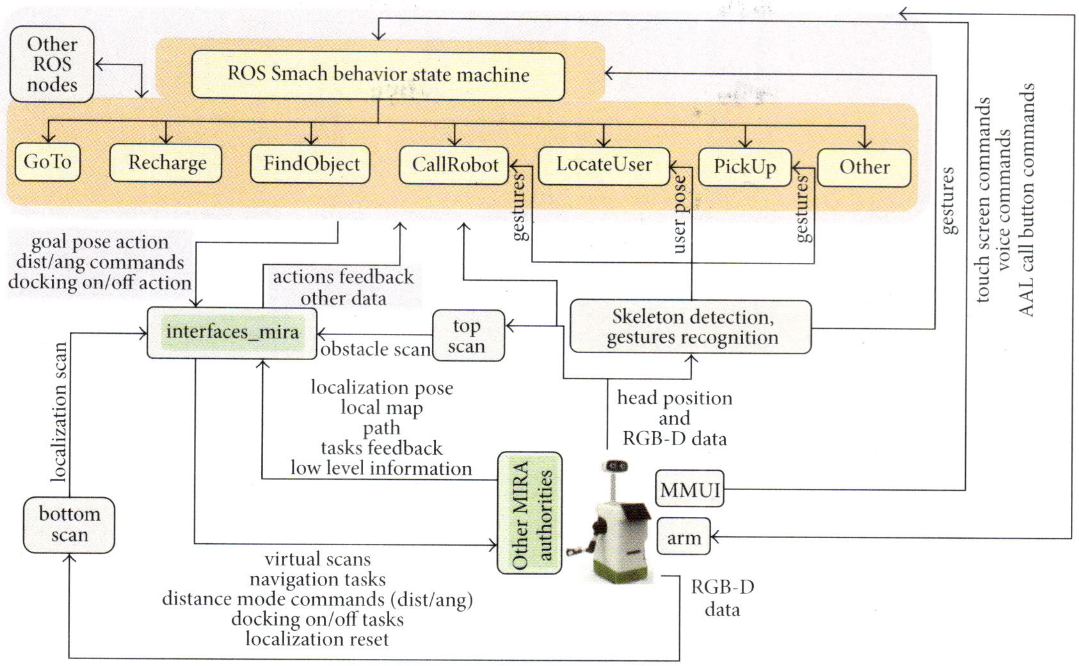

FIGURE 4: Hobbit behavior architecture.

In the following, we discuss the overall behavior coordination architecture and how the Mutual Care concept was implemented and go into detail of some of the building blocks necessary to construct the 11 scenarios. We further present the methods we developed to realize the goals of the project while respecting the limits set by the low-cost approach of our robots.

4.1. Behavior Coordination Architecture. Following the scenario descriptions, as defined by our specialists in elderly care, their implementation and the execution followed a script-based approach. A state-machine framework, SMACH (http://wiki.ros.org/smach), was therefore chosen to handle the behavior execution for all high-level codes.

An overview of the implemented architecture is shown in Figure 4. The top structure in this architecture is the PuppetMaster, which handles the decision-making outside of any scenario execution, where it can start, preempt, and restart any sub-state-machines. For this, it collects the input from those ROS nodes that handle gesture and speech recognition, text input via touchscreen, emergency detection (fallen and falling person detection, emergency button on the robot itself, and emergency gesture), and scheduled commands that need to be executed at a specific time of the day. The PuppetMaster delegates the actual scenario behavior execution to the sub-state-machines, which only rely on the input data needed for the current scenario. Each of these sub-state-machines corresponds to one of the scenarios designed to assist the users in their daily lives. As we needed to deal with many different commands with different execution priorities, it was necessary to ensure that every part of the execution of the state-machines can safely be interrupted without the risk of lingering in an undefined state. Particularly in situations when the arm of the robot was moving, it was necessary to be able to bring it into a position in which it would be safe to perform other tasks. The movement of the robot within the environment would have been unsafe if the arm would still stick out of the footprint of the robot itself. The priorities of the commands were defined with respect to the safety of the user, so that emergency situations can always preempt a possibly running state-machine, regardless of the state the system is currently in.

4.2. RGB-D Based Navigation in Home Environments. Autonomous navigation in user's homes, especially with low-cost RGB-D sensors, is a challenging aspect of care mobile robots. These RGB-D sensors pose additional challenges for safe navigation [18, 20–22]. The reduced field of view, the blind detection area, and the short maximum range of this kind of sensors provides limited information about the robot's surroundings. If the robot, for example, turns around in a narrow corridor, it might happen that the walls are already too close to be observed while turning, leading to increased localization uncertainty. In order to prevent such cases, we defined *no-go* areas around walls in narrow passages, preventing the robot from navigating too close to walls in the first place. For obstacle avoidance, the head is tilted down during navigation, so that the head camera partially compensates for the blind spot of the bottom camera. If obstacles are detected, they are remembered for a certain time in the robot's local map. However, a suitable trade-off had to be found for the decay rate. On one hand, the robot must be able to avoid persisting obstacles, but, on the other hand, it should not be blocked for too long when an obstacle in front of it (e.g., a walking person) is removed.

While localization methods generally assume that features of the environment can be detected, this assumption

Figure 5: Risky areas to be avoided. Obstacles like high shelves or stairs may not be perceived by Hobbit's sensor setup.

Figure 6: Examples of installed ramps to overcome door thresholds.

does not hold for the used RGB-D cameras with limited range and long corridors. In this situation, according to the detected features, the robot could be anywhere along the parallel walls, which can cause problems in cases where the robot should enter a room after driving along in such a corridor. When entering a room, it is especially important that the robot be correctly localized in the transversal direction to the doorway and that the doorway be approached from the front, so accurately driving through doors located on one side of a corridor is much more difficult than through doors located at the beginning or at the end of a corridor. In order to approach doors from the front, avoiding getting too close to the corner sides, a useful strategy for wide enough places is adding *no-go* areas at sides of a doorway entrance or at sharp corners. This way, it is possible to have safer navigation behavior in wide areas while keeping the ability to go through narrower areas. This provides more flexibility than methods with fixed security margins for the whole operational area.

No-go areas were also useful to avoid potentially dangerous and restricted areas and rooms. A few examples are shown in Figure 5. Areas with cables and thin obstacles on the floor and very narrow rooms (usually kitchens), where a nonholonomic robot as Hobbit cannot maneuver, were also avoided. However, it is worth noting that *no-go* areas are only useful as long as overall localization is precise enough. Other challenging situations were caused by thresholds and bumps on the floor and carpets. To overcome thresholds, we tested commercial and homemade ramps (Figure 6). After testing different configurations and finding proper incline limits, the robot was usually able to pass thresholds. Problems with standard planning methods, for example, when a new plan caused the robot to turn while driving on a ramp, were observed. A situation-dependent direct motion control instead of a plan-based approach can reduce the risk during such situations.

In order to facilitate the tasks to be carried out in the home environment, the concept of using rooms and labeled places inside the rooms (locations) was applied. The rooms are manually defined, such that spatial ambiguity is not a problem. Also, the geometry of the defined rooms does not have to be very precise with respect to the map, as long as the rooms contain all the places of interest that the user wants to label. Places are learned by tele-operating the robot to specific locations and the subsequent association of places to rooms operates automatically, based on the crossing number algorithm to detect whether a point lies inside a generic polygon [23]. Figure 7 shows several examples of rooms and places defined in the user trials for different tasks.

FIGURE 7: Rooms and places defined in two real apartments in Vienna.

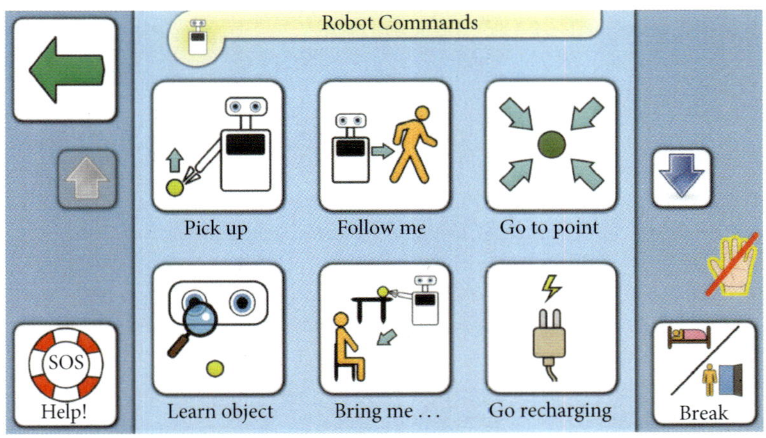

FIGURE 8: GUI of Hobbit showing one of the menu pages for robot commands. The strike-through hand on the right side indicates that the gesture input modality is disabled currently. A similar indicator was used for the speech input.

4.3. Multimodal Interaction Between the User and the Robot. The Hobbit robot deploys an improved version of the multimodal user interface (MMUI) used on Hobbit PT1. Generally speaking, the MMUI is a framework containing the following main building blocks: a Graphical User Interface (GUI) with touch, Automatic Speech Recognition (ASR), Text to Speech (TTS), and Gesture Recognition Interface (GRI). The MMUI provides emergency call features, web services (e.g., weather, news, RSS feed, and social media), control of robotic functions, and entertainment features. Compared to PT1, the graphical design of the GUI (Figure 8) was modified to better meet the user's needs. Graphical indicators on the GUI for showing current availability of GRI and ASR were iteratively improved.

During PT1 trials, we found that most of the users did not use the option of extending the MMUI to a comfortable ergonomic position for them. Therefore the mounting of the touchscreen was changed to a fixed position on Hobbit. Additionally, while the PT1 robot approached the user from the front, the Hobbit robot approaches the user from the right or left side while seated, which is more positively experienced by the user [24]. This offers the additional advantage that the robot is close enough for the user to interact via the touchscreen, while at the same time does not invade the personal space of the user (limiting her/his movement space or restricting other activities such as watching TV). Hobbit makes use of the MMUI to combine the advantages of the various user interaction modalities [25]. The touchscreen has strengths such as intuitiveness, reliability, and flexibility for multiple users in different sitting positions but requires a rather narrow distance between user and robot (Figure 9). ASR allows a larger distance and can also be used when no free hands are available, but it has the disadvantage of being influenced by the ambient noise level, which may reduce recognition performance significantly. GRI allows a wider distance between the robot and user and also works in noisy environments, but it only succeeds when the user is in the field of view of the robot. The interaction with Hobbit always depends on the distance between the user and Hobbit. It can be done through a wireless call button (far from other rooms), ASR and GRI (2 m to 3 m), and touchscreen (arm length, see Figure 9).

The ASR of Hobbit is speaker-independent, continuous, and available in four languages: English, German, Swedish,

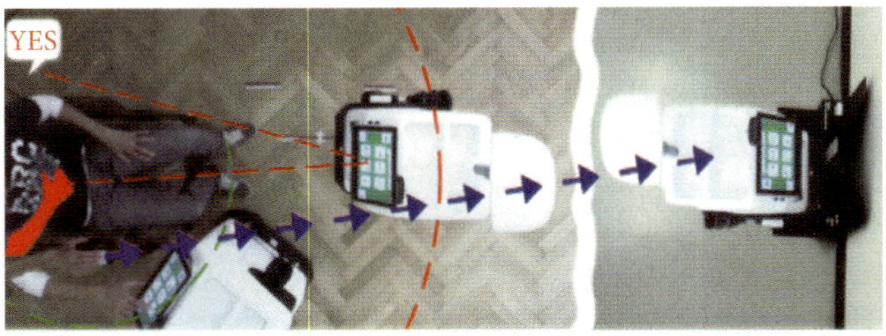

FIGURE 9: Different interaction distances between user and Hobbit seen from a ceiling camera. Short range: touch; middle range: speech and gesture; long range: wireless call button.

and Greek. Contemporary ASR systems work well for different applications, as long as the microphone is not moved far from the speaker's mouth. The latter case is called distant or far-field ASR and shows a significant drop in performance, which is mainly due to three different types of distortion [26]: (a) background noise, (b) echo and reverberation, and (c) other types of distortions, for example, room modes or the orientation of the speaker's head. For distant ASR, currently no off-the-shelf solution exists, but acceptable error rates can be achieved for distances up to 3 m by careful tuning of the audio components and the ASR engine [27]. An interface to a cloud based calendar was introduced, allowing PU and SU of Hobbit to access and partly to also edit events and reminders.

Despite the known difficulties with speech recognition in the far field and the local dialects of the users, the ASR of Hobbit worked as expected. The ASR was activated all over the Hobbit user trials, but the performance rate was commented on by users as necessary to be improved. The same was observed for the GRI. Eventually, the touchscreen as input modality was used most often by the majority of users, followed by speech and gesture. Touch was used more than twice as often as it was the case with ASR. Additionally, many users did not wait until the robot had completed its own speech output before starting to give a speech command which reduced the recognition rate. Considering these lessons learned, the aims for future work on the ASR are twofold: improving the performance of the ASR and providing better indication when the MMUI is listening to spoken commands and when it is not. The aspect of using two different variants for text messages from the robot to the user was taken over from Hobbit PT1. Based on other researches, it can be concluded that using different text variants does have an influence, for example, by increasing users' impression of interacting with a (more) vivid system. Some users demanded additional ASR commands, for example, *right, left, forward, reverse*, and *stop* in addition to *come closer*, as they would like to position (move) the robot with the help of voice commands or a remote control.

4.4. Person Detection and Tracking. To serve as building block for components like activity recognition [28] and natural human-robot communication [19, 29] as well as specialized functions like the fitness application [30], we developed a human body detection and tracking solution. Person detection and tracking in home environments is a challenging problem because of its high dimensionality and the appearance variability of the tracked person. A challenging aspect of the problem in Hobbit-related scenarios is that elderly users spend a considerable amount of time sitting in various types of chairs or couches. Therefore, human detection and tracking should consider human body figures that do not stand out from their background. On the contrary, they may interact with cluttered scenes, exhibiting severe partial occlusions. Additionally, the method needs to be capable of detecting a user's body while standing or walking based on frontal, back, or side views.

The adopted solution [31] enables 3D part-based, full/upper body detection and tracking of multiple humans based on the depth data acquired by the RGB-D sensor. The 3D positions and orientations for all joints of the skeletal model (full or upper body) relative to the depth sensor are computed for each time stamp. A conventional face detection algorithm [32] is also integrated using the color data stream of the sensor to facilitate human detection in case the face of the user is visible by the sensor. The proposed method has a number of beneficial properties that are summarized as follows: (1) performs accurate markerless 3D tracking of the human body that requires no training data, (2) requires simple inexpensive sensory apparatus (RGB-D camera), (3) exhibits robustness in a number of challenging conditions (illumination changes, environment clutter, camera motion, etc.), (4) has a high tolerance with respect to variations in human body dimensions, clothing, and so forth, (5) performs automatic human detection and automatic tracking initialization, thus recovering easily from possible tracking failures, (6) handles self-occlusions among body parts or occlusions due to obstacles/furniture and so forth, and (7) achieves real-time performance on a conventional computer. Indicative results of the method are illustrated in Figure 10.

4.5. Gesture Recognition. A vision-based gestural interface was developed to enrich the multimodal user interface of Hobbit in addition to speech and touch modalities. This enables natural interaction between the user and the robot by recognizing a predefined set of gestures performed by the user using her/his hands and arms. Gestures can be of varying complexity and their recognition is also affected by the scene context, actions that are performed in the foreground or the background at the same time, and by preceding and/or following actions. Moreover, gestures are often culture-specific, providing additional evidence to substantiate the interesting as well as challenging nature of the problem.

For Hobbit, existing upper body gestures/postures as used on PT1 had to be replaced with more intuitive hand/finger-based gestures that can be performed more easily by elderly users while sitting or standing. We redesigned the gestural vocabulary for Hobbit that now consists of six hand gestures that convey messages of fundamental importance in the context of human-robot dialogue. Aiming at natural, easy-to-memorize means of interaction, users have identified gestures consisting of both static and dynamic hand configurations that involve different scales of observation (from arms to fingers) and exhibit intrinsic ambiguities. Recognition needs to be performed in continuous video streams containing

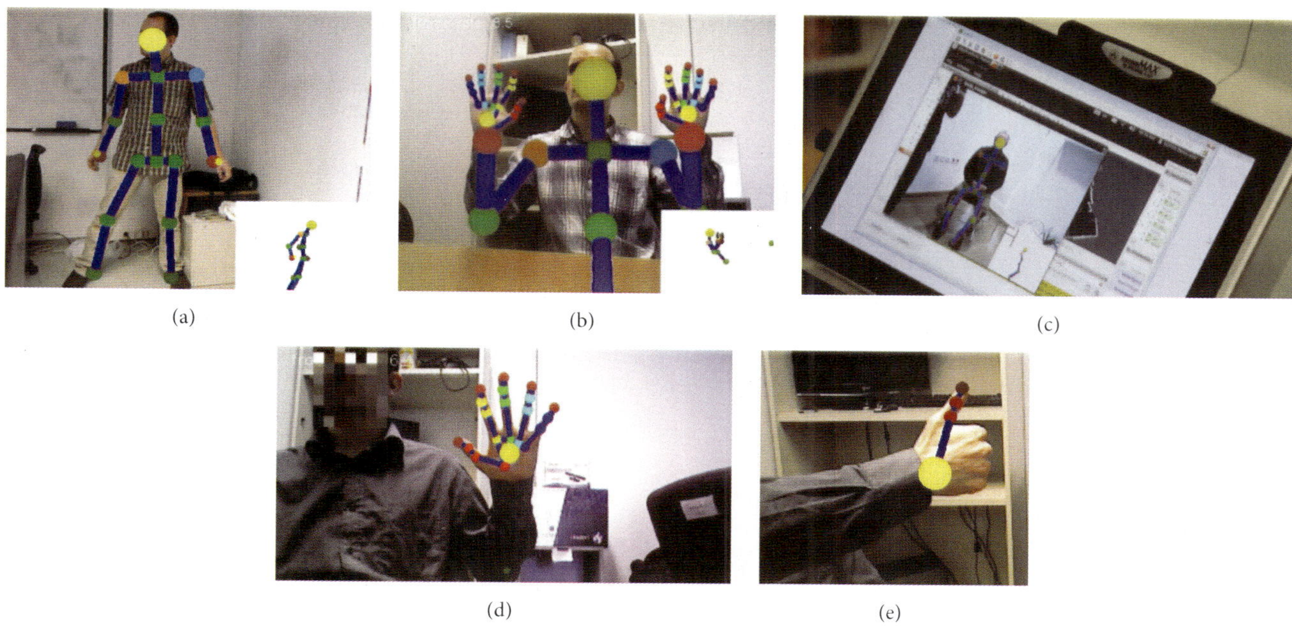

FIGURE 10: Qualitative results of the 3D skeletal model-based person detection and tracking method. (a) Full model of a standing user. (b) Upper body (including hands and fingers) of a sitting user. (c) Full model of a sitting user. ((d) and (e)) Hand and finger detection supporting the gesture recognition framework (see Section 4.5).

other irrelevant actions. All the above need to be achieved by analyzing information acquired by a possibly moving RGB-D camera in cluttered environments with considerable light variations.

The proposed framework for gesture recognition [19, 29] consists of a complete system that detects and tracks arms, hands, and fingers and performs spatiotemporal segmentation and recognition of the set of predefined gestures, based on data acquired by the head camera of the robot. Thus, the gesture recognition component is integrated with the human detection and tracking module (see Section 4.4). At a higher level, hand posture models are defined and serve as building blocks to recognize gestures based on the temporal evolution of the detected postures. The 3D detection and tracking of hands and fingers relies on depth data acquired by the head camera of Hobbit, geometrical primitives, and minimum spanning tree features of the observed structure of the scene in order to classify foreground and background and further discriminate between hand and nonhand structures in the foreground. Upon detection of the hand (palm and fingers), the trajectories of their 3D positions across time are analyzed to achieve recognition of hand postures and gestures (Table 1). The last column describes the assignment of the chosen physical movements to robot commands. The performance of the developed method has been tested not only by users acquainted with technology but also by elderly users [19] (see Figure 11). Those tests formed a very good basis for fine-tuning several algorithmic details towards delivering a robust and efficient hand gesture recognition component. The performance of the final component was tested during field trials achieving high performance according to the evaluation results.

4.6. Fall Detection. According to the assessed user needs and the results of PT1 laboratory studies [17], a top-priority and prominent functionality of Hobbit regards fall prevention and fall detection. We hereby describe a relevant vision-based component that enables a patrolling robot to (a) perform fall detection and (b) detect a user lying on the floor. We focused mostly on the second scenario, as observing a user falling in the field of view of an autonomous assistive robot is of very low probability. The proposed vision-based emergency detection mechanism consists of three modes, each of which initiates an emergency handling routine upon successful recognition of the emergency situation:

(1) Detection of a falling user in case the fall occurs while the body is observable by the head camera of the robot

(2) Detection of a fallen user who is lying on the floor while the robot is navigating/patrolling

(3) Recognition of the emergency (help) gesture that can be performed by a sitting or standing user via the gesture recognition interface of Hobbit (see Figure 11, middle)

The methodology for (1) regards a simple classifier trained on the statistics of the 3D position and velocity of the observed human body joints acquired by the person detection and tracking component. For (2), once the general assumption, the fact that the human's head is above the rest of the body, does no longer hold true, an alternative, simple, yet effective approach to the problem has been adopted. This capitalizes on calibrated depth and thermal visual data acquired from two different sensors that are available on the head of Hobbit. More specifically, depth data from both cameras of the robot

Table 1: Set of hand/arm postures/gestures considered for the gestural interface of Hobbit.

User command	Upper body gesture/posture	Robot command	Related scenarios/tasks
Yes	Thumb up-palm closed	Positive response to confirmation dialogues. YES gesture	All (1 m to 2 m distance to robot)
No	Close palm, waving with index finger up	Negative response to confirmation dialogues	All (1 m to 2 m distance to robot)
Come closer	Bend the elbow of one arm repeatedly towards the platform and the body	Reposition the platform closer to the sitting user	All (1 m to 2 m distance to robot)
Cancel task	Both open palms towards the robot	Terminate an on-going robot behavior/task	All
Pointing	Extend one arm and point in 3D space towards an object (lying on the floor)	Detect and grasp the object of interest towards the pointing 3D direction	Pick up an (unknown) object from the floor
Reward	Open palm facing towards the robot and circular movement (at least one complete circle is needed)	Rewards the robot for an accomplished action/task	Approach the user
Emergency	Cross hands pose (normal-range interaction)	Emergency detection, initiated by the user	Emergency detection

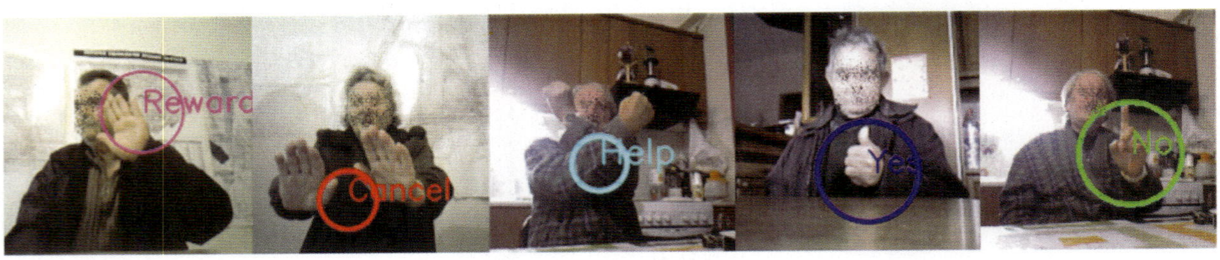

Figure 11: Snapshots of Hobbit users performing gestures during lab trials. The recognition results are superimposed as text and a circle on the images indicating the location and the name of the recognized gesture (taken from [19]).

(head and base) are acquired and analyzed while observing the floor area in front of the robot. Figure 12 illustrates sample results of the fallen user detection component. In Figure 12(a), the upper part illustrates the color frame captured by the head camera of the robot that is titled down towards the floor, while navigating. In the bottom image, the viewpoint of the bottom camera is illustrated, after the estimation of the 3D floor plane has been performed.

The methodology for vision-based emergency detection of case (3) refers to successful recognition of the emergency "Help me," based on the gesture and posture recognition module, as described in Section 4.5. The developed component is constantly running in the background within the robot's behavior coordination framework, while the robot is active during all robot tasks, except from object detection and recognition tasks.

4.7. Approaching the User. Specific behavior coordination was developed so that the robot could approach the user in a more flexible and effective way compared to standard existing methods. Using fixed predefined positions can be sufficient in certain scenarios, but it often presents limitations in real-world conditions [22]. The approach we developed incorporates user detection and interaction (Section 4.4), remembered obstacles and discrete motion for coming closer to the user with better, and adaptive positioning.

First, a safe position to move to is obtained from the local map and the robot moves there. Secondly, the user communicates to the robot whether it should move even closer or not in any of the three available modes (speech, touch, or gesture). Finally, the robot moves closer by a fixed distance of 0.15 m for a maximum of three times if the user wishes. This gives the users more control over final distance adjustments. A more detailed description of this novel approach will be published elsewhere.

4.8. User Following. As the head camera is not available for observing the full body of a user during navigation (obstacle detection), we designed a new approach [33] to localize a user by observing its lower body part, mainly the legs, based on RGB-D sensory data acquired by the bottom camera of the platform.

The proposed method is able to track moving objects such as humans, estimate camera ego-motion, and perform map construction based on visual input provided by a single RGB-D camera that is rigidly attached to a moving platform. The moving objects in the environment are assumed to move on a planar floor. The first step is to segment the static background from the moving foreground by selecting a small number of points of interest whose 3D positions are estimated directly from the sensory information. The camera motion is computed by fitting those points to a progressively built model of the environment. A 3D point may not match the current version of the map either because it is a noise contaminated observation or because it belongs to a moving object or because it belongs to a structure

(a) (b) (c)

FIGURE 12: Vision-based emergency detection of a fallen user lying on the floor. The upper and lower middle images show the captured frame from the head and bottom cameras, respectively. The green dots mark a found skeleton within the search area (green and blue rectangles). (a–c) No human, no detection; person lying on the floor, correct detection; volumetric data from the head's depth and temperature sensor are in conflict with the volumetric date provided by the bottom depth sensor.

(a) (b) (c) (d)

FIGURE 13: (a–d) User points to an object on the floor and Hobbit drives to a point from where it can be picked up and moves the arm to a position to grasp it. The object is lifted and the check if grasp was successful is performed: the object is moved forward to check if something has changed at the previous position of the object on the floor. If successful, the object is placed on the tray on top of the robot.

attached to the static environment that is observed for the first time. A classification mechanism is used to perform this disambiguation. Additionally, the method estimates the camera (ego) motion and the motion of the tracked objects in a coordinate system that is attached to the static environment (robotic platform). In essence, our hypothesis is that a pair of segmented and tracked objects of specific size/width that move independently side-by-side at the same distance and direction in the field of view of a moving RGB-D camera correspond to user's legs being followed by the robot with high probability. The method provides the 3D position of user's legs with respect to the moving or static robotic platform. Other moving objects in the environment are filtered out or can be provided to an obstacle avoidance mechanism as moving obstacles, thus facilitating safe navigation of the robot.

4.9. Pick Up Objects from the Floor. To reduce the risk of falling, Hobbit was designed to be able to pick up unknown objects from the floor. Figure 13 shows the steps of the "Pick up object" task. The user starts the command and points at the object on the floor. If the pointing gesture is recognized, the robot navigates to a position from where it could observe the object. At this position, the robot looks at the approximate position of the object. Hobbit then makes fine adjustments to position itself at a location from where grasping is possible. If it is safe to grasp the object, the robot executes the arm trajectory and subsequently checks if the grasp was successful and will try to do so a second time if it was not.

Several autonomous mobile robots have been developed to fetch and deliver objects to people [34–38]. None of these publications evaluate their robot grasping from floor, and none evaluate the process of approaching an object and

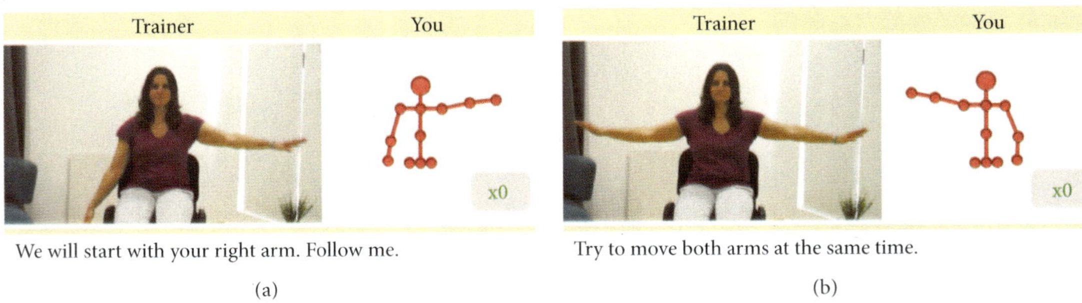

FIGURE 14: (a) Avatar mirroring the trainer's movement proved easier for users to follow. (b) Correction suggested by the system to the user.

grasping it as a combined action. Detection of the user and recognition of a pointing gesture were performed using the work presented in [19, 31]. Checks are performed to rule out unintentional or wrong pointing gestures and to enhance the accuracy of the detected pointing gesture.

A plausibility check tests if the pointing gesture is pointing towards the floor. To guarantee an exact position of the robot to bring the arm in a position where the gripper can approach the object in a straight line before closing, the accurate movement to the grasping position can be done as a relative movement to the object instead of using the global navigation. This is a crucial step as the region, in which the head camera is able to perceive objects and where the 6-DoF arm is able to perform a movement straight down to the floor without changing gripper orientation, is limited to $15 \times 10\,\text{cm}^2$. For calculating grasps, we use the method of Height Accumulated Features [39]. These features reduce the complexity of a perceived point cloud input, increase the value of given information, and hence enable the use of machine learning for grasp detection of unknown objects in cluttered and noncluttered scenes.

4.10. Fitness Application. The fitness application was introduced as a feature to the Hobbit robot after the PT1 trials and was made available during the PT2 trials for evaluation. The motivation behind this application comes from the fact that physical activity can have a significant positive impact on the maintenance or even on the improvement of motor skills, balance, and general physical well-being of elderly people, which in turn can lower the risk of falls in the long run. Based on feedback from the Community and Active Ageing Center of the municipality of Heraklion, Greece, the following requirements were produced. The exercises must (1) be easy to learn, (2) target different joints and muscles, (3) provide appropriate feedback to the user, (4) keep the user engaged while providing enough breaks, and (5) be designed to be performed from a seated position.

Based on these requirements and feedback from test users, we developed an application including three difficulty levels and seven different exercises. The user interface consisted of a split view of a video recording of the actual trainer performing each exercise on the left side and an avatar figure depicting the user's movement while executing the instructed exercise on the right side as shown in Figure 14. This side-to-side viewing setup allowed the user to compare his or her movements to those of the trainer. The bottom part of the interface was allocated for the instructions at the beginning of each exercise and also for any feedback and guidance to the user when needed. The design and development of the fitness application are described in more detail in [30]. The fitness application was explained to the participants of the trials by the facilitator at the initial introduction of the system during the installation day. The participants could access the application if desired at any time. Almost all users tried the fitness application at least once with some using it multiple times during the three-week evaluation period. From the comments received during the mid-term and end-of-trial interviews, it can be concluded that the overall concept of having the fitness program as a feature of the robot received positive marks by many of users as far as its usefulness and importance are concerned. However, most users who tried it out said that they would have liked it to be more challenging and to offer a larger variety of exercise routines with various challenging levels to choose from.

5. Field Trials

We conducted field trials in the households of 18 PU with 5 Hobbit robots in Austria, Greece, and Sweden. The trials lasted ~21 days for each household, resulting in a total of 371 days. During this time, the robots were placed in the homes of 18 older adults living on their own, where users could use and explore the robot on a 24/7 basis. Detailed results of the trials will be published elsewhere; preliminary results can be found in [40] (a first analysis only of the robot log data without any cross-analysis to the other data collected) and in [41] (a first overview on the methodological challenges faced during the field trails).

The trial sample consisted of 16 female and 2 male PU; their age ranged from 75 to 90 years ($M = 79.67$). All PU were living alone, either in flats (13 participants) or in houses. In adherence with inclusion criteria set by the research consortium, all participants had fallen in the last two years or were worried about falling and had moderate impairments in at least one of the areas of mobility, vision, and hearing. 15 PU had some form of multiple impairments. Furthermore, all participants had sufficient mental capacity to understand the project and give consent. In terms of technology experience, 50.0% of the PU stated that they were using a computer every day, 44.45% stated that they were never using a computer or used it less than once a week, and only one participant used a computer two to three times a week.

Before the actual trials, the PU were surveyed to make sure that they matched the criteria for inclusion and to discuss possible necessary changes to their home environments for the trials (e.g., removing carpets and covering mirrors). After an informed consent was signed, the robot was brought into the home and the technical setup took place. After this setup, a representative from the elderly care facility explained the study procedure and the robot functionalities to the PU in an individual open-ended manner. Afterwards, a manual was left within the household in case participants wanted to look up a functionality during the 21 days. All users experienced two behavioral roles of the robot. The robot was set to *device-mode* until day 11 when it was switched to *companion-mode* (i.e., Mutual Care). The real-world environment in which the field tests took place bears certain challenges, such as unforeseen changes in the environment and uncontrollable settings. Assessment by means of qualitative interviews and questionnaires took place at four stages of each trial: before trial, midterm, end of trial, and after trial (i.e., one week after the trial had ended). Moreover, log data was automatically recorded by the robot during the whole trial duration. The field trial methodology is comparable to similar studies (e.g. [42]).

The field trials revealed that several functions of the robot lack stability over time. Those technical issues certainly influenced the evaluation of the system because a reliable working technical system is a prerequisite for positive user experience. We tried to minimize potential negative feelings due to potential malfunctioning by informing our users that a prototype of a robot is a very complex technical system that might malfunction. Additionally, they were given the phone number of the facilitator who was available for them around the clock, 7 days per week, for immediate support. However, malfunctions certainly had an influence on subjects' answers during the assessments and may have attracted attention with the result that the subtle behavioral changes introduced by the switch from *device-mode* to *companion-mode* may have been shifted out of the attentional focus. Availability of commands was equally distributed across the two phases of *Mutual Care* as can be seen in Table 2. Please note that unavailability or malfunctioning of functions in one but not the other mode (unequal distribution of functionality) would have led to a bias within the evaluation. Table 2 gives an overview of the functional status across all PU during the field trials. It is based on the combination of (i) a check of the robot's features by the facilitator during the preassessment, midterm assessment, and end-of-trial assessments, (ii) protocols of the calls of the users because they had a problem with the robot, and (iii) analysis of the log data by technical partners.

The Hobbit field trials marked the first time an autonomous, multifunctional service robot, able to manipulate objects, was put into the domestic environment of older adults for a duration of multiple weeks. Our field trials provided insight into how the elderly used the Hobbit robot and which functionalities they deemed useful for themselves and how the robot influenced their daily life. Furthermore, we could show that it is in principal feasible to support elderly with a low-cost, autonomous service robot controlled by a rather simple behavior coordination system.

6. Lessons Learned

Based on all the insights gained from developing and testing Hobbit in the field, we can summarize the following recommendations for fellow researchers in the area of socially assistive robots for enabling independent living for older adults in domestic environments.

6.1. Robot Behavior Coordination. The developed behavior control based on a state-machine proved to be very useful and allowed us to implement many extensions in a short time. A close interconnection with the user was therefore helpful. In the following, we present our main lessons learned regarding the implementation of the robot behavior.

6.1.1. Transparency. Actions and their effects need to be communicated in a clear fashion so that the robot's presented functionality can be fully understood by the user. Users reported missing or nonworking functionality (e.g., reminders not being delivered to them and patrol not being executed). Most of these reported issues were caused by the fact that the users did not understand the technical interdependencies between robot functions. For example, if a command was not available due to a certain internal state of the robot, the user was not aware of this and did not understand the shown behavior of the robot. These functional relations need to be made explicit and stated more clearly to the users.

6.1.2. Legibility. The log data and conversations with participants revealed that the robot needs to communicate its intentions. For instance, when the robot proactively moved out of its charging station, the user was not always aware what was going to happen next. When they did not understand what the robot was doing, they canceled the robot's action, effectively stopping part of the robot's benefit to them. To work around this, a robot needs to clearly state the reason of its action and which goal it is trying to achieve when performing an autonomously started task.

6.1.3. Contradictory Commands. Log data presented an interesting effect while interacting with the touchscreen. When moving the hand towards the touchscreen on the robot, the gesture recognition system detected the movement of the hand as the *come closer* gesture, shortly followed by a command from the touch input on the GUI. We could replicate this behavior later on in our internal tests in the lab. A simple solution for such contradictions of commands is to simply wait for a short period of time (less than 0.2 seconds) before a gesture close to the robot is processed by the behavior coordination system to wait for a possibly following touch input.

6.1.4. Transparency of Task Interdependencies. The interviews revealed that the interdependencies between the tasks were not clear to the user; the best example was the learn-and-bring-object task. As described, for the bring-object task, the object first had to be learned so that it can be found in the apartment. However, this fact needs to be remembered by

Table 2: System reliability across 18 PU.

(a)

muc mode	Statistics	Call Hobbit	Come closer	Stop Hobbit	Emergency	Pick up object	Teach a new object	Bring object to user	Calendar reminders	Follow me	Move to location
Device	Days total	226	226	226	226	226	226	226	226	226	226
	Days of introduction	31	31	31	31	31	31	31	31	31	31
	Days switched off	55	55	55	55	55	55	55	55	55	55
	Days in use	140	140	140	140	140	140	140	140	140	140
	Days when feature was not working	14	13	13	19	84	12	79	49	105	15
	Days when feature was only partially working	23	20	22	44	47	116	62	83	13	32
Companion	Days total	148	148	148	148	148	148	148	148	148	148
	Days switched off	20	20	20	20	20	20	20	20	20	20
	Days in use	128	128	128	128	128	128	128	128	128	128
	Days when feature was not working	14	10	8	12	83	17	85	54	92	9
	Days when feature was only partially working	25	16	17	38	40	95	43	64	18	29
Device	Working over days in use	81.79%	83.57%	82.86%	70.71%	23.21%	50.00%	21.43%	35.36%	20.36%	77.86%
Companion	Working over days in use	79.30%	85.94%	87.11%	75.78%	19.53%	49.61%	16.80%	32.81%	21.09%	81.64%

(b)

muc mode	Statistics	Go recharge	Take a break	Telephone	Information	Surprise me	Entertainment audio	Entertainment games	Entertainment fitness	Reward
Device	Days in total	226	226	226	226	226	226	226	226	226
	Days of introduction	31	31	31	31	31	31	31	31	31
	Days feature was disabled	55	55	55	55	55	55	55	55	55
	Days of feature in use	140	140	140	140	140	140	140	140	140
	Days when feature was not working	19	16	19	11	11	11	23	20	11
	Days when feature was only partially working	20	6	22	9	23	7	20	27	6
Companion	Days in total	148	148	148	148	148	148	148	148	148
	Days feature was disabled	20	20	20	20	20	20	20	20	20
	Days of feature in use	128	128	128	128	128	128	128	128	128
	Days when feature was not working	10	8	22	8	8	7	26	19	14
	Days when feature was only partially working	33	9	16	7	23	8	11	23	7
Device	Working over days in use	79.29%	86.43%	78.57%	88.93%	83.93%	89.64%	76.43%	76.07%	90.00%
Companion	Working over days in use	79.30%	90.23%	76.56%	91.02%	84.77%	91.41%	75.39%	76.17%	86.33%

the user and as this is often not the case, users wanted to ask Hobbit to bring them an object even though it had not learned any objects before. In this specific case, the problem could be easily fixed by only offering the task "bring object" when an object was actually learned beforehand (e.g., the task could be greyed out in the MMUI).

6.1.5. Full Integration without External Programs. The handling of user input and output must be fully integrated with the rest of the robot's software architecture to be able to handle interruptions and continuations of interaction between the user and the robot. The user interface on the tablet computer (MMUI) incorporated multiple external programs (e.g., Flash games, speech recognition, and the fitness functionality). As those were not directly integrated, the behavior coordination was not aware about their current state, leading to multiple interaction issues with users. For example, a game would be exiting when a command with higher priority (e.g., emergency from fall detection) would start the emergency scenario. External programs need to be included in a way that makes it possible to suspend and resume their execution at any time.

6.1.6. Avoiding Loops. Reviewing the log data revealed that the behavior coordination system could be trapped in a loop without a way to continue the desired behavior execution. The behavior coordination needs to provide a fallback solution in case of a seemingly endless loop in any part of the behavior. The behavior coordination communicates with the MMUI in a way that does not provide immediate feedback over the same channels of communication. Due to timing issues, it occurred that a reply was lost between the communicating partners (i.e., the fact that the robot stopped speech output). From there on, the behavior coordination was in a state that should not be reached and was unable to continue program execution in the desired manner. Thus, the communication structures should always have a fallback solution to continue execution as well as the feedback data on the same channels to prevent such a stop in a scenario.

6.2. Human-Robot Interaction with the MMUI. The interaction with the user was based on a multimodal user interface that was perceived as easy to use during our field trials. While touch input turned out to be the most reliable modality, speech and gesture interaction was highly welcome. Many of the entertainment functions of the MMUI relied on Internet connectivity. Many users either were not interested in some UI features which therefore should be removed or asked for special configuration of the preferred features (e.g., selection of entertainment). The main way the user was able to communicate remotely with Hobbit was with the use of physical switches (call buttons) placed at several fixed places inside the house of the user. The user had to physically go to the designated switch spot and press the switch for the robot to approach her/him. A smartphone/tablet application could be developed to allow a better remote communication experience with the robot.

6.2.1. Internet Connectivity. Internet connectivity was not reliable depending on location and time. While in most countries Internet (line-based or mobile) coverage is no problem in general, local availability and quality vary significantly, which makes Internet-based services difficult to implement for technically unaware users. The integration of rich Internet-based content into the interaction therefore lacks usability in case of intermittent connectivity.

6.2.2. Graphical User Interface. The GUI could be personalized by the user for increased comfort during interaction. This, however, shows the need for localized content to be available. As the setup phase during the trials showed that PU are likely not aware what content is available, some (remote) support and knowledge from SU are necessary for the configuration of the user interface.

6.2.3. Speech Recognition. Field trials showed that speech recognition is still not working well for many users. Despite the overall acceptable recognition rate that varies largely from user to user and from language to language and that is based on the environment and distance, users often do not support the needs of current ASR technology for clearly expressed and separated commands in normal voice. The Sweet-Home project once more emphasizes the findings from the DiRHA 2 project that practical speech recognition for old people in the home environment is still a major challenge by itself [43]. However, our ASR provided a positively experienced natural input channel when used in a multimodal HRI, where the touchscreen with its GUI provides a reliably working base.

6.2.4. Smarthome Integration. The setup phase during the field trials showed that the integration into smarthome environments can be beneficial. Field trials showed that context awareness and adaptations highly impact the acceptance of the robot. Imagined features could be automatic on/off of the light or the stove or adjusting the proactively level of the robot based on the user's mood.

6.2.5. Remote End User Control. Reflecting on the field trial indicates that a potential valuable extension of the interaction modalities would be a remote control of the robot, for instance, on a smartphone enabling PU but also maybe SU to control the robot from outside the home. Potential useful scenarios could be to send the robot to the docking station or to patrol the flat and search for an object or the PU or the SU video calling the PU.

6.3. Implementation of Mutual Care Behavior. In the beginning of the trials, we implemented Mutual Care in such a fashion that in the companion mode the robot offers to return the favor after every interaction with the user. This was done in order to guarantee that the users would notice the difference between the modes during the interaction. The positive fact was that users noticed the changes. However, they were soon very annoyed by the robot. Consequently, we changed this implementation during the running trials. The return of favor frequency was reduced; it was no longer offered after the commands *Recharge batteries, Go to, Call button*, and *Surprise*. Further feedback from the second and third Austrian and the second and third Swedish users led to further reduction of the return a favor frequency to offering it only after the following three commands:

(1) Pick up command (favor: Hobbit offers music: *I'd like to return the favor. I like music. Shall I play some music for you?*)

(2) Learn object command (favor: Hobbit offers to play a game (suitable because the user is already sitting down): *I'd like to return the favor. Do you want to play a game?*)

(3) Reward command (favor: Hobbit offers to surprise the user: *I'd like to return the favor. I like surprises. Do you want a surprise?*)

However, as the interviews showed, these behavioral changes were no longer recognized by the users. Similarly, the differences in proactivity and presence were not reflectively noticed by the users, but the changes in dialogue were noticed.

6.3.1. Help Situations. For the development of Mutual Care behavior in completely autonomous scenarios, which helping situations the robot can really identify in order to ask for help and how the robot can notice that it actively recovered through the help have to be considered.

6.3.2. Design of Neediness. In the interviews, PU reflected that they did not really recognize that the robot needed their input to continue its task. For Mutual Care, the *need of help* seems to be essential. For future version of the robot, how to design the *neediness* needs to be considered. This could be achieved with facial expressions, sounds, or movements. Also for behaviors such as presence and proactivity, the robot could say after an interaction: *"I would prefer staying with you in your room"* or proactivity (e.g., *"I would like to spend more time with you"* before offering an activity). This would give a better explanation of the robot's behavior to the user and an expected raise of acceptance.

7. Conclusions

In this article, we presented the second prototypical implementation of the Hobbit robot, a socially assistive service robot. We presented the main functionality it provided, as well as the behavior coordination that enabled autonomous interaction with the robot in real private homes. Hobbit is designed especially for fall detection and prevention, providing various tasks (e.g., picking up objects from the floor, patrolling through the flat, and employing reminder functionalities), and supports multimodal interaction for different impairment levels. We focused on the development of a service robot for older adults, which has the potential to promote aging in the home and to postpone the need to move to a care facility. Within the field trials, we reached the desirable long-term goal that a mobile service robot with manipulation capabilities enters real homes of older adults and showed its usefulness and potential to support independent living for elderly users.

To conclude, we believe that methods, results, and lessons learned presented in this article constitute valuable knowledge for fellow researchers in the field of assistive service robotics and serve as a stepping stone towards developing affordable care robots for the aging population.

Conflicts of Interest

The authors declare that there are no conflicts of interest.

Acknowledgments

This research has received funding from the European Community's Seventh Framework Programme (FP7/2007–2013) (Grant Agreement no. 288146, Hobbit).

References

[1] P. Dario, P. F. M. J. Verschure, T. Prescott et al., "Robot companions for citizens," *Procedia Computer Science*, vol. 7, pp. 47–51, 2011.

[2] J. M. Beer, C.-A. Smarr, T. L. Chen et al., "The domesticated robot: Design guidelines for assisting older adults to age in place," in *Proceedings of the 7th Annual ACM/IEEE International Conference on Human-Robot Interaction, (HRI '12)*, pp. 335–342, USA, March 2012.

[3] D. Fischinger, P. Einramhof, K. Papoutsakis et al., "Hobbit - The Mutual Care Robot," in *Assistance and Service Robotics in a Human Environment Workshop in conjunction with IEEE/RSJ International Conference on Intelligent Robots and Systems*, 2013.

[4] J. Bohren, R. B. Rusu, E. G. Jones et al., "Towards autonomous robotic butlers: Lessons learned with the PR2," in *Proceedings of the 2011 IEEE International Conference on Robotics and Automation, (ICRA '11)*, pp. 5568–5575, China, May 2011.

[5] Y. Demiris and B. Khadhouri, "Hierarchical attentive multiple models for execution and recognition of actions," *Robotics and Autonomous Systems*, vol. 54, no. 5, pp. 361–369, 2006.

[6] T. Carlson and Y. Demiris, "Collaborative control for a robotic wheelchair: evaluation of performance, attention, and workload," *IEEE Transactions on Systems, Man, and Cybernetics, Part B: Cybernetics*, vol. 42, no. 3, pp. 876–888, 2012.

[7] M. Cashmore, M. Fox, D. Long et al., "Rosplan: Planning in the robot operating system," in *Proceedings of the 25th International Conference on Automated Planning and Scheduling, (ICAPS '15)*, pp. 333–341, June 2015.

[8] M. Mansouri and F. Pecora, "More knowledge on the table: Planning with space, time and resources for robots," in *Proceedings of the 2014 IEEE International Conference on Robotics and Automation, (ICRA '14)*, pp. 647–654, China, June 2014.

[9] S. Goetze, S. Fischer, N. Moritz, J. E. Appell, and F. Wallhoff, "Multimodal Human-Machine Interaction for Service Robots in Home-Care Environments," in *Proceedings of the 1st Workshop on Speech and Multimodal Interaction in Assistive Environments*, 2012.

[10] H. S. Koppula, A. Jain, and A. Saxena, "Anticipatory planning for human-robot teams," *Springer Tracts in Advanced Robotics*, vol. 109, pp. 453–470, 2016.

[11] P. Kwang-Hyun and Z. Zenn Bien, "Intelligent sweet home for assisting the elderly and the handicapped," in *Independent Living for Persons with Disabilities and Elderly People: ICOST*, p. 151, 2003.

[12] T. Fukuda, P. Di, F. Chen et al., "Advanced service robotics for human assistance and support," in *Proceedings of the International Conference on Advanced Computer Science and Information Systems, (ICACSIS '11)*, pp. 25–30, December 2011.

[13] D.-J. Kim, R. Lovelett, and A. Behal, "An empirical study with simulated ADL tasks using a vision-guided assistive robot arm," in *Proceedings of the 2009 IEEE International Conference on Rehabilitation Robotics, (ICORR '09)*, pp. 504–509, Japan, June 2009.

[14] M. B. Hong, S. J. Kim, T. Um, and K. Kim, "KULEX: An ADL power-assistance demonstration," in *Proceedings of the 10th International Conference on Ubiquitous obots and Ambient Intelligence, (URAI '13)*, pp. 542–544, November 2013.

[15] L. Lammer, A. Huber, A. Weiss, and M. Vincze, "Mutual Care: How older adults react when they should help their care Robot," in *Proceedings of the 3rd international symposim on New Frontiers in Human-Robot interaction*, April 2014.

[16] T. Körtner, A. Schmid, D. Batko-Klein et al., "How social robots make older users really feel well - A method to assess users' concepts of a social robotic assistant," in *Proceedings of the International Conference on Social Robotics*, vol. 7621, pp. 138–147, Springer, 2012.

[17] D. Fischinger, P. Einramhof, K. Papoutsakis et al., "Hobbit, a care robot supporting independent living at home: First prototype and lessons learned," *Robotics and Autonomous Systems*, vol. 75, pp. 60–78, 2016.

[18] P. De La Puente, M. Bajones, P. Einramhof, D. Wolf, D. Fischinger, and M. Vincze, "RGB-D sensor setup for multiple tasks of home robots and experimental results," in *Proceedings of the IEEE/RSJ International Conference on Intelligent Robots and Systems, (IROS '14)*, pp. 2587–2594, USA, September 2014.

[19] D. Michel, K. Papoutsakis, and A. A. Argyros, "Gesture recognition supporting the interaction of humans with socially assistive robots," in *Advances in Visual Computing*, vol. 8887 of *Lecture Notes in Computer Science*, pp. 793–804, Springer International Publishing, Berlin, Germany, 2014.

[20] J. González-Jiménez, J. Ruiz-Sarmiento, and C. Galindo, "Improving 2D Reactive Navigators with Kinect," in *Proceeding of the 10th International Conference on Informatics in Control, Automation and Robotics (ICINCO '13)*, 2013.

[21] M. Jalobeanu, G. Shirakyan, G. Parent, H. Kikkeri, B. Peasley, and A. Feniello, "Reliable kinect-based navigation in large indoor environments," in *Proceedings of the 2015 IEEE International Conference on Robotics and Automation, (ICRA '15)*, pp. 495–502, USA, May 2015.

[22] P. De La Puente, M. Bajones, Reuther. C., D. Fischinger, Wolf. D., and M. Vincze, "Experiences with rgb-d based navigation in real home robotic trials," in *Austrian Robotics Workshop (ARW)*, 2016.

[23] M. Shimrat, "Algorithm 112: Position of point relative to polygon," *Communications of the ACM*, vol. 5, no. 8, p. 434, 1962.

[24] M. L. Walters, K. Dautenhahn, S. N. Woods, and K. L. Koay, "Robotic etiquette: Results from user studies involving a fetch and carry task," in *Proceedings of the HRI 2007: 2007 ACM/IEEE Conference on Human-Robot Interaction - Robot as Team Member*, pp. 317–324, USA, March 2007.

[25] P. Mayer, C. Beck, and P. Panek, "Examples of multimodal user interfaces for socially assistive robots in ambient assisted living environments," in *Proceedings of the 3rd IEEE International Conference on Cognitive Infocommunications, (CogInfoCom '12)*, pp. 401–406, Slovakia, December 2012.

[26] M. Wölfel and J. McDonough, *Distant Speech Recognition*, John Wiley & Sons, 2009.

[27] P. Panek and P. Mayer, "Challenges in adopting speech control for assistive robots," in *Ambient Assisted Living*, Advanced Technologies and Societal Change, pp. 3–14, Springer International Publishing, Berlin, Germany, 2015.

[28] D. I. Kosmopoulos, K. Papoutsakis, and A. A. Argyros, "Online segmentation and classification of modeled actions performed in the context of unmodeled ones," in *Proceedings of the 25th British Machine Vision Conference, (BMVC '14)*, pp. 1–12, September 2014.

[29] D. Michel and K. Papoutsakis, "Gesture Recognition Apparatuses," *Methods and Systems for Human-Machine Interaction*, 2016.

[30] M. Foukarakis, I. Adami, D. Ioannidi et al., "A robot-based application for physical exercise training," in *Proceedings of the 2nd International Conference on Information and Communication Technologies for Ageing Well and e-Health, (ICT4AWE '16)*, pp. 45–52, April 2016.

[31] D. Michel and A. Argyros, "Apparatuses, methods and systems for recovering a 3-dimensional skeletal model of the human body, 2016".

[32] P. Viola and M. J. Jones, "Robust real-time face detection," *International Journal of Computer Vision*, vol. 57, no. 2, pp. 137–154, 2004.

[33] P. Panteleris and A. A. Argyros, "Vision-based SLAM and moving objects tracking for the perceptual support of a smart walker platform," *Computer Vision-ECCV 2014 Workshops*, vol. 8927, pp. 407–423, 2014.

[34] A. Remazeilles, C. Leroux, and G. Chalubert, "SAM: A robotic butler for handicapped people," in *Proceedings of the 17th IEEE International Symposium on Robot and Human Interactive Communication, RO-MAN*, pp. 315–321, Germany, August 2008.

[35] S. S. Srinivasa, D. Ferguson, C. J. Helfrich et al., "HERB: A home exploring robotic butler," *Autonomous Robots*, vol. 28, no. 1, pp. 5–20, 2010.

[36] B. Graf, U. Reiser, M. Hägele, K. Mauz, and P. Klein, "Robotic home assistant care-O-bot 3—product vision and innovation platform," in *Proceedings of the IEEE Workshop on Advanced Robotics and Its Social Impacts (ARSO '09)*, pp. 139–144, Tokyo, Japan, November 2009.

[37] D. Fischinger and M. Vincze, "Empty the basket - A shape based learning approach for grasping piles of unknown objects," in *Proceedings of the 25th IEEE/RSJ International Conference on Robotics and Intelligent Systems, (IROS '12)*, pp. 2051–2057, Portugal, October 2012.

[38] D. Fischinger and M. Vincze, "Shape based learning for grasping novel objects in cluttered scenes," in *Proceedings of the 10th IFAC Symposium on Robot Control, (SYROCO '12)*, pp. 787–792, Croatia, September 2012.

[39] D. Fischinger, A. Weiss, and M. Vincze, "Learning grasps with topographic features," *International Journal of Robotics Research*, vol. 34, no. 9, pp. 1167–1194, 2015.

[40] M. Bajones, A. Weiss, and M. Vincze, "Log data analysis of long-term household trials: Lessons learned and pitfalls. 2016. Workshop on Challenges and best practices to study HRI in natural interaction settings," in *Proceeding of the International Conference on Human-Robot Interaction*, 2016.

[41] J. Pripfl, T. Körtner, D. Batko-Klein et al., "Results of a real world trial with a mobile social service robot for older adults," in *Proceedings of the 11th Annual ACM/IEEE International*

Conference on Human-Robot Interaction, (HRI '16), pp. 497-498, IEEE Press, New Zealand, March 2016.

[42] F. Palumbo, D. La Rosa, E. Ferro et al., "Reliability and human factors in ambient assisted living environments," *Journal of Reliable Intelligent Environments*, vol. 3, no. 3, pp. 139–157, 2017.

[43] M. Vacher, F. Portet, A. Fleury, and N. Noury, "Development of audio sensing technology for ambient assisted living: Applications and challenges," *International Journal of E-Health and Medical Communications*, vol. 2, no. 1, pp. 35–54, 2011.

2

Hierarchical Sliding Mode Algorithm for Athlete Robot Walking

Van Dong Hai Nguyen,[1] Xuan-Dung Huynh,[1] Minh-Tam Nguyen,[1] Ionel Cristian Vladu,[2] and Mircea Ivanescu[2]

[1] *Ho Chi Minh City University of Technology and Education (HCMUTE), 01 Vo Van Ngan Street, Thu Duc District, Ho Chi Minh City, Vietnam*
[2] *University of Craiova (UCV), 13 Alexandru Ioan Cuza Street, 200585 Craiova, Romania*

Correspondence should be addressed to Van Dong Hai Nguyen; hainvd@hcmute.edu.vn

Academic Editor: Yangmin Li

Dynamic equations and the control law for a class of robots with elastic underactuated MIMO system of legs, athlete Robot, are discussed in this paper. The dynamic equations are determined by Euler-Lagrange method. A new method based on hierarchical sliding mode for controlling postures is also introduced. Genetic algorithm is applied to design the oscillator for robot motion. Then, a hierarchical sliding mode controller is implemented to control basic posture of athlete robot stepping. Successful simulation results show the motion of athlete robot.

1. Introduction

Two-legged robot is an interesting topic which lasts for long time [1–3]. One of the classical forms of this type, humanoid robot, which only has solid links, can move or balance through ZMP method [4–6]. Anyway, this method is only appropriate with MIMO system which has the same number of inputs and outputs. One of the other disadvantages of ZMP is that the robot moves very slowly and unflexibly. In order to improve the flexibility of robot and energy saving capabilities, some authors [7–13] put elastic components for smooth dynamic motion. By these researches [7–13], robot is flexible when the feet of robot still exist. One approach suggested [14, 15] is to replace the leg and foot by an elastic leg. This approach makes the robot become the underactuated MIMO system, called athlete robot (AR). The AR robot study is based on the idea of elastic legs for disabled people (Figures 1 and 2). In the case of AR, solid legs are replaced by elastic legs and two torques are substituted by one torque (Figure 3). Ryuma Niiyama and his colleagues studied this class of robots but their efforts were focused on the experimental results and real biomechanical structure. The dynamic equations and control algorithm were not analyzed. This paper will present the dynamic equation of AR.

Classical methods of ZMP control for humanoid robot [4–6] become useless for ARs due to structure of AR as a MIMO underactuated system. Some researches [16–18] simplified the complicated structure of two-legged robot into simple form: spring-load inverted pendulum (SLIP). Anyway, the SLIP model is not completely equivalent to the former MIMO model. Hence, the controller which is based on the former complicated model will be more reliable. Beside the opinion of control methods for MIMO underactuated nonlinear system [19], this paper also presents new idea that three proportional-controllers (P-controllers) are designed to transform AR to a SIMO system. Hence, it will be more convenient to use multiple control methods for SIMO underactuated system [20–23]. A basic method of hierarchical sliding mode (HSM) control algorithm for SIMO system, which was presented by Qian et al. [20], can balance a SIMO system. Other related works [21–23] used HSM to balance a specific SIMO system. Qian used mathematical methods to prove the stability of sliding surfaces of each layer in [20]. Despite the remarkable contribution in [20], there is still a boundary that control parameters "should" be in. Therefore, based on that boundary, genetic algorithm (GA) is used in our paper to find the appropriate control parameters.

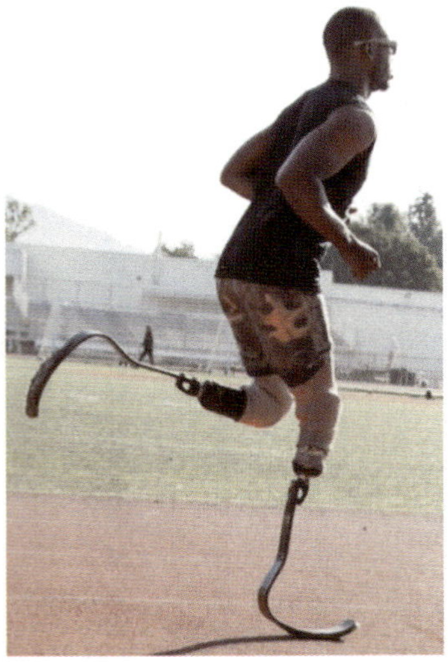

FIGURE 1: Motion of disabled people with elastic legs.

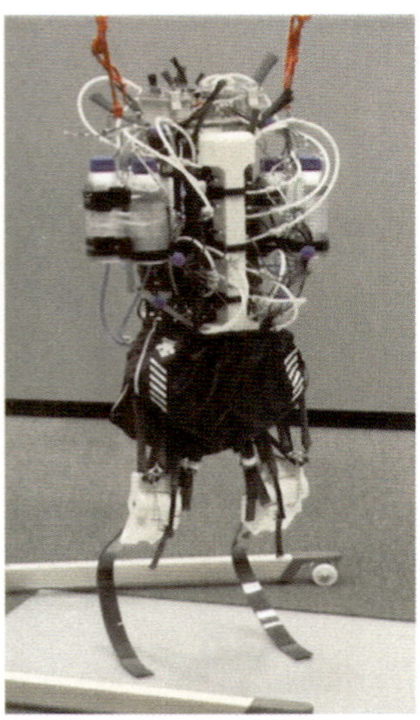

FIGURE 2: AR, University of Tokyo, Japan [14].

Also, GA is also used to design the prescribed trajectory of motion that defines a step of robot.

The paper concludes five sections. The dynamic equations of AR are generated in Section 2. Section 3 infers the mathematical transformation from MIMO underactuated nonlinear structure of AR into a SIMO one. Section 3 also presents application of HSM controller for that robot. Section 4 introduces simulation results. A conclusion in Section 5 ends the paper.

2. Mathematical Model

The elastic legs are designed to be able to self-balance when there is not external force on them (Figure 4) and to accumulate elastic energy for motion. We introduce the following notations:

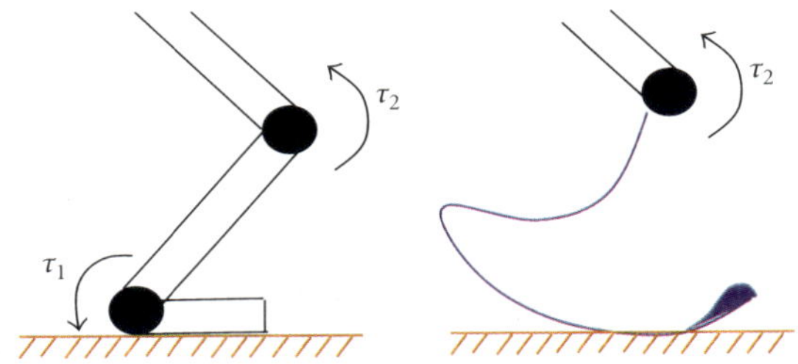

FIGURE 3: Correlation between leg of solid robot and elastic leg of AR.

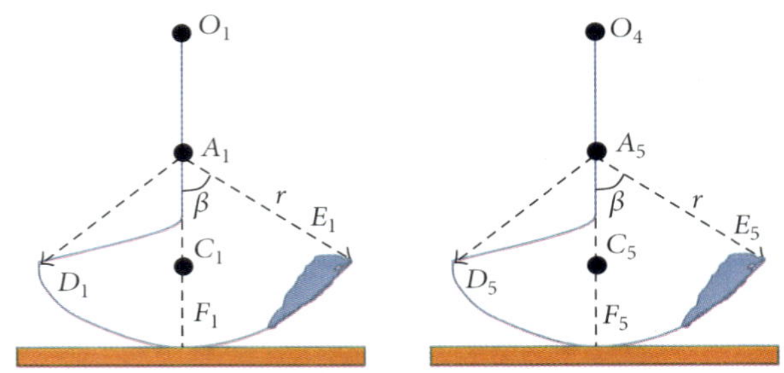

FIGURE 4: Self-balanced position of elastic legs.

(i) C_i: mass center of link i $(i = 1, 5)$

(ii) O_i: (with $(i = 1, 2, 3, 4)$) connecting point of link i and link $i + 1$

(iii) A_i: center of the curved part of elastic leg $(i = 1, 5)$

(iv) D_i, E_i: two-edge points of elastic legs with coordinates: (x_{Di}, y_{Di}) and (x_{Ei}, y_{Ei}) $(i = 1, 5)$

(v) F_i: middle point of elastic legs with coordinate: (x_{Fi}, y_{Fi}) $(i = 1, 5)$

(vi) $\beta = \widehat{E_i A_i C_i}$: angle of curved part of elastic legs

Main parameters of AR are described in Figure 5 as follows:

(i) m_i: mass of i $(i = 1, 2, 3, 4, 5)$; $m_1 = m_5$; $m_2 = m_4$ (kg)

(ii) l_i: length of link i $(i = 2, 3, 4)$; $l_2 = l_4$; $l_3 = 0$ (m)

(iii) α_1: angle between link 1 and vertical axis, obtaining $\alpha_1 = \widehat{C_1 O_1 y'}$ (rad)

(iv) α_i: angle between next link and link i $(i = 2, 3, 4, 5)$ (rad)

(v) I_i: inertial moment of link i (kgm2)

(vi) r: radius of hunch part (m)

(vii) k: rotational spring coefficient (Nm/rad)

Behavior of elastic leg is determined by Castigliano's Theorem [24] which provides a good tool for analyzing forces on curved components. AR can be regarded as an equivalent inverted pendulum in Figure 6 where the elastic legs are equivalent to springs.

Consider Figure 6(b) that describes the equivalent model of AR's leg. The strain potential energy of system can be defined as

$$P_{\text{strain}} = \frac{1}{2} k \alpha_1^2. \tag{1}$$

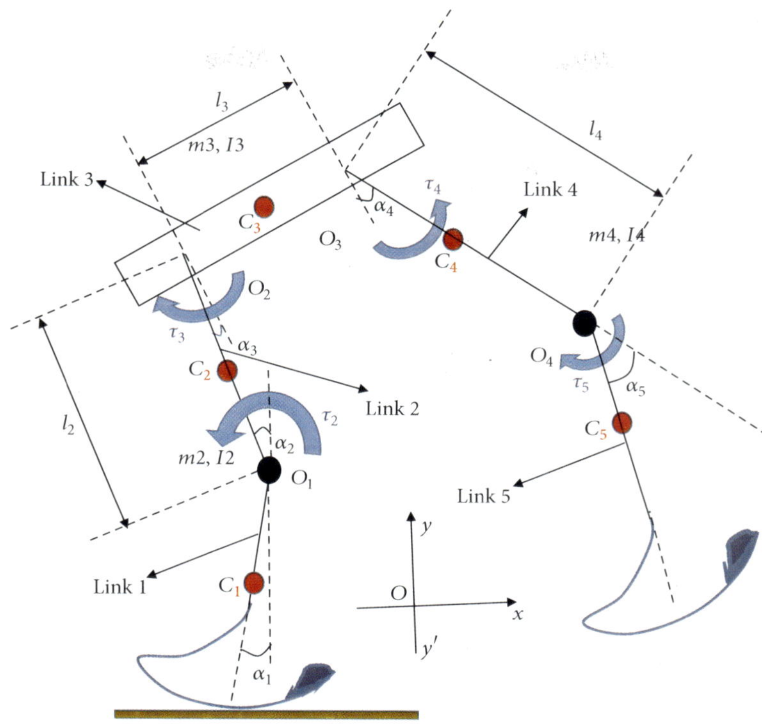

FIGURE 5: Mathematical model of AR.

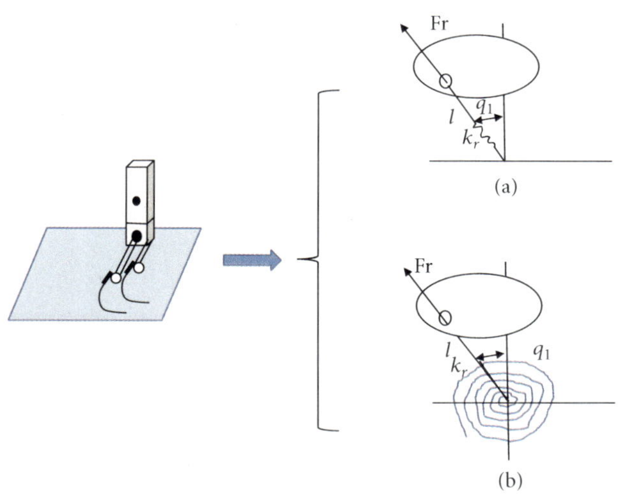

FIGURE 6: Equivalent pendulum model: (a) as linear spring and (b) as rotational spring.

Total potential energy of system is

$$P = g\sum_{i=1}^{5} m_i y_{Ci} + P_{\text{strain}}. \tag{2}$$

Kinetic energy of system is

$$K = \sum_{i=1}^{5}\left[\frac{m_i}{2}\left(\dot{x}_{Ci}^2 + \dot{y}_{Ci}^2\right) + \frac{I_i}{2}\dot{\alpha}_i^2\right]. \tag{3}$$

Lagrange operator is

$$L = K - P. \tag{4}$$

By using Euler-Lagrange method, dynamic equations will be

$$\begin{aligned} \frac{\partial L}{\partial \alpha_i} - \frac{d}{dt}\left(\frac{\partial L}{\partial \dot{\alpha}_i}\right) &= \tau_i \quad (i = 2, 3, \ldots, 5) \\ \frac{\partial L}{\partial \alpha_1} - \frac{d}{dt}\left(\frac{\partial L}{\partial \dot{\alpha}_1}\right) &= 0. \end{aligned} \tag{5}$$

From (5), the dynamic model can be written as

$$J(\alpha_i)\ddot{\alpha} = F(\alpha_i, \dot{\alpha}_i) + \tau, \tag{6}$$

where $\alpha = [\alpha_1 \ \alpha_2 \ \cdots \ \alpha_5]^T$, $\tau = [0 \ \tau_2 \ \tau_3 \ \tau_4 \ \tau_5]^T$, and the matrices J, F are calculated by MATLAB/Simulink simulation. New variables q_i are defined to simplify forms of equation (Figure 7).

Relation of variables α_i and q_i is inferred in

$$\begin{aligned} \alpha_1 &= q_1; \\ \alpha_2 &= q_1 - q_2; \\ \alpha_3 &= q_2 + q_3; \\ \alpha_4 &= -q_3 - q_4; \\ \alpha_5 &= q_5 - q_4 \end{aligned}$$

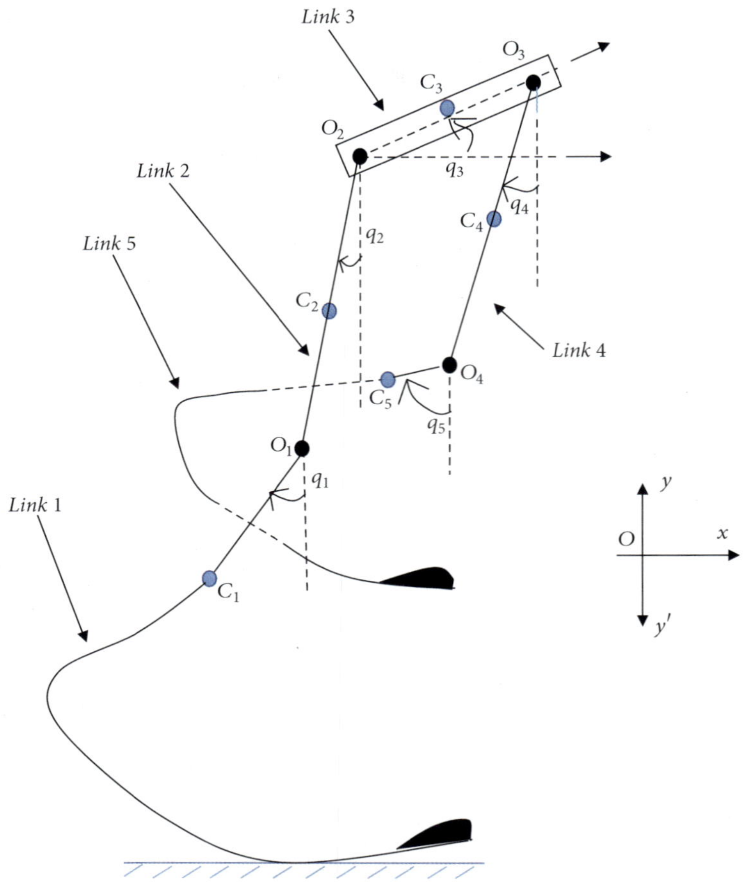

FIGURE 7: Model of AR with variables are referenced to the vertical axis.

$$q_1 = \alpha_1;$$
$$q_2 = q_1 - \alpha_2;$$
$$q_3 = \alpha_3 - q_2;$$
$$q_4 = -q_3 - \alpha_4;$$
$$q_5 = q_4 + \alpha_5. \quad (7)$$

Also, denote by α_{id} and q_{id} the reference signal of variables α_i and q_i. Consider $q_0 = \text{const} > 0$ the angle between one leg and the vertical axis when both legs touch the ground. Sample-time T is defined as period time of a step of robot

Many authors [25–27] proposed several solutions for designing the oscillators for the motion of two-legged walking robots. Their methods are very complex and based on intuition and developed by simulations and experiments. In this paper, an algorithm is proposed based on GA.

We consider the reference trajectories of q_1, q_2, q_3, q_4, and q_5 as $q_{1d} = q_0 = \text{const}$, $q_{2d} = q_0 = \text{const}$, $q_{3d} = 0$, q_{4d}, q_{5d}. Reference trajectories of q_4 and q_5 are described in Figure 9.

Coordinates (x_{Aj}, y_{Aj}) and (x_{Bj}, y_{Bj}) have to be selected through GA.

3. Control Algorithm

In order to use HSC algorithms, new variables will be defined as

$$e_i = \alpha_i - \alpha_{id} \quad (i = 1, 2, \ldots, 5), \quad (8)$$

where $\alpha_d = [\alpha_{1d} \ \alpha_{2d} \ \cdots \ \alpha_{5d}]^T$, $e = [e_1 \ e_2 \ \cdots \ e_5]^T$.

The dynamic model (6) is rewritten as

$$\ddot{e} = J^{-1}F + J^{-1}[0 \ 0 \ \tau_3 \ \tau_4 \ \tau_5]^T - \ddot{\alpha}_d + J^{-1}[0 \ \tau_2 \ 0 \ 0 \ 0]^T. \quad (9)$$

A simply proportional controller for links 3, 4, and 5 is proposed

$$\tau_i = K_{pi}e_i \quad (i = 3, 4, 5). \quad (10)$$

Also, define

$$f = [f_1 \ f_2 \ \cdots \ f_5]^T = J^{-1}F + J^{-1}[0 \ 0 \ \tau_3 \ \tau_4 \ \tau_5]^T - \ddot{\alpha}_d, \quad (11)$$
$$b = [b_1 \ b_2 \ \cdots \ b_5]^T,$$

where b is the second column of matrix J^{-1}.

Substitute (10) and (11) into (9), and obtain

$$\ddot{e} = f + b\tau_2. \tag{12}$$

By using the conventional change of control inputs (10), the dynamic model (6) is transformed in SIMO models in (12). The AR is a five-order system and it is impossible to calculate directly J^{-1} with just unknown variables due to its complexity and limitation of simulating software (MATLAB/Simulink) (only 25000 characters can appear on MATLAB window). Hence, exact dynamic equations cannot be described visibly. Anyway, simulation process has to be implemented following Figure 10.

From (8) to (12), the error model will be

$$\ddot{e}_i = f_i + b_i\tau_2 \quad (i = 1, 2, \ldots, 5). \tag{13}$$

Equation (13) has the form of equation of a SIMO system. Therefore, a solution of using controller which is suitable for high-order SIMO system can be considered. This controller stabilizes variables $e_i \xrightarrow{t\to\infty} 0$. This leads to $\alpha_i \xrightarrow{t\to\infty} \alpha_{id}$.

Hierarchical sliding mode of structure of hierarchical sliding surfaces is shown in Figure 10.

From the description in Figure 11, sliding surfaces are chosen

$$s_i = c_i e_i + \dot{e}_i \quad (i = 1, 2, \ldots, 5) \tag{14}$$

$$S_i = a_{i-1} S_{i-1} + s_i \quad (i = 1, 2, \ldots, 5), \tag{15}$$

where $a_{i-1} = \text{const}$ and $a_0 = S_0 = 0$.

$$S_i = \sum_{r=1}^{i} \left(\prod_{j=r}^{i} a_j \right) s_r \quad (i = 1, 2, \ldots, 5). \tag{16}$$

Derivative (16) with respect to time yields

$$\dot{S}_i = \sum_{r=1}^{i} \left(\prod_{j=r}^{i} a_j \right) \dot{s}_r \quad (i = 1, 2, \ldots, 5). \tag{17}$$

From (14) and (15), we deduce that the ith layer SMC comprises the information of $(i-1)$th, $(i-2)$th, ..., 1st layer, and subsystem layer. Hence, define the ith layer sliding mode control law as

$$u_i = u_{i-1} + u_{\text{eq}i} + u_{\text{sw}i} \quad (i = 1, 2, \ldots, 5). \tag{18}$$

Here, $\tau_0 = 0$. And $u_{\text{sw}i}$, $u_{\text{eq}i}$ is switching and equivalent control law for ith layer. Let

$$u_{\text{eq}i} = -\frac{c_i \dot{e}_i + f_i}{b_i} \quad (i = 1, 2, \ldots, 5) \tag{19}$$

Define Lyapunov function for ith layer as

$$V_i = \frac{S_i^2}{2} \quad (i = 1, 2, \ldots, 5) \tag{20}$$

Derive (20), and from (17), we obtain

$$\left(\dot{V}_i =\right) S_i \dot{S}_i = S_i \left[\sum_{r=1}^{5} \left(\prod_{j=r}^{5} a_j \right) \dot{s}_r \right] \quad (i = 1, 2, \ldots, 5). \tag{21}$$

Derive (14), and substitute (14) and (18) into new results which is obtained by deriving (14), which yields

$$\begin{aligned} \dot{S}_i &= \sum_{l=1}^{i} \left[\sum_{\substack{r=1 \\ r\neq l}}^{i} \left(\prod_{j=r}^{i} a_j \right) b_r \right] u_{\text{eq}l} \\ &\quad + \sum_{l=1}^{i} \left[\sum_{r=1}^{i} \left(\prod_{j=r}^{i} a_j \right) b_r \right] u_{\text{sw}l} \quad (i = 1, 2, \ldots, 5). \end{aligned} \tag{22}$$

By considering stability of ith layer sliding surface, let

$$\dot{S}_i = -k_i S_i - \eta_i \operatorname{sgn} S_i \quad (\text{where } k_i > 0,\ \eta_i > 0) \tag{23}$$

From (22) and (23), switching control law of ith layer can be obtained as

$$\begin{aligned} u_{\text{sw}i} &= -\sum_{l=1}^{i-1} u_{\text{sw}l} \\ &\quad - \frac{\left[\sum_{l=1}^{i=1} \left[\sum_{r=1,r\neq l}^{i} \left(\prod_{j=r}^{i} a_j \right) b_r \right] u_{\text{eq}l} - \left(k_i S_i + \eta_i \operatorname{sgn} S_i \right) \right]}{\sum_{r=1}^{i} \left(\prod_{j=r}^{i} a_j \right) b_r} \\ &\quad (i = 1, 2, \ldots, 5), \end{aligned} \tag{24}$$

and the final control law for controller is selected as

$$\begin{aligned} \tau_2 &= u_5 \\ &= \frac{\left[\sum_{r=1}^{5} \left(\prod_{j=r}^{5} a_j \right) b_r u_{\text{eq}r} - \left(k_5 S_5 + \eta_5 \operatorname{sgn} S_5 \right) \right]}{\sum_{r=1}^{5} \left(\prod_{j=r}^{5} a_j \right) b_r}. \end{aligned} \tag{25}$$

From general results in [20], we obtain Theorems 1 and 2 below.

Theorem 1. *Consider that equations of error of system (9) are described in (13). If control law is chosen as (25) and ith sliding surfaces are identified as in (16), then S_i is asymptotically stabilized.*

Proof. The Lyapunov function of ith layer is chosen as in (20). From (23), we obtain

$$\dot{V}_i = S_i \dot{S}_i = S_i \left(-k_i S_i - \eta_i \operatorname{sgn} S_i \right) \quad (i = 1, 2, \ldots, 5). \tag{26}$$

Integrating two sides of formula (21), we obtain

$$\int_0^t \dot{V}_i d\tau = \int_0^t \left(-k_i S_i^2 - \eta_i \left| S_i \right| \right) d\tau \quad (i = 1, 2, \ldots, 5). \tag{27}$$

Consider

$$\begin{aligned} V_i(0) &= V_i(t) + \int_0^t \left(k_i S_i^2 + \eta_i \left| S_i \right| \right) d\tau \\ &\geq \int_0^t \left(k_i S_i^2 + \eta_i \left| S_i \right| \right) d\tau \quad (i = 1, 2, \ldots, 5). \end{aligned} \tag{28}$$

Therefore, we obtain

$$\begin{aligned} &\lim_{t\to\infty} \int_0^t \left(k_i S_i^2 + \eta_i \left| S_i \right| \right) d\tau \leq V_i(0) < \infty \\ &\quad (i = 1, 2, \ldots, 5). \end{aligned} \tag{29}$$

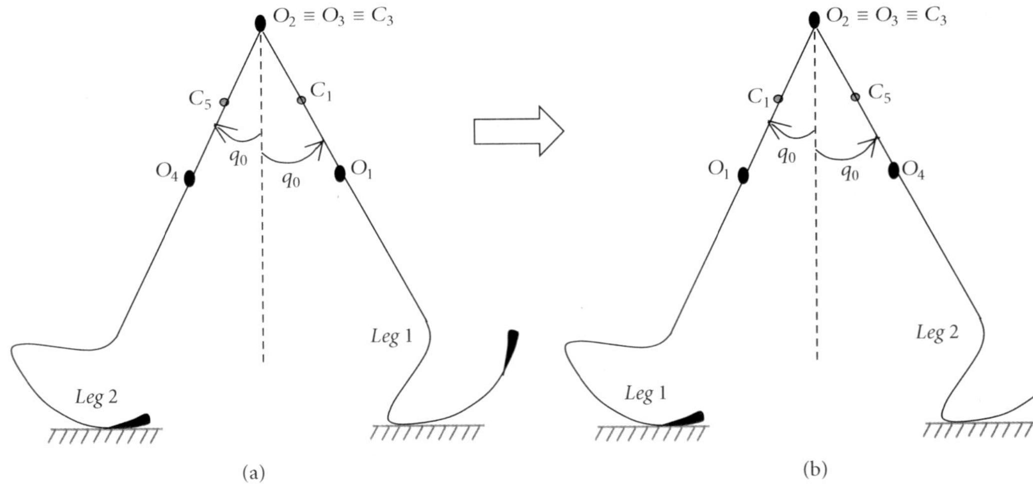

FIGURE 8: Position of two legs. (a) Before the step and (b) after the step.

According to Barbalat lemma, there exists

$$\lim_{t\to\infty} \left(k_i S_i^2 + \eta_i \left|S_i\right|\right) d\tau \le V_i(0) < \infty \quad (i = 1, 2, \ldots, 5). \tag{30}$$

From (30), it means that $\lim_{t\to\infty} S_i = 0$. Hence, the ith sliding surface of S_i is asymptotically stable.

Theorem 2. *Consider that equations of error of system (9) are described in (13). If control law is chosen as (25) and ith subsystem sliding surfaces are identified as in (14), then s_i is asymptotically stabilized.*

Proof. Assume s_i is not asymptotically stabilized when conditions in Theorem 2 are satisfied

$$\lim_{t\to\infty} s_i \ne 0 \quad (i = 1, 2, \ldots, 5). \tag{31}$$

Hence, from (16), we obtain

$$\lim_{t\to\infty} S_i = \lim_{t\to\infty} \sum_{r=1}^{i} \left(\prod_{j=r}^{i} a_j\right) s_r = \sum_{r=1}^{i} \left(\prod_{j=r}^{i} a_j\right) \lim_{t\to\infty} s_r \ne 0 \quad (i = 1, 2, \ldots, 5). \tag{32}$$

Equation (32) contradicts Theorem 1. Therefore, we obtain

$$\lim_{t\to\infty} s_i = 0 \quad (i = 1, 2, \ldots, 5) \tag{33}$$

and s_i is asymptotically stabilized.

Theorems 1 and 2 prove the stability of sliding surface. However, the stability of e_i is not guaranteed. Therefore, searching algorithm, such as genetic algorithm (GA), is a solution to find appropriate control parameters in (25). In this case, GA is defined by population of 40 individuals. Each individual contains 24 chromosomes which include values of the following:

(i) Control parameters: $a_1, a_2, a_3, a_4, a_5, c_1, c_2, c_3, c_4, c_5, \eta_5, k_5, K_{p1}, K_{p2}, K_{p3}$

(ii) Trajectory parameters: $x_{A4}, x_{A5}, x_{B4}, x_{B5}, y_{A4}, y_{A5}, y_{B4}, y_{B5}$

(iii) Half-period time of a motion cycle (time for a step of AR): T (if AR moves fast, then T is chosen in the range: $0\,(s) \le T \le 2\,(s)$)

The individuals chosen for crossover process have to satisfy the following conditions:

(i) All y_{Di}, y_{Ei}, and y_{Fi} $(i = 1, 5)$ are positive for $t = 0 \to T$.

(ii) $0 \le x_{Ai} \le x_{Bi} \le T$ $(i = 4, 5)$.

Fitness function of GA is defined as

$$J = \frac{1}{T} \sum_{i=1}^{5} \zeta_i^2 \quad (i = 1, 2, \ldots, 5), \tag{34}$$

where $\zeta_i = q_i - q_{di}$.

In GA process, the fitness function of GA is designed to evaluate the results after each loop of searching. If each link tracks trajectory, then $\zeta_i \to 0$ and $\sum_{i=1}^{5} \zeta_i^2 \to 0$. However, sample time of step T is also chosen through GA. Only $\sum_{i=1}^{5} \zeta_i^2$ cannot represent the ability of tracking trajectory in a period of time. Therefore, fitness function should be selected as $J = (1/T) \sum_{i=1}^{5} \zeta_i^2$. The detected result which generates smaller J from (34) will be considered as better.

The GA process is shown in Figure 12.

In Figure 12, firstly, 40 random couples were created. In each individual in couples, the structure of gene concludes encoded parameters of control system and reference trajectories (control parameters: $a_1, a_2, a_3, a_4, a_5, c_1, c_2, c_3, c_4, c_5, \eta_5, k_5, K_{p1}, K_{p2}, K_{p3}, x_{A4}, x_{A5}, x_{B4}, x_{B5}, y_{A4}, y_{A5}, y_{B4}, y_{B5}$). We denote a constant value $J_{\min}$ to check if the result is best or not. $J_{\min}$ will be updated if there is better value and the result that makes $J < J_{\min}$ will be stored as the best result. After

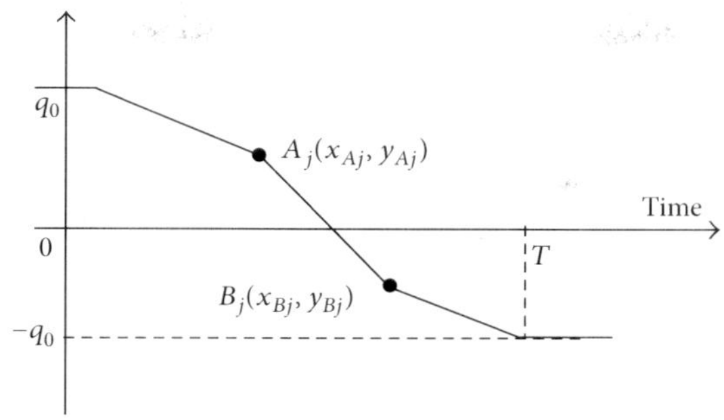

FIGURE 9: Reference trajectories q_{jd} ($j = 4, 5$).

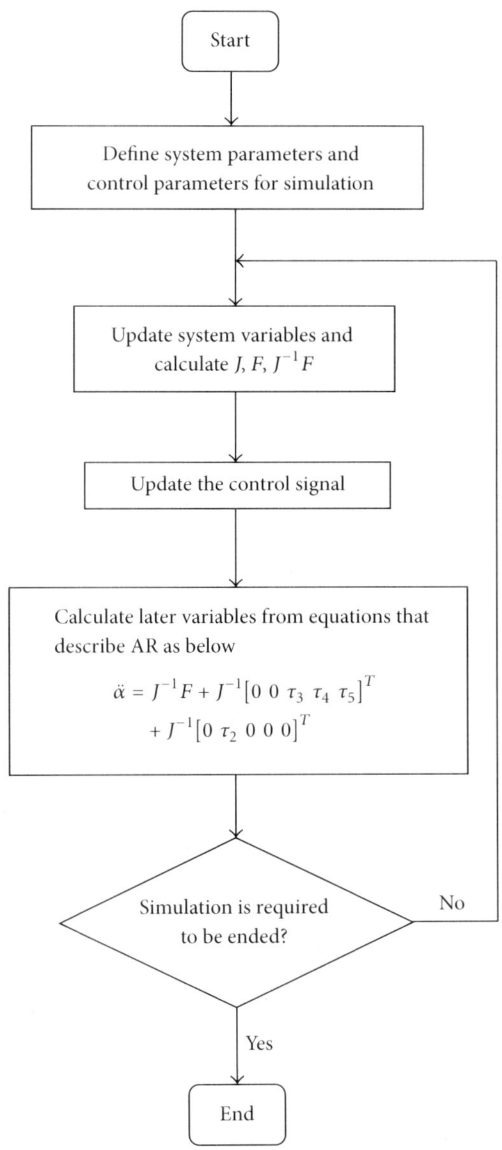

FIGURE 10: Simulation process.

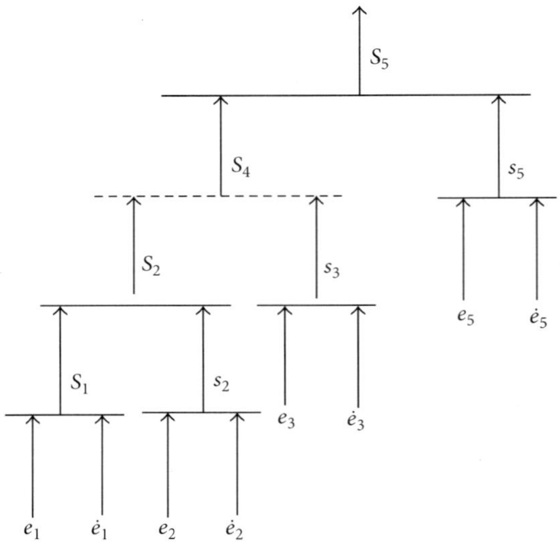

FIGURE 11: Hierarchical sliding surfaces structure for system which has one control input and five output variables.

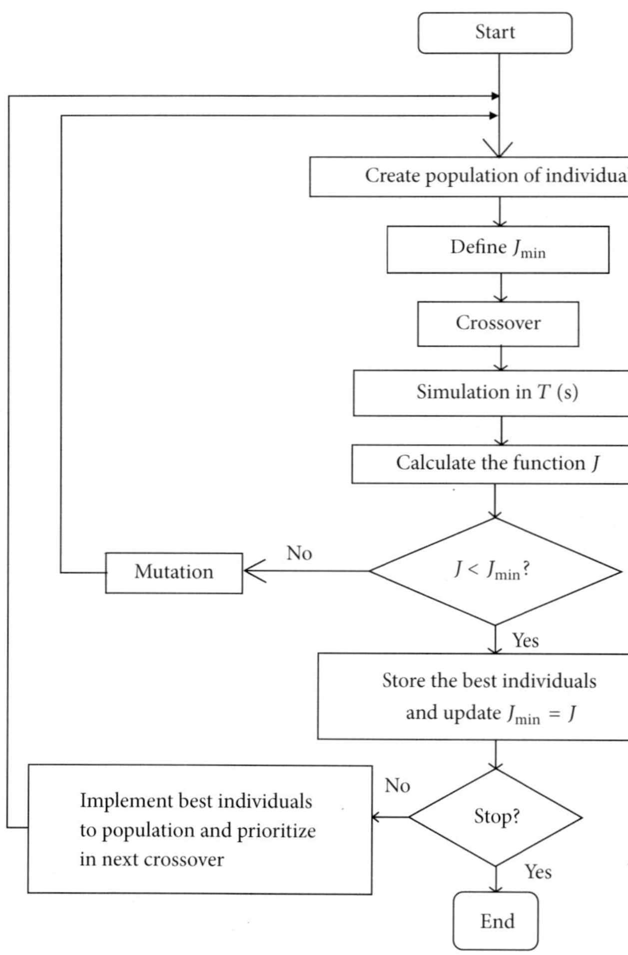

FIGURE 12: Flowchart of genetic calculation process.

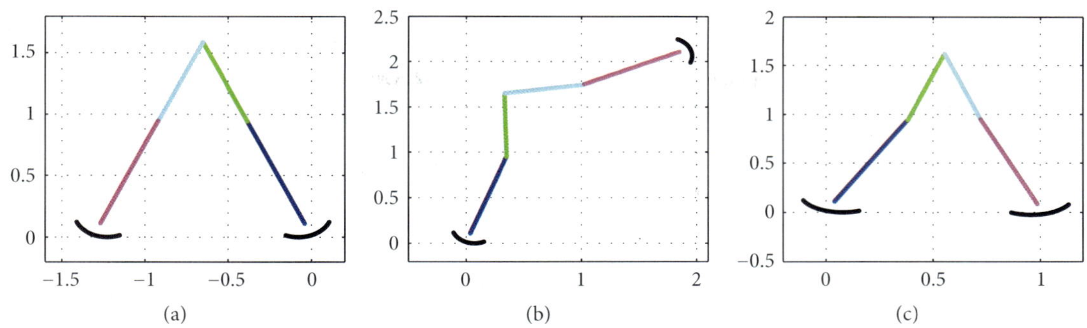

FIGURE 13: AR at representative time (a) $t = 0$ (s), (b) $t = T = 1.01$ (s), and (c) $t = 1.12$ (s). Number 1; curved part of elastic part of leg 2; number 2: linear part of elastic part of leg 2; number 3: link 4; number 4: link 2; number 5: linear part of elastic part of leg 1; number 6: curved part of elastic part of leg 1.

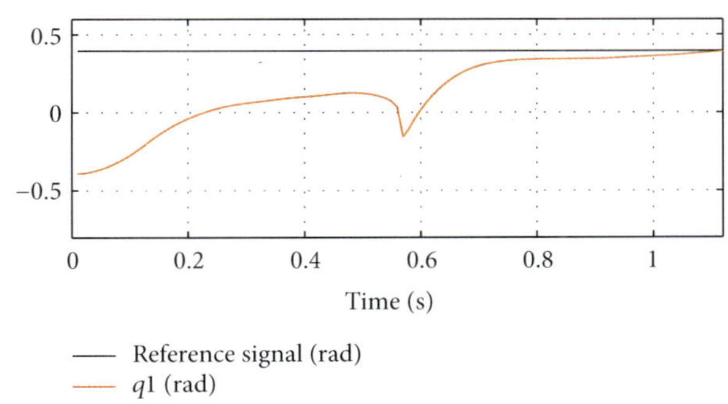

FIGURE 14: Trajectory of q_1 (rad).

crossover couples, 80 new individuals are created. By checking the function J of each individual after each simulation, better individuals have more priority to become couples and have more offspring in next generation. Worse individuals have less priority to give their gene to next generation. The loop will continue until we want to stop the program.

4. Simulation Results

System parameters are selected as follows:

$\gamma_1 = 0.1$ (m); $\gamma_2 = 1$ (m); $r = 0.2$ (m); $\gamma_3 = r - \gamma_1$; $\gamma_4 = \gamma_2 - r$; $m_1 = 0.01$ (kg); $m_2 = 0.1$ (kg); $m_3 = 0.5$ (kg); $I_1 = 0.001$ (kgm^2); $I_2 = 0.02$ (kgm^2); $I_3 = 0.1$ (kgm^2); $l_3 = 0$ (m); $l_2 = 0.7$ (m); $g = 9.81$ (N/kgm^2); $k = 0.2$ (N/kgm^2); $T = 1.01$ (s); $q_0 = \pi/8$ (rad).

Parameters of HSM controller in (25) are found by GA method as follows:

$a_1 = 9.4$; $a_2 = 6.94$; $a_3 = 4.77$; $a_4 = 1.67$; $a_5 = 1.37$; $c_1 = 6.9$; $c_2 = 9.43$; $c_3 = 3.02$; $c_4 = 9.94$; $c_5 = 4.18$; $\eta_5 = 6.88$; $k_5 = 9.83$; $K_{p1} = 76.5$; $K_{p2} = 43.3$; $K_{p3} = 81.4$.

Parameters of reference trajectories in Figure 9 are determined as follows:

$x_{A4} = 0.1213$ (m); $y_{A4} = -0.3456$ (rad); $x_{B4} = 0.2526$ (m); $y_{B4} = -0.5341$ (rad); $x_{A5} = 0.3536$ (m); $y_{A5} = 1.5080$ (rad); $x_{B5} = 0.9394$ (m); $y_{B5} = -1.4923$ (rad).

Initial values of variables at $t = 0$ (s) are specified in Figure 8. Motions of AR under HSM controller are described in Figures 13 and 19.

From initial position (Figure 13(a)), after a period time T, AR moves to a new position (Figure 13(b)). An additional time $t_{\text{add}} = 0.11$ (s) is needed to achieve the exact position of a final sequence. This additional time can be improved by GA techniques. Period time of a step which is T will be $T + t_{\text{add}}$. Trajectories of each link variables q_i are shown in Figures 14–18.

The walking motion of AR for a single step is shown in Figure 19.

From Figures 14 to 18, after 1.12 s, AR finishes a step. However, in the operating process, q_i does not track well the reference trajectories but the final positions are still close to the reference trajectories. Figures 13 and 19 also consolidate that reasoning. The motion of each link affects others significantly. Only motion of link 5 is less affected (Figure 24). Because links 3, 4, and 5 are controlled directly by P-controllers for each link (from (10)), they follow trajectories (Figures 22–24) more closely than link 1 or link 2 does. Anyway, the vibration exists in these figures due to the effect from other links. The response of link 1 and link 2 is not good

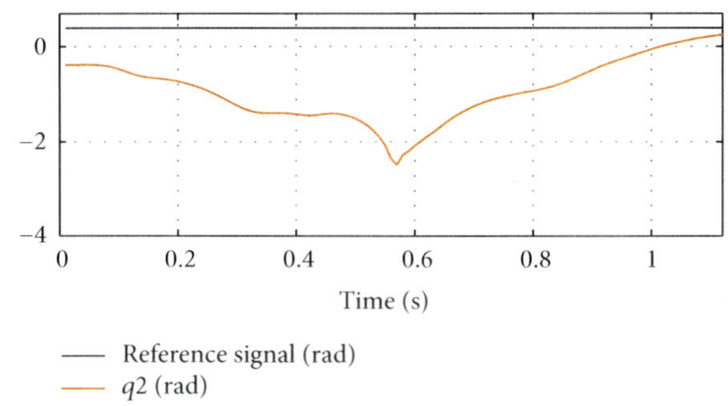

Figure 15: Trajectory of q_2 (rad).

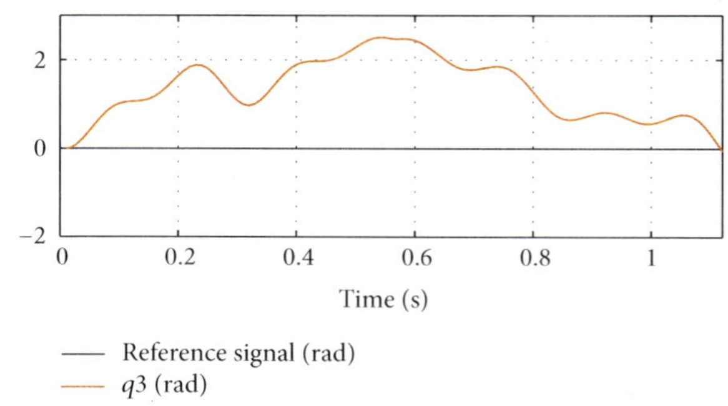

Figure 16: Trajectory of q_3 (rad).

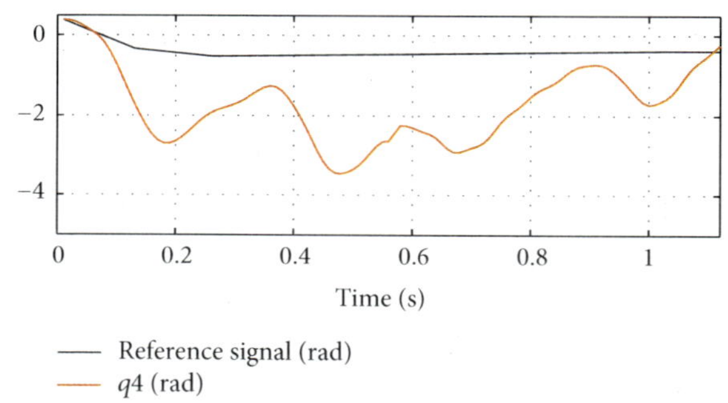

Figure 17: Trajectory of q_4 (rad).

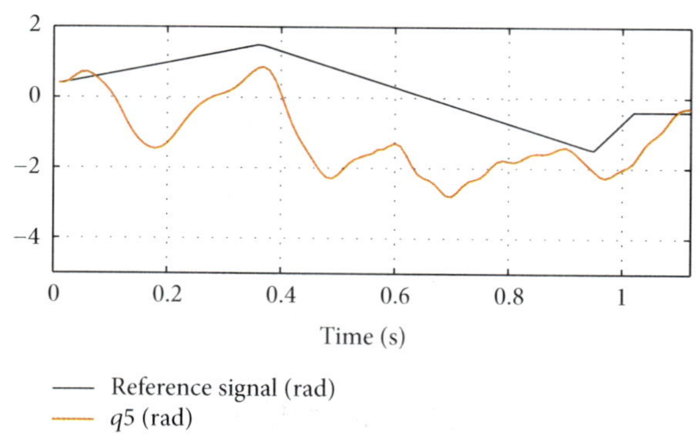

Figure 18: Trajectory of q_5 (rad).

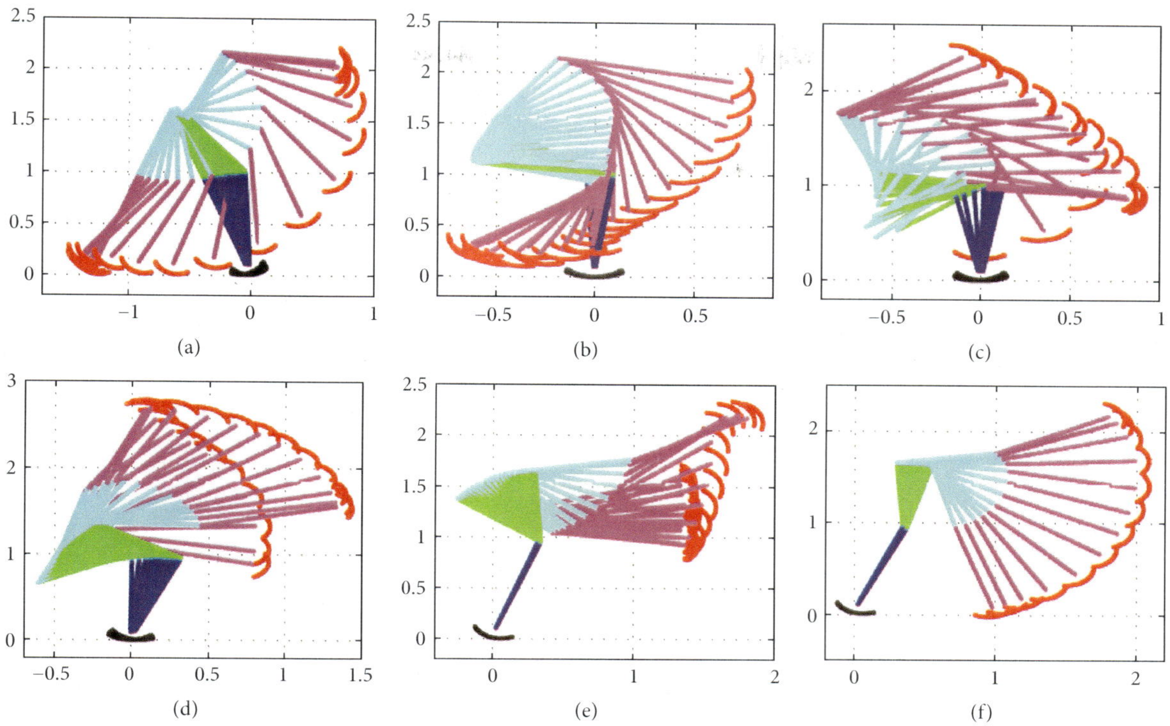

FIGURE 19: Motion description of AR. (a) From 0 s to 0.2 s. (b) From 0.2 s to 0.4 s. (c) From 0.4 s to 0.6 s. (d) From 0.6 s to 0.8 s. (e) From 0.8 s to 1 s. (f) From 1 s to 1.12 s.

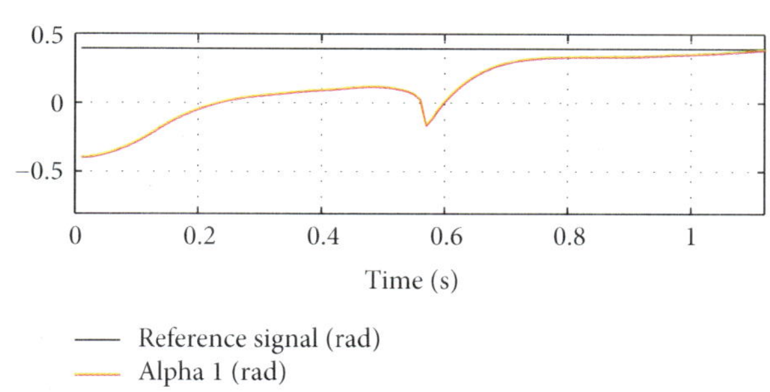

FIGURE 20: Trajectory of α_1 (rad).

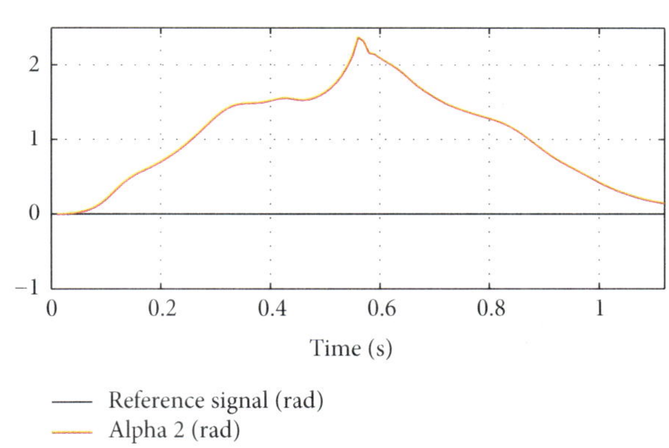

FIGURE 21: Trajectory of α_2 (rad).

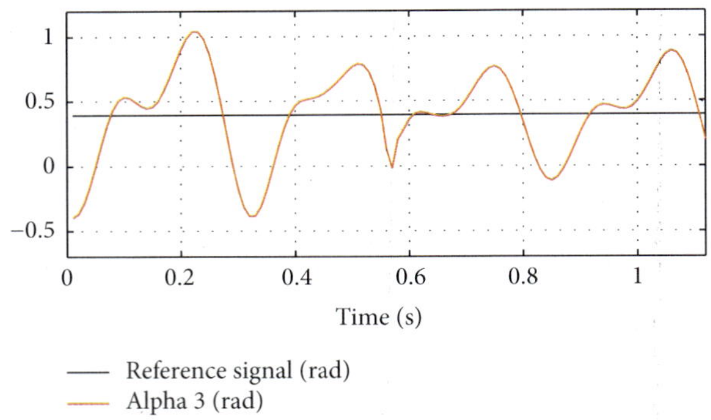

FIGURE 22: Trajectory of α_3 (rad).

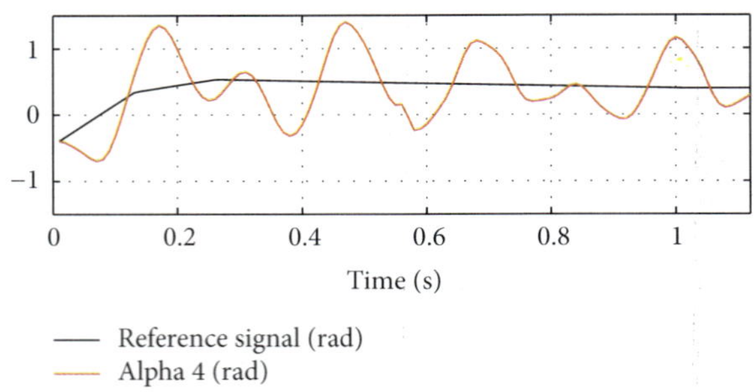

FIGURE 23: Trajectory of α_4 (rad).

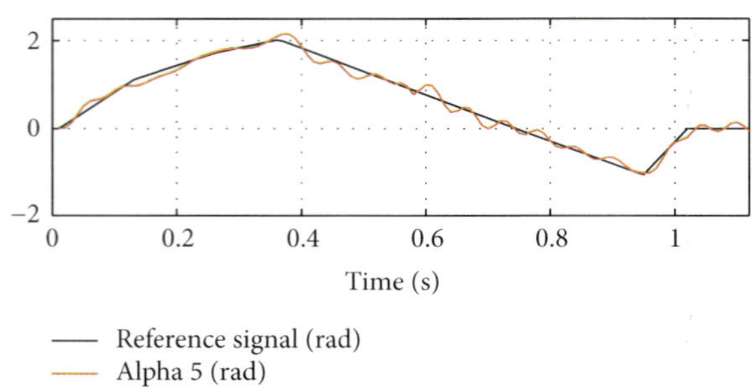

FIGURE 24: Trajectory of α_5 (rad).

at first but HSM signal finally leads links 1 and 2 to reference trajectories at the end of period of the step (Figures 20 and 21).

5. Conclusion

In the paper, the authors represent method of generating dynamic equations of AR with elastic legs through Euler-Lagrange. Due to the complexity of structure, dynamic equations cannot be visibly shown. But, simulation can still be implemented after each loop of simulation by general matrix form of AR. A method of using three P-controllers is introduced to transform MIMO nonlinear underactuated form of AR into a SIMO system. Then, the authors also propose HSM controller, which was well implemented for SIMO system, for motion of AR in one-step period. Although sliding surfaces in that method were proved to work well, a mathematical problem is not completely guaranteed. This led to difficulty in choosing exactly control parameters. GA is proposed to solve that problem. Along with selecting the acceptable control parameters, GA is also successfully used to design the reference trajectories for motion of each link. The success of this approach in control AR robot is consolidated through simulation.

Conflicts of Interest

The authors declare that there are no conflicts of interest regarding the publication of this paper.

References

[1] V. V. Beletzky, "Nonlinear Effects in Dynamics of Controlled Two-legged Walking," in *Proceedings of the Nonlinear Dynamics in Engineering Systems, International Union of Theoretical and Applied Mechanics*, pp. 17–26, Springer, 1990.

[2] H. Qiang, Y. Kazuhito, K. Shuuji et al., "Planning Walking Patterns for a Biped Robot," *IEEE Transactions on Robotics and Automation*, vol. 17, no. 3, pp. 280–289, 2001.

[3] W. Sujan, B. Amin, S. Nathan, and S. Madhavan, "Bipedal Walking—A Developmental Design," in *Proceedings of the International Symposium on Robotics and Intelligent Sensors, Procedia 41*, pp. 1016–1021, Elsevier, 2012.

[4] L. Qinghua, A. Takanishi, and I. Kato, "A Biped Walking Robot having a ZMP Measurement System using Universal Force-moment Sensors," in *Proceedings of the Intelligent Robots and System*, pp. 1568–1573, IEEE, 1991.

[5] S. N. Napoleon and S. Mitsuji, "Balance control analysis of humanoid robot based on ZMP feedback control," in *Proceedings of the 2002 IEEE/RSJ International Conference on Intelligent Robots and Systems*, pp. 2437–2442, IEEE, October 2002.

[6] D. W. Kim, N.-H. Kim, and G.-T. Park, "ZMP based neural network inspired humanoid robot control," *Nonlinear Dynamics*, vol. 67, no. 1, pp. 793–806, 2012.

[7] J. A. Smith and A. Seyfarth, "Elastic Leg Function in a Bipedal Walking Robot," *Journal of Biomechanics, Page*, vol. 40, Supplement 2, p. S306, 2007.

[8] F. Iida, Y. Minekawa, J. Rummel, and A. Seyfarth, "Toward a human-like biped robot with compliant legs," *Robotics and Autonomous Systems*, vol. 57, no. 2, pp. 139–144, 2009.

[9] C. K. Toh, M. Xie, H. Yang, G. Zhang, Q. P. Bui, and B. Tian, "Flexible Foot Design for Biped Walking on Uneven Terrain<," in *Proceeding of International Conference on Intelligent Robotics and Applications (ICIRA 2010), Part of the Lecture Notes in Computer Science book series*, vol. 6424, pp. 442–452, Springer, 2010.

[10] B.-H. Kim, "Work analysis of compliant leg mechanisms for bipedal walking robots," *International Journal of Advanced Robotic Systems*, vol. 10, article no. 334, 2013.

[11] A. Ming, K. Sato, R. Sato, E. Kazama, I. Miyamoto, and M. Shimojo, "Development of robot leg composed of parallel linkage and elastic spring for dynamic locomotion," in *Proceedings of the 2015 IEEE International Conference on Information and Automation, ICIA 2015 - In conjunction with 2015 IEEE International Conference on Automation and Logistics*, pp. 38–43, August 2015.

[12] D. Lakatos, A. Albu-Schäffer, C. Rode, and F. Loeffl, "Dynamic bipedal walking by controlling only the equilibrium of intrinsic elasticities," in *Proceedings of the 16th IEEE-RAS International*

Conference on Humanoid Robots, Humanoids 2016, vol. 2016, pp. 1282–1289, IEEE, November 2016.

[13] M. A. Sharbafi, C. Rode, S. Kurowski et al., "A new biarticular actuator design facilitates control of leg function in BioBiped3," *Bioinspiration & Biomimetics*, vol. 11, no. 4, Article ID 046003, 2016.

[14] N. Ryuma, N. Satoshi, and K. Yasuo, "Athlete Robot with Applied Human Muscle Activation Patterns for Bipedal Running," in *Proceeding of International Conference on Humanoid Robots (Humanoids)*, IEEE, 2010.

[15] N. Ryuma, N. Satoshi, and K. Yasuo, "Biomechanical Approach to Open-Loop Bipedal Running with a Musculoskeletal Athlete Robot," *Journal of Advanced Robotics*, vol. 26, no. 3-4, pp. 445–452, 2012.

[16] G. Garofalo, C. Ott, and A. Albu-Schäffer, "Walking control of fully actuated robots based on the bipedal SLIP model," in *Proceedings of International Conference on Robotics and Automation (ICRA)*, vol. 2012, pp. 1456–1463, IEEE.

[17] Y. Liu, P. M. Wensing, D. E. Orin, and Y. F. Zheng, "Dynamic walking in a humanoid robot based on a 3D Actuated Dual-SLIP model," in *Proceedings of the 2015 IEEE International Conference on Robotics and Automation, ICRA 2015*, pp. 5710–5717, May 2015.

[18] M. Shabazi, R. Babuska, and G. A. D. Lopes, "Unified Modeling and Control of Walking and Running on Inverted Pendulum," *IEEE Transactions of Robotics*, vol. 32, no. 5, pp. 1178–1195, 2016.

[19] A. Kulkarni and A. Kumar, "Adaptive hierarchical control for a class of MIMO uncertain underactuated systems," in *Proceedings of the 5th IEEE International Conference on Computational Intelligence and Computing Research, IEEE ICCIC 2014*, pp. 305–308, December 2014.

[20] D. Qian, J. Yi, and D. Zhao, "Hierarchical sliding mode control for a class of simo under-actuated systems," *Control and Cybernetics*, vol. 37, no. 1, pp. 159–175, 2008.

[21] D. Qian and J. Yi, *Book of Hierarchical Sliding Mode Control for Under-actuated Cranes: Design, Analysis and Simulation*, Springer, 2015.

[22] F. Baklouti, S. Aloui, and A. Chaari, "Adaptive Fuzzy Sliding Mode Tracking Control of Uncertain Underactuated Nonlinear Systems: A Comparative Study," *Journal of Control Science and Engineering*, vol. 2016, Article ID 9283103, 12 pages, 2016.

[23] K. Rsetam, Z. Cao, and Z. Man, "Hierarchical sliding mode control applied to a single-link flexible joint robot manipulator," in *Proceedings of the 2016 International Conference on Advanced Mechatronic Systems, ICAMechS 2016*, pp. 476–481, December 2016.

[24] A. Castigliano and E. S. Andrews, *Elastic Stresses in Structures*, Cambridge University Press, Cambridge, UK, 2014.

[25] T. Zielinska, "Gait Rhythm Generators of a Two-Legged Walking Machine," in *Proceedings of the ROMANSY 11, Series of International Centre for Mechanical Sciences*, vol. 318, pp. 165–171, Springer, 1997.

[26] G. Endo, J. Morimoto, J. Nakanishi, and G. Cheng, "An empirical exploration of a neural oscillator for biped locomotion control," in *Proceedings of the 2004 IEEE International Conference on Robotics and Automation*, pp. 3036–3042, IEEE, May 2004.

[27] D. Hein, M. Hild, and R. Berger, "Evolution of Biped Walking Using Neural Oscillators and Physical Simulation," in *Part of the Lecture Notes in Computer Science book series (LNCS)*, vol. 5001, pp. 433–440, Springer, 2007.

3

On Indexicality, Direction of Arrival of Sound Sources, and Human-Robot Interaction

Ivan Meza, Caleb Rascon, Gibran Fuentes, and Luis A. Pineda

Instituto de Investigaciones en Matematicas Aplicadas y en Systemas, Universidad Nacional Autonoma de Mexico, Ciudad Universitaria, Coyocan, 04510 Mexico City, MEX, Mexico

Correspondence should be addressed to Ivan Meza; ivanvladimir@turing.iimas.unam.mx

Academic Editor: Gordon R. Pennock

We present the use of direction of arrival (DOA) of sound sources as an index during the interaction between humans and service robots. These indices follow the notion defined by the theory of interpretation of signs by Peirce. This notion establishes a strong physical relation between signs (DOAs) and objects being signified in specific contexts. With this in mind, we have modeled the call at a distance to a robot as indexical in nature. These indices can be later interpreted as the position of the user and the user herself/himself. The relation between the call and the emitter is formalized in our framework of development of service robots based on the SitLog programming language. In particular, we create a set of behaviours based on direction of arrival information to be used in the programming of tasks for service robots. Based on these behaviours, we have implemented four tasks which heavily rely on them: following a person, taking attendance of a class, playing Marco-Polo, and acting as a waiter in a restaurant.

1. Introduction

Knowing the origin of a sound source is an important skill for a robot. Usually, this skill is associated with survival behaviours in which this knowledge is used to react to natural predators or eminent danger. However, this skill also plays a key role during interaction, for instance, in calling a waiter over. At first, knowing the direction of arrival (DOA) of a sound source may seem too basic to be an element in the interaction; however, as we will show, this information can be meaningful so that the robot can react in a proper manner and enhance the interaction.

We consider the direction of arrival information as an index in the sense proposed by Peirce's sign theory [1, 2]. In this theory, there are three types of signs given their relation with the represented object. These types are *icon*, *index*, and *symbol*. *Icons* reflect qualitative features of the object, for instance, pictures or drawings of the object. *Indices* have an existential or physical connection with the objects in a specific context; for instance, a dark cloud is a sign for rain. On the other hand, *symbols* have a stable interpretation based on a convention that connects them to the object, for instance, a logo of a product. This also holds for spoken communication; onomatopoeic words, for instance, are iconic since they resemble what they represent (e.g., *chop*), pronouns are indices since they "point" to the object they represent, and nouns are symbolic since they are conventional and detached from the object they represent. In the context of this classification, DOA information has an indexical quality; it is an index of the position of the object which emits the sound and also of the object itself, since it holds a strong connection between the emitter of a sound (object) and the direction of arrival (sign). In this work, we exploit the use of this index to support the interaction between a robot and the users.

The present work of indexical DOAs is formalized and implemented in the SitLog programming language that we have developed and used to program our service robot [3, 4]. SitLog defines a set of behaviours which range from simple (one skill) to composed (more than one skill) behaviours. These behaviours are the basic blocks used to program our robot at a higher level. Examples of simple behaviours are *walk*, which takes the robot from its current position to a destination, and *ask*, which prompts the user with a question and waits for the spoken answer. An example of a composed behaviour is *searching for an object* in different locations, since it relies on other basic behaviours such as *walk* and *see object*. In essence, a behaviour has to be (1) portable, so it can be used in different tasks, (2) compositional, so by coupling it with different behaviours we can create more complex behaviours or program a task, and (3) able to handle potential failures (e.g., not arriving to its destination, not listening to an answer from the user, or not finding the object).

In order to model the indexicality of the DOAs and incorporate it into the interaction capabilities of our robot, we propose a simple behaviour to handle DOAs and interpret them as indices. This supports the interaction by allowing calls at a distance. In this case, the user or users can bring the attention of the robot to a specific area in which the sound originates. This behaviour can also be used in combination with the *walk* behaviour to allow interruptions during walking such as calling a waiter over. Additionally, the behaviour also supports the verification of the number of sources during a conversation. In particular, this can be used in combination with the *ask* behaviour to ensure that when the robot is listening to an answer, there is only one person talking.

This paper is organized as follows: Section 2 presents previous work in which sound information is part of a task for a service robot. Section 3 reviews indexicality as interpretation procedure. Section 4 presents the framework we use to interpret DOAs signs as indices. Section 5 describes the manner in which the DOA behaviour resolves a DOA index. Section 6 shows the four tasks on which the indices from sound sources are used to direct the interaction. Section 7 presents a discussion about our proposal and findings.

2. Previous Work

The use of sound source localization (SSL) in robotics is a blooming field; Okuno and Nakadai and Argentieri et al., present reviews of the main methods and their use with different types of robots [6–8]. Since its application to robotics, SSL has been promoted as a main skill for robots. Brooks et al. proposed it as a basic skill for interaction in the Cog project [9]. The robot SIG was proposed as an experimental setting for the RoboCup competition [10]; and the robot Spartacus participated in the 2005 AAAI Mobile Robot Challenge implementing SSL as a part of its skill set [11].

Although the interaction goal is a driving force in the field, there is a great effort on developing a robust SSL module. With this goal in mind, the preferred setting is the use of an array of microphones on board of the robot. A many-microphone solution provides good performance given its redundancy. In [12], Valin et al. propose an 8-microphone 3D array to enhance speech recognition and as a preamble for DOA estimation. More microphones are possible. Hara et al. use two arrays of 8 microphones (16 in total) [13]. However, minimal systems with 2 microphones are also possible, such as the one presented in [14]. A common setting, which we follow, is to use 3 microphones [15, 16]. The best approach and configuration are still an open question. Badali et al. present an evaluation of the main approaches over an 8-microphone array for mobile robots [17].

Source localization can be reactive as a modality of the interaction as presented in works that deal with following a conversation [18–21]; or it can be an essential part of the interaction, such as using sound localization to control the navigation of the robot. In particular, Teachasrisaksakul et al. propose a system that follows a person through a room by her or his voice [22]. On the other hand, Li et al. demonstrated a robot which plays *hide and seek*, using visual and audio modalities in the interaction [23]. These works follow a similar approach on how to incorporate sound source location as a part of the interaction as the one we propose. However, since only one user is present in their work, its treatment as an index is trivial. The DOA sign is directly interpreted as the user; there are no conflicts regarding who or what the sign could represent.

Given the progress in the field, there has been an effort directed at capturing a more complicated interaction. For instance, Fransen et al. present a multimodal robot which follows the instructions among two users to reach a target [24]. Nakadai et al. present a robot which is able to judge rock-paper-scissors sound games [25]. The Quizmaster robot plays a game in which it asks a question to up to four participants and uses SSL and source separation to decide who answered it first [26]. Do et al. present a robot which together with a caregiver log and detect the origin of certain sounds [27]. At this level of complexity of the interaction, DOA signs are being indirectly used as indices: the DOAs become indices which signify directly the users and this information is used to disambiguate who calls or wins. In [27], the DOAs and a category associated with them signify types of events. At this level, the multiple sources make a method to assign the DOA to the right entity necessary. As we will see in this work, the benefit of considering a DOA as an index in the right context allows us to use indexical resolution to perform this disambiguation in the cases that more than one DOA competes. In this work, we formalize this resolution similarly to other reference resolution mechanisms on robots, such as deictic references [28, 29]. This differs from other approaches in which DOAs and multimodal information complement each other for disambiguation [30].

3. Indexical Expressions

An interactive agent such as a robot has to understand the intention of its users or have expectations about the events that can occur in the environment in order to produce an adequate response. Considering the theory of signs by Pierce, we can formulate the goal of "understanding" as providing a sign or a set of signs to identify the signified object or objects [1, 2, 31]. In order to denote the object, the robot builds a representation of it. For example, if the user commands *robot go to the kitchen*, the sequence of symbols *robot*, *go*, *to*, *the*, and *kitchen* get translated into the representation $go(robot, kitchen)$, with which the robot can establish the signified objects (itself, the kitchen, and the action) and act accordingly. A similar mechanism can be used for the interpretation of iconic signs; for instance, when the robot sees a table and identifies on it a bottle of juice, it will

TABLE 1: Examples of indexical expressions and their resolution.

Sign	Representation	Constraint	Resolved
This is juice + pointing gesture	$\lambda x.name(x : pose, juice)$	Pose of object	$name(object, juice)$
Take it	$\lambda x.take(x : object)$	Object	$take(juice)$

generate the representation $on(juice, table)$, which denotes the corresponding juice object fully.

When the set of signs of the interaction includes indexical signs, the representation is not fully specified by the signs since the index has to be linked in an extra resolution step to the denoted object. For instance, if the user states the command *robot come here*, the representation for this utterance is underspecified as $\lambda h.go(robot, h)$. In order to fully specify this expression, it has to be analysed in relation to the context of the task, in which the position of the user can be inferred. For example, if the user is located in the kitchen, the expression should be resolved to $go(robot, kitchen)$.

In this scheme, indexical signs produce an underspecified representation at first which should be later resolved in the presence of contextual information. Furthermore, indices provide additional constraints on the type of object which can resolve the expression; for instance, in the previous example, the destination of the robot has to be resolved into a valid position. In order to account for this phenomenon, inspired by the treatment of indices in [32], we propose the following formulation for an index in lambda calculus in which we use the symbol ":" to signal a constraint α over a certain variable x:

$$\lambda \alpha Q x. (Q @ x : \alpha). \tag{1}$$

Indices defined in this form receive as parameters the constraint α and the underspecified predicate Q; the effect of applying such arguments is the binding of the constrained variable $x : \alpha$ into the expression Q. In this case, the symbol @ signifies a functional application (i.e., β reduction in lambda calculus.) If we apply the index in our previous example $Q = \lambda h.go(robot, h)$ and $\alpha = user_{pos}$, the following representation is produced: $\lambda h.go(robot, h : user_{pos})$. It is still underspecified, but it has constrained the variable h to be the position of the user. A second process of resolution has to identify from the context the fact that the user is in the kitchen.

Table 1 shows two examples of indexical signs using this formulation with a robot. For these examples, consider the following context: the user and the robot are in front of a table with an orange juice placed on it. The first illustrated example is a multimodal expression: the user says something and points to an object. The word *this* in conjunction with the pointing gesture define an index which imposes a spatial constraint on the object being signified by both. In this case, the constraint is the pose of the object (position and orientation extracted from the visual system). This constraint is used to identify the object in the scene, which in the example resolves into the bottle of juice. The second example shows an indexical reference with the word *it* which also adds a constraint [33]. In this case, the underspecified variable has to be resolved to an object in the real world, the bottle of juice again. Notice that we have invoked a spatial and deictic resolution; however, we have not specified a particular mechanism to perform this type of resolutions. The topic of effective mechanisms for the resolution of indexical expressions continues to be studied in different fields [34–36], and which mechanism is applied depends on the type of expression and the type of task being solved. In this work, we define the mechanism for the resolution of indexical DOA expressions and the conditions in which it can be applied.

4. Dialogue Models and Robot

In our robot, the coordination among modules during the execution of a task is done by the execution of dialogue models. These dialogue models are written using the SitLog programming language for service robots. Dialogue models are composed of a set of *situations* which have associated a tuple of *expectations*, *actions*, and next *situations*. In a typical interaction cycle, the robot is in a given situation in which it has certain expectations; if the arriving interpretation matches some expectation, the robot proceeds to perform the associated actions and moves to another situation. In this framework, performing a task consists in traversing a dialogue model while the modules provide the interpretations and perform the actions defined by the dialogue model. Additionally, in SitLog, it is possible to define the main elements as functions that dynamically construct the structure of the dialogue model. It is also possible to call another dialogue model from a particular situation. Figure 1 depicts these cases:

(a) It shows a typical SitLog arc in which a situation S_i has one expectation α; if this is satisfied, the set of actions β are performed and the arc arrives to the situation S_j.

(b) It exemplifies the case in which the elements are defined by functions and the expectation will be defined by the evaluation of function f, actions by g, and the next situation by h. In particular, these properties make it possible to program dynamic dialogue models which change as the task is executed.

(c) It shows the case for *recursive* situations in which a situation calls an entire dialogue model for execution, and, when finished, it uses the name of the last situation as its own expectation. Algorithm 1 presents the SitLog code for these cases.

SitLog at its core is framework-agnostic; one could implement different frameworks such as subsumption [37], reactive [38], or simple state machine architectures. At our current laboratory, we have developed a full library of behaviours and tasks based on our interaction cycle which is a high level cognitive architecture (IOCA [4, 39]). Associated with

```
[ id ==> s_i,
  arcs ==> [ alpha : betha => s_j]
]
[ id ==> s_i,
  arcs ==> [ f(x) : g(y) => h(z) ]
]
[ id ==> s_i,
  embeded_dm ==> r(m)
  arcs ==> [ alpha : betha => s_j ]
]
```

ALGORITHM 1: Example of code for SitLog: (a) typical SitLog arc, (b) functional arc, and (c) recursive situation.

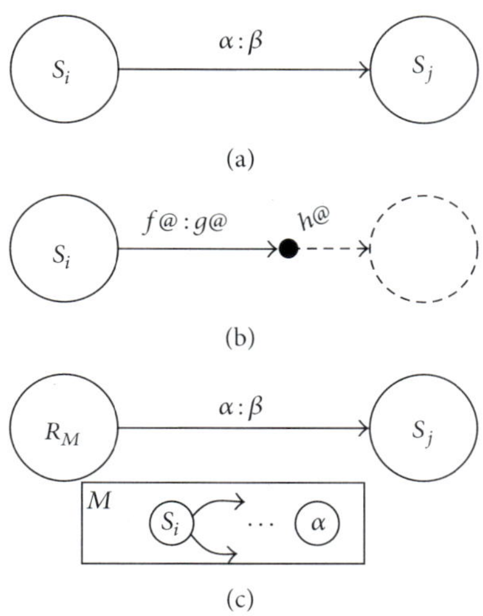

FIGURE 1: Examples of situations, expectations, and actions. (a) Typical SitLog arc, (b) functional arc, and (c) recursive situation.

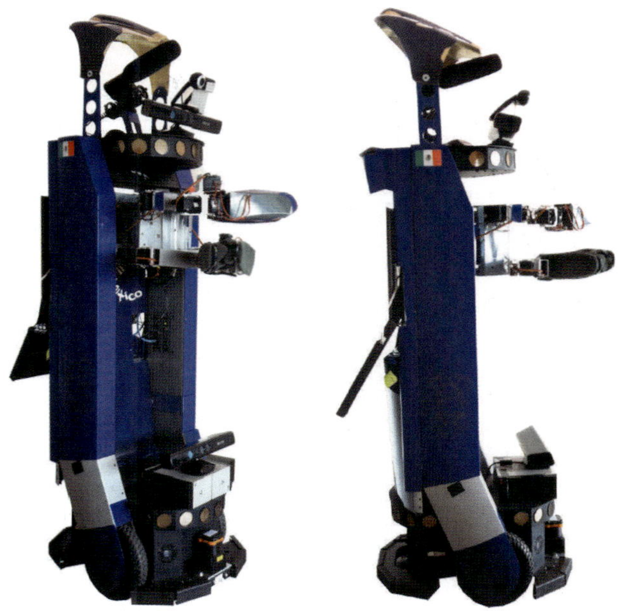

FIGURE 2: The Golem-II+ robot.

the architecture, we have implemented several skills such as vision, language [40], sound [5], navigation, manipulation, and movement. Using this framework, we have programmed the Golem-II+ robot (depicted in Figure 2 [41]). Table 2 summarizes the main capabilities, modules, and hardware that compose the robot. Within this framework and the current version of the hardware, we have implemented several tasks such as following a person, introducing itself, searching for objects, acting as a waiter in a dinner party or a restaurant [42], guarding a museum, and playing a match memory game (for examples of the tasks, visit http://golem.iimas.unam.mx/).

4.1. Multiple-DOA Estimation System. The DOA skill is performed by the multiple-DOA estimation system. This is based on previous work that focuses on a small, lightweight hardware setup that is able to estimate more DOAs than the amount of microphones employed [5, 43].

From the hardware standpoint (see Table 2, sound perception section), the system uses a 2-dimensional 3-microphone array (see Figure 4) going through a 4-channel USB interface. From the software standpoint, the architecture of this can be seen in Figure 3. The system is divided into three parts.

Audio Acquisition. This is the underlying module that feeds audio data to the next part of the system. It is based on the Jack Audio Connection Toolkit [44], which provides high resolution data (48 kHz, in our case) at real-time speeds with a very high tolerance in the amount of microphones it manages and a relative small resource requirement. This captured audio passes through a VAD which activates the localization.

Initial Single-DOA Estimation. For each time window, each microphone pair is used to estimate a signal delay using a cross-correlation vector (CCV) method (see Figure 3, single-DOA estimation block). Each delay is used to calculate two preliminary DOAs (to consider the back-to-front mirroring issue of 1-dimensional arrays pose). Using a basic search method, the most *coherent* set of 3 DOAs (1 from each microphone pair) is found, using a proposed coherency metric (see Figure 5). If the coherency of that set is above a certain threshold, the DOA of the most perpendicular pair towards the source is proposed as its DOA; this pair is chosen to avoid nonlinearities in the relation between the delay and the resulting DOA. Figure 5 illustrates this stage.

Multiple-DOA Tracking. Because the previous part carries DOA estimation near real-time speeds, it is able to estimate the DOA from one source in small time windows, even in instances where there are two or more simultaneous sound sources (see Figure 3, multi-DOA tracking block). However, the occurrence of 1-sound source time windows in such instances is stochastic. To this effect, when a DOA is proposed from the previous part of the system, a clustering method is

TABLE 2: Capabilities, modules, and hardware used on the Golem-II+ robot (IH: in-house developed).

Module	Hardware	Software libraries
	Dialogue management	
Dialogue manager	—	SitLog
Knowledge-base	—	Prolog
	Vision	
Object recognition	Flea 3 camera	MOPED
Person recognition	WebCam	OpenCV
Person tracking	Kinect	OpenNI
Gestures recognition	Kinect	OpenNI
	Language	
Speech recognition	Directional Mic	PocketSphinx
Speech synthesiser	Speakers	Festival TTS
Understanding	—	GF grammar, IH parser
	Sound perception	
DOA system	3 omnidirectional mics	—
Volume monitor	Directional mic	Jack, IH library
Volume monitor	M-Audio Fast Track interface	Jack, IH library
	Navigation, manipulation, and neck	
Navigation	Robotic base, laser	Player
Manipulation	IH robotic arm	Dynamixel RoboPlus
Neck movement	IH neck	Dynamixel RoboPlus

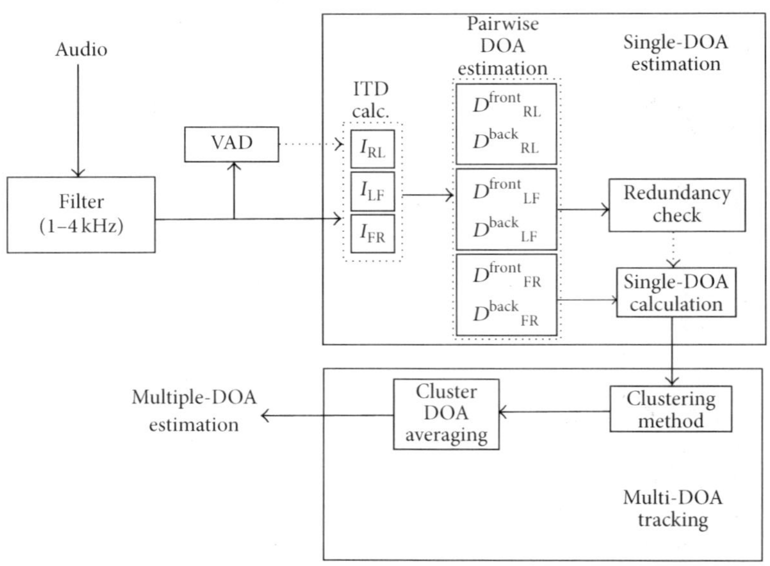

FIGURE 3: Architecture of the multiple-DOA estimation system.

employed to group several nearby DOAs into one or more estimated sound sources directions.

Figure 6 shows the tracking of three speakers during 30 seconds; each of the speakers is separated by 120 degrees. As it can be seen, this tracking is quite effective, as it could localize each of the found sources. In this example, the system had a 69% precision and 60% recall performance at the frame level. Table 3 presents an evaluation of the whole system. Although it could be thought that this performance would not be enough for HRI interactions, since it is at the frame label (i.e., 100 ms), in an interaction setting when turns are being taken, this performance is more than enough to catch the interaction of a user. An extensive evaluation of this module can be consulted in [43].

5. DOA Behaviour

The goal of the DOA behaviour is to transform DOA measurements into a reference to an object in the context of the task. When the DOA system detects a source, this

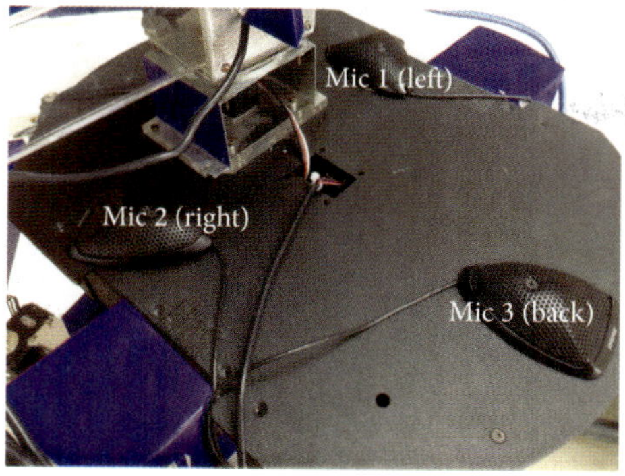

FIGURE 4: Microphones of the audio localization system in the robot Golem-II+.

TABLE 3: Multiple-DOA estimation system performance.

Minimum distance of speaker	30 cm
Maximum distance of speaker	5 m
Maximum number of simultaneous speakers	4
Response time	s
Range (minimum angle between speakers)	10
Coverage	360
*F*1-*score* 1 speaker	100.00%
*F*1-*score* 2 speakers	78.79%
*F*1-*score* 3 speakers	72.34%
*F*1-*score* 4 speakers	61.02%

sign has a strong relation with the emitter of the sound. The DOA behaviour takes advantage of this strong relation to resolve the object being "pointed at" by imposing a spatial constraint on the possible sources. The spatial constraint consists of the relative position a of the source x given the current robot position (i.e., $\lambda a\ x.rel\_pos(x, a)$). At this point, we only have the information that something is at position a; however, since we consider the DOA an index, we can use our formulation (1) for indices so that we obtain the following reduction:

$$\lambda x.rel\_pos\left(x : a, a\right). \tag{2}$$

Although we are able to constrain the object, so far, we have not resolved the index to the right referent; to achieve this, we need to use the contextual information. The angle of the DOA is used to identify potential objects given the context. Algorithm 2 is in charge of this resolution stage by analysing which objects on the context satisfy the DOA constraint.

The whole mechanism is encapsulated in a behaviour which is programmed as a simple dialogue model (see Figure 7). This dialogue model can process multiple DOAs at once; for each one, it will try to resolve the object being refered. If successful, it creates an *ok* situation associated with a list of referred objects, otherwise, it reaches an *error* situation. The proposed DOA behaviour supports interaction in two modalities:

(i) Unrestricted call.

(ii) Contextualized call.

```
Require: Pred, DOAs, Context
Ensure: Pred(listObjects, listDOAs)
  listObjects ← []
  listDOAs ← []
  for doa ← DOAs do
    for id, α ← Context do
      if doa α then
        listObjects.push(id)
        listDOAs.push(doa)
      end if
    end for
  end for
```

ALGORITHM 2: Implementation algorithm for the interpretation of DOAS indices.

The first case corresponds to the scenario in which the position of the user cannot be expected/predicted beforehand; in this case, the description of the context C is empty meaning that only one user is present (Figure 8(a) depicts the robot in such situation); that is, the indices are directly interpreted as the user the robot is having the interaction with. In the second case, the robot has an expectation of the direction from which it will be called and it uses this expectation to identify the caller. For instance, Figure 8(b) depicts the robot having a conversation with two users; with the information of their position, it can discard a third user which is not part of the conversation.

So far, we have assumed that the robot is not moving, but the user should be able to call the robot while it is moving. In order to tackle this situation, we created a composed behaviour which allows the robot to be walking and listening for possible calls at a distance. Figure 9 shows the dialogue model of this behaviour. The call for such behaviour is $walkdoa(D, C)$; D represents the destination and C the description of the spatial context. First, the behaviour starts with the action of *walking* to D and, while doing so, the robot polls for possible DOAs sources in C. If there are none, it continues walking; otherwise, it ends the dialogue with a situation called *interrupted* which contains the information relevant to such interruption. The spatial context gets updated in each call with the current position of the robot. Notice that this behaviour reuses the simple DOA behaviour (situation $R_{doas(C)}$).

For the alternative case, when the user moves, the DOA signal imposes an extra constraint into the index. This constraint is that the DOA at a relative position comes from the same sound source as the previous DOA and this extra information is provided by the tracking stage of our multiple-DOA estimation system. Thanks to this constraint, we can use the same mechanism we have outlined so far to handle this case. In combination with the walking and listening behaviour, they can candle the scenario in which both user and robot move.

Additionally, we also propose to use the DOA behaviour to verify the number of sound sources in a given moment. In particular, we have created a composed behaviour between

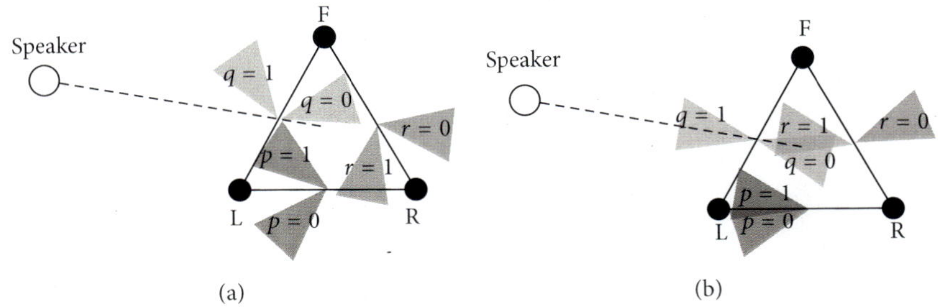

FIGURE 5: (a) Incoherent set of measures. (b) Coherent set of measures (taken from [5]).

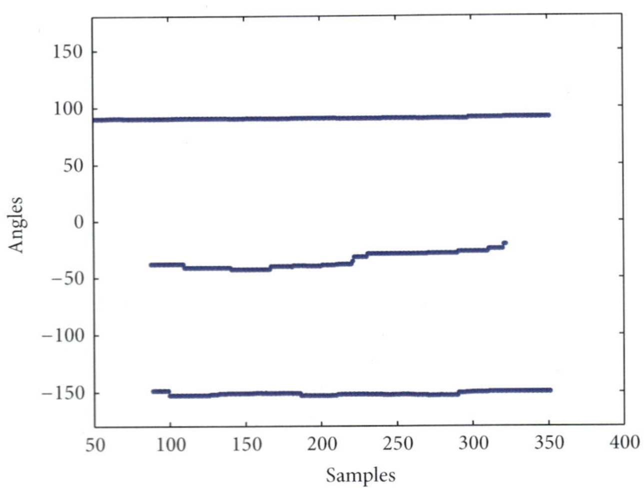

FIGURE 6: DOAs for three speakers at 90, −30, and −150 degrees.

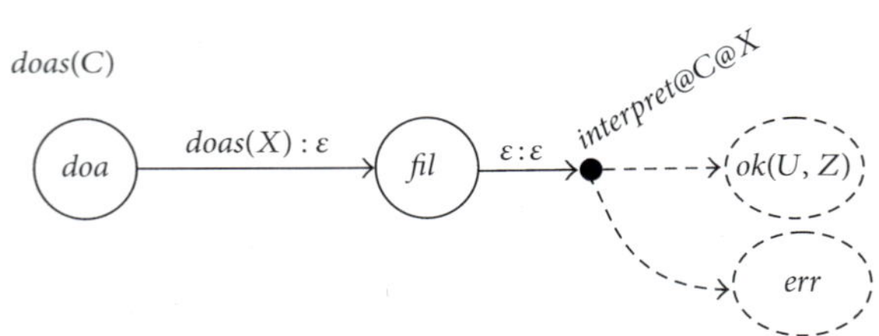

FIGURE 7: Dialogue model for simple DOA behaviour.

the ask and DOA behaviours which take advantage of this situation. Figure 10 shows the dialogue model for such behaviour. After asking a question P and listening to the answer from the user, the robot checks how many indices were recovered (i.e., the number of elements in Z): if it is only one, it could proceed with the interaction; if they are more than one, the robot infers that more than one user is talking, so it produces an error which has to be handled by the robot. Similarly to the walking behaviour, this behaviour reuses the simple DOA behaviour.

6. Implemented Tasks

The DOA behaviours have been used to program the following tasks with our robot Golem-II+ (videos of these tasks can be watched at https://youtube/Q6prwIjoDnE?list=PL4EiERt__u4faJoxHFlM5EMwhEoNH4NSc).

Following a Person. In this task, the robot follows a person by using the visual tracking system (kinect based). The robot tries to keep a distance of 1 m; while the user is moving, the robot tries to catch it at a safe speed. In case the robot loses him, it asks the user to call it so the robot could infer which direction to look for. Once positioned on the right direction, it uses vision to continue the tracking it uses. When lost, the user can be in any direction and inside a ratio of 5 m and the robot would identify where it is. If the user is not found, it will insist to be called and try again to locate him (for a demo of the full system, see supplementary video *following a person.mp4* in Supplementary Material available online at http://dx.doi.org/10.1155/2016/3081048).

Taking Attendance. In this task, the robot takes attendance in a class by calling the names of the members of the class and for each call it expects an auditory response. When the direction of the response is identified, it faces towards that direction, checks if a person is there by asking to wave, and gets closer to verify the identity of the student. The number of students is limited by the vision system which only can see up to four different persons. They have to be located in a line; at the moment, the task does not consider the case of multiple rows. We take advantage of this setting and we specify a spatial context in which students are only in front of the robot (for a demo of the full system, see supplementary video *taking attendance of a class.mp4*).

Playing Marco-Polo. In this task, the robot plays the role of looking for the users by enunciating *Marco* and expecting one or more *Polo* responses. When the direction of a response is identified, it moves towards that direction if possible. When advancing to this direction, if the laser sensor recognizes that there is someone in front at close distance, it assumes it has won the game; otherwise, it continues calling *Marco* until the user responds. The robot has n tries to catch the user; if not, it gives up and loses the game. The laser hypotheses that there is someone in front if it detects a discontinuity in the reading. In this setting, the players can be located at any direction in a ratio of 5 m (for a demo of the full system, see supplementary video *playing Marco-Polo.mp4*).

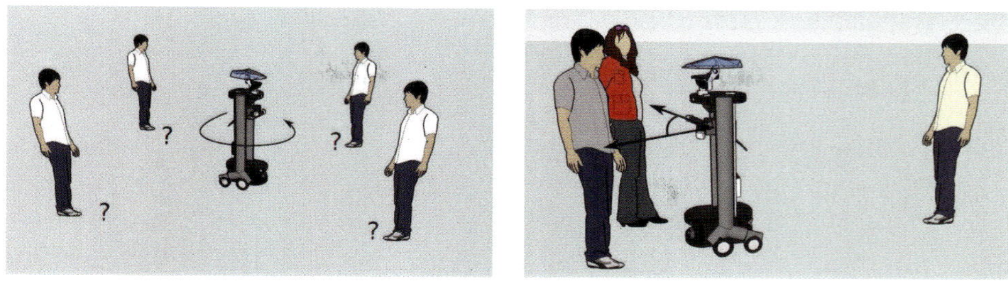

(a) (b)

FIGURE 8: Examples of calls at a distance: unrestricted, one user not at predictable position, and contextualized, multiple users at predictable position.

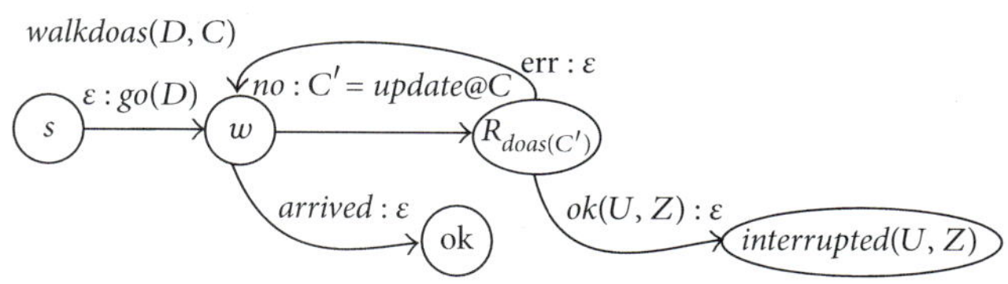

FIGURE 9: Dialogue model for walk DOA behaviour.

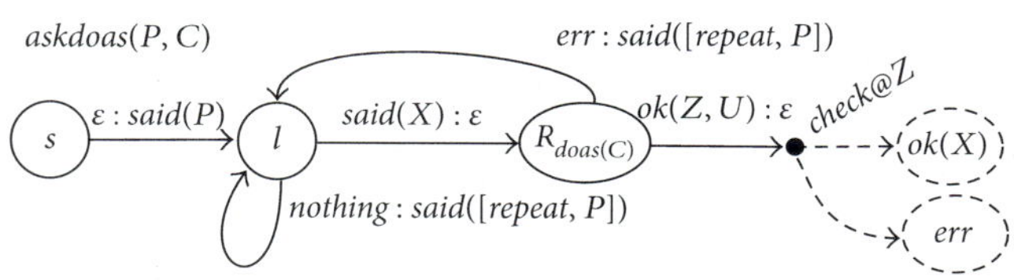

FIGURE 10: Dialogue model for ask DOA behaviour.

Waiter. In the previous examples, the DOA behaviour relies on the robot's initiative. This task pushes the behaviour into a mixed-initiative strategy. In this case, the robot is a waiter and it waits for calls from the users located in the restaurant tables. Once a table calls at a distance, the robot will walk to the table to take the orders of the clients. However, while walking, it will listen for other calls from different tables; if this happens, it will let the users in the corresponding table know that it will be there as soon as it finishes with the table it is currently walking to. While taking the order, the robot directs the interaction if more than one client talks at the same time. Once it has the order, it proceeds to bring it by taking the drinks/meals from a designated pickup location in the kitchen area or by asking the chef directly for them. This task is limited by the maximum number of sources that could be simultaneously detected (4). In this case, the maximum number of tables to attend would be 4 with a maximum number of 4 clients in each table; this have to be in a ratio of 5 m (for a demo of the whole system, see supplementary video *waiter in a restaurant.mp4*).

We hypothesize that the success of the interaction in each of the tasks is possible because of the use of the DOA information as an index which is fully interpreted in the context of each task. These contexts make the physical relation between the sign (the call) and the object (the user or users) evident. See Table 4 for a summary of the main elements involved in the resolution of the DOA indices in each of the tasks and detailed as follows.

In the case of the *following a person* task, it is until the robot has lost the user and addresses her or him that the response is interpreted as the user and the direction where it could look for its user; this is a case of unconstrained call in which the spatial context is empty. Without context, the robot is able to establish the relative position of the user.

In the case of the *taking attendance* task, the protocol for the interaction is quite standard. The robot calls the name of the student, and she or he has to respond to such call. Such response is interpreted as the presence of the student and its relative position is established. Here, we can constrain the calls to be in front of the robot.

In the case of the *playing Marco-Polo* task, the rules define the context on which the call is interpreted. These rules specify that when the robot says *Marco*, the users have to respond with *Polo*; the index of such response can be interpreted as the relative position of a specific player and it can direct the robot to look in that area. This task also uses an unconstrained call. In fact, the user is not even required to say *Polo*, since the objective of the game is that the robot "catches" one of the users via sound alone; therefore, the DOA of any sound may be considered as an index; this DOA will refer to the user.

In the case of the *waiter* task, the spatial context is defined by the arrangement of the tables in the restaurant. This task uses this spatial context to interpret which table calls for the attention of the robot. The constraint triggered by the DOA information helps identify the calling table, and it also establishes its relative position. The task uses both the simple version and the walking version of the DOA behaviour to resolve the indices. Additionally, it uses the ask version of the DOA behaviour when taking the order. The information provided by the ask version is used by the robot to face the clients while taking their orders and to direct the interaction by asking the users to talk one at the time.

6.1. Application of the DOA Behaviour in Tasks. Figure 11 shows excerpts from the dialogue model of the *following a person*, *taking attendance*, and *playing Marco-Polo* tasks. Given the information provided by the DOA behaviour, the *following a person* and *playing Marco-Polo* tasks cannot

TABLE 4: Characteristics of the tasks using DOA behaviours.

	Following a person	Taking attendance
Agents	Person being followed	Students
Robot goal	Track a person	Verify students present
Users goals	Being followed	Be in the class
DOA behaviour	Call at a distance	Call at a distance
Situation	User lost	Student name called
Sign	*Here*	*Present*
Expression	$\lambda x.rel\_pos(x : a, a)$	$\lambda x.rel_p os(x : a, a)$
Resolution	$rel_p os(user, a)$	$rel_p os(name, a)$
Interpretation	Target and its position	Presence of student and position
	Playing Marco-Polo	Waiter
Agents	Players	Clients
Robot goal	Catch a player	Take and deliver orders
Users goals	Not being caught	Receive orders
DOA behaviour	Call at a distance	Call at a distance Verification
Situation	Say *Marco*	Robot doing nothing or asking order
Sign	*Polo*	*Waiter*
Expression	$\lambda x.rel\_pos(x : a, a)$	$\lambda x_a.rel_p os(x : a, a)$
Resolution	$rel\_pos(player, a)$	$rel_p os(table, a)$
Interpretation	Player and its position	Table to order and client talking

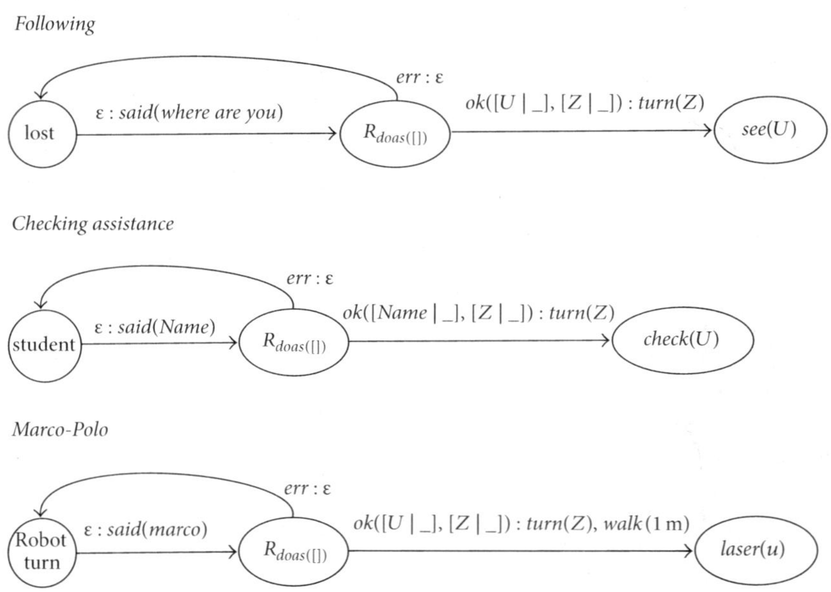

FIGURE 11: The DOA behaviour used on the *following a person*, *taking attendance*, and *playing Marco-Polo*.

identify the user; it is as if he or she would have only said *I am here*. The response of the robot to this interaction is not verbal but an action; in both cases, it turns towards the direction of the call. In case of an error in the detection of the DOA, the robot will try to recover. In the *following a person* task, the person tracker will trigger an error when no one is found and it will ask again for a call by the user. In the *playing Marco-Polo* task, the robot will repeat *Marco* and wait for the answer, continuing with the game. Because of the nature of the *take attendance* task, the interpretation of the index signals a particular student even in an underspecified context because the robot knows the name of the student being called, and the response is attributed to her or him. It is as if he or she would have responded *me X, I am here*. In this task, the robot tries to prevent an error in the DOA location and it checks if there is an actual person in that direction, but in case it is not, it will keep calling the name of the student.

Figure 12 shows the DOA behaviours for the *waiter* task. This task uses the three behaviours based on the DOA skill. First, it uses the DOA behaviour but with a defined context.

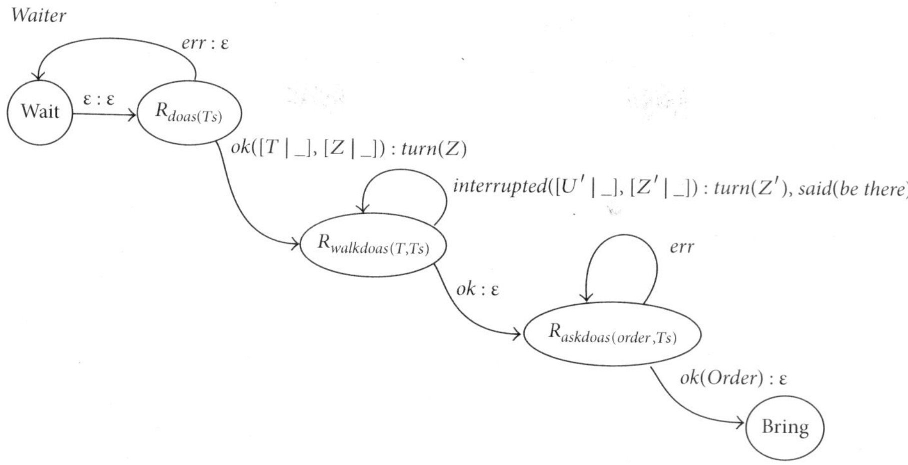

FIGURE 12: The DOA behaviours used on the *waiter* task.

The interpretation of the index will depend on the spatial context and it will define which table calls the robot. In this case, it is as if the client would have said *here table X needs something*. In the task, an error has a larger effect on the interaction; if the robot mistakenly approaches a table that did not call for it, it will offer to take the order to which the clients there could refuse or if the table is empty the robot will leave. However, the table which really wanted to ask for an order could call the robot while walking, so that when the robot is done with the erroneous table, it will approach the correct one. During such movements, the task uses the walking DOA behaviour. Finally, when arriving to a table to ask an order, the task uses the ask DOA behaviour, and if more than one index is returned during the response, it will let the clients know that only one user is supposed to talk at a time. If an error on the detection of DOAs occurs during this part of the task, it will create a confusing situation on which the robot will face clients which are not there; however, if no client answers, the robot will assume the "ghost" client does not want anything and will carry on with the rest of the order taking and delivery.

6.2. History of Demonstration and Evaluation of the Tasks. The *following a person, playing Marco-Polo,* and *waiter* tasks have been repetitively demonstrated in our lab to the general public and in the RoboCup@Home competition [45, 46]. The experience with these demonstrations during the competition has been positive. In 2012, the *following a person* task was demonstrated in the German Open competition, where the demonstration allowed the team to obtain the third place, and in the Mexican Tournament of Robotics national competition, where the robot won the first place. In 2013, the *waiter* task was presented in the first stage of the RoboCup competition and it was awarded the Innovation Award of the @Home league that year. Additionally, that same year, a demonstration of the robot playing Marco-Polo with another robot was attempted but technical problems prevented the execution of the demo during competition, but it was successfully carried out to the general public. In all these cases, we have observed a positive appreciation of the integration of sound localization in the interaction between humans and robots. In particular, the *waiter* task has been extensively evaluated [42]. This evaluation was done with final users (30) which were asked to place an order to the robot. 60% of the users did not have previous experience with a service robot. We found that, when asking for an order, clients repeatedly called at a distance (100%) and fully completed the task (90%). And even though it was not common that two users talked at the same time (40%), when they did, the robot was able to direct the interaction.

In the case of the *taking attendance* task, this was programmed by students during a three-week summer internship. The students had no previous knowledge of our development framework. During the internship, they learned the framework, developed the idea, and implemented it on the robot. Other four teams of students developed other four different tasks. The abstraction of the skill as an element of the SitLog language made it relatively easy to directly implement the task and reach an appropriate performance in a short time span.

7. Discussion

In this work, we proposed the use of direction of arrival of sound sources as indices. In many Human-Robot Interaction tasks, the position of objects is essential for the execution of the task, for instance, losing someone while following him or her. Under this perspective, DOAs are second-class citizens since they do not provide a position but a direction. However, treating DOAs as indices allows the robot to link such directions to objects in the spatial context that leads to the correct interpretation and continuation of the task, guessing that the lost one is a direction. This consideration permits modeling calls at a distance to a robot which is a desirable way to interact with an interactive agent. In order to demonstrate this concept, we implemented a set of behaviours that depend on DOAs to be interpreted as indices and tested them in four tasks: following a person, taking the attendance of a class, playing Marco-Polo, and acting as a waiter in a restaurant.

The spatial context, or a lack of it, defines aspects of the type of interaction on the task. When there is no available spatial context, such as when playing Marco-Polo, any index is considered by the robot as a possible participant. However, when a spatial context is defined, for instance, in attending the tables in a restaurant, it is possible to keep track of more than one user. It is also possible to have a mixed-initiative interaction in which the robot waits for the call, rather than visiting each table and asking if they want anything. The indexicality of the DOAs allows the robot to react to the direction of the user and continue with the interaction, not necessarily in the same modality: in *playing Marco-Polo,* the robot moves in the direction of the index; in *taking the attendance* and *following a person* tasks, the robot confirms the presence of the user by visual methods in the direction of the index; in the *waiter* task, the robot faces the tables which correspond to the indices. These three possible aspects of the interaction when using indexical DOAs (multiuser, mixed-initiative, and multimodal) are an example of the richness of the use of calls at a distance in an interaction.

Additionally, we have showed that the DOA behaviour complements the verbal interaction. In particular, knowing that there is more than one person talking when the robot is listening is a good indication that the interaction may not be good. With the indexical information, the robot can take control of the interaction and direct it by asking each user to speak one at a time, preventing an error in the communication.

We were able to implement these tasks since the multiple-DOA estimation system is light and robust up to four sound sources. However, this system and other robotic systems impose some restrictions on the interaction. In the case of following a person, it can only follow one person at a close distance (1 m); in the case of taking attendance, the students have to be aligned; In our test, there were three students; in the case of playing Marco-Polo and waiting tables, the users, player and tables, have to be in a ratio of 5 m. In all these cases, the maximum number of speakers could have been four. However, under this condition, the performance of the system was the poorest and had an effect on the interaction; our evaluation shows that one user for playing Marco-Polo and two users for table in the waiter case were the limit to achieve good performance. At this point, another major problem is possibility of hijacking the attention of the robot. This is related to the fact that when the spatial context is not defined, the robot automatically interprets the DOA as its target. This is exemplified with the Marco-Polo game. At the moment, our implementation of the task takes advantage of this situation and an index is interpreted as a possible user which becomes the target to approach. However, this situation can provoke the fact that the robot switches targets constantly rather than targeting one which might be a better strategy to win the game. In order to account for this situation, we have identified that the DOA behaviour could be complemented with a speaker verification stage in which the identity of who is responding with *Polo* would be verified. This complement is persistent with our proposal since it becomes an extra constraint in the referred object. Other tasks can also take advantage of this, as such verification can be part of the *following a person, taking attendance*, and *waiter* tasks. In the future, we will explore complementing the DOA behaviour with such verification and its consequences on the interaction.

Competing Interests

The authors declare that there are no competing interests regarding the publication of this paper.

Acknowledgments

The authors thank the support of CONACYT through Projects 81965 and 178673, PAPIIT-UNAM through Project IN107513, and ICYTDF through Project PICCO12-024.

References

[1] C. Peirce, N. Houser, and C. Kloesel, *The Essential Peirce: Selected Philosophical Writings*, vol. 1 of *The Essential Peirce*, Indiana University Press, 1992.

[2] C. Peirce, N. Houser, C. Kloesel, and P. E. Project, *The Essential Peirce: Selected Philosophical Writings*, vol. 2 of *The Essential Peirce*, Indiana University Press, 1998.

[3] L. A. Pineda, L. Salinas, I. V. Meza, C. Rascon, and G. Fuentes, "Sit log: A programming language for service robot task," *International Journal of Advanced Robotic Systems*, vol. 10, pp. 1–12, 2013.

[4] L. Pineda, A. Rodrguez, G. Fuentes, C. Rascn, and I. Meza, "Concept and functional structure of a service robot," *International Journal of Advanced Robotic Systems*, vol. 6, pp. 1–15, 2015.

[5] C. Rascon and L. Pineda, "Multiple direction-of-arrival estimation for a mobile robotic platform with small hardware setup," in *IAENG Transactions on Engineering Technologies*, H. K. Kim, S.-I. Ao, M. A. Amouzegar, and B. B. Rieger, Eds., vol. 247 of *Lecture Notes in Electrical Engineering*, pp. 209–223, Springer, Amsterdam, Netherlands, 2014.

[6] S. Argentieri, A. Portello, M. Bernard, P. Danes, and B. Gas, "Binaural systems in robotics," in *The Technology of Binaural Listening*, pp. 225–253, Springer, Berlin, Germany, 2013.

[7] H. G. Okuno and K. Nakadai, "Robot audition: its rise and perspectives," in *Proceedings of the IEEE International Conference on Acoustics, Speech and Signal Processing (ICASSP '15)*, pp. 5610–5614, South Brisbane, Australia, April 2015.

[8] S. Argentieri, P. Danès, and P. Souères, "A survey on sound source localization in robotics: from binaural to array processing methods," *Computer Speech & Language*, vol. 34, no. 1, pp. 87–112, 2015.

[9] R. A. Brooks, C. Breazeal, M. Marjanović, B. Scassellati, and M. M. Williamson, "The cog project: building a humanoid robot," in *Computation for Metaphors, Analogy, and Agents*, vol. 1562 of *Lecture Notes in Computer Science*, pp. 52–87, Springer, Berlin, Germany, 1999.

[10] H. Kitano, H. G. Okuno, K. Nakadai, T. Sabisch, and T. Matsui, "Design and architecture of sig the humanoid: an experimental platform for integrated perception in robocup humanoid challenge," in *Proceedings of the IEEE/RSJ International Conference on Intelligent Robots and Systems (IROS '00)*, vol. 1, pp. 181–190, IEEE, Takamatsu, Japan, 2000.

[11] F. Michaud, C. Côté, D. Létourneau et al., "Spartacus attending the 2005 AAAI conference," *Autonomous Robots*, vol. 22, no. 4, pp. 369–383, 2007.

[12] J.-M. Valin, S. Yamamoto, J. Rouat, F. Michaud, K. Nakadai, and H. G. Okuno, "Robust recognition of simultaneous speech by a mobile robot," *IEEE Transactions on Robotics*, vol. 23, no. 4, pp. 742–752, 2007.

[13] I. Hara, F. Asano, H. Asoh et al., "Robust speech interface based on audio and video information fusion for humanoid HRP-2," in *Proceedings of the IEEE/RSJ International Conference on Intelligent Robots and Systems (IROS '04)*, vol. 3, pp. 2404–2410, September-October 2004.

[14] J. C. Murray, H. R. Erwin, and S. Wermter, "Robotic sound-source localisation architecture using cross-correlation and recurrent neural networks," *Neural Networks*, vol. 22, no. 2, pp. 173–189, 2009.

[15] H.-D. Kim, J.-S. Choi, and M. Kim, "Human-robot interaction in real environments by audio-visual integration," *International Journal of Control, Automation and Systems*, vol. 5, no. 1, pp. 61–69, 2007.

[16] B.-C. Park, K.-D. Ban, K.-C. Kwak, and H.-S. Yoon, "Sound source localization based on audio-visual information for intelligent service robots," in *Proceedings of the 8th International Symposium on Advanced Intelligent Systems (ISIS '07)*, pp. 364–367, Sokcho, South Korea, 2007.

[17] A. Badali, J.-M. Valin, F. Michaud, and P. Aarabi, "Evaluating real-time audio localization algorithms for artificial audition in robotics," in *Proceedings of the IEEE/RSJ International Conference on Intelligent Robots and Systems (IROS '09)*, pp. 2033–2038, St. Louis, Miss, USA, October 2009.

[18] D. Bohus and E. Horvitz, "Facilitating multiparty dialog with gaze, gesture, and speech," in *Proceedings of the International Conference on Multimodal Interfaces and the Workshop on Machine Learning for Multimodal Interaction (ICMI-MLMI '10)*, pp. 5:1–5:8, Beijing, China, November 2010.

[19] K. Nakadai, H. G. Okuno, and H. Kitano, "Real-time sound source localization and separation for robot audition," in *Proceedings of the IEEE International Conference on Spoken Language Processing (ICSLP '02)*, pp. 193–196, Denver, Colo, USA, September 2002.

[20] V. M. Trifa, A. Koene, J. Morén, and G. Cheng, "Real-time acoustic source localization in noisy environments for human-robot multimodal interaction," in *Proceedings of the 16th IEEE International Conference on Robot and Human Interactive Communication (RO-MAN '07)*, pp. 393–398, August 2007.

[21] J. G. Trafton, M. D. Bugajska, B. R. Fransen, and R. M. Ratwani, "Integrating vision and audition within a cognitive architecture to track conversations," in *Proceedings of the 3rd ACM/IEEE International Conference on Human-Robot Interaction (HRI '08)*, pp. 201–208, ACM, March 2008.

[22] K. Teachasrisaksakul, N. Iemcha-Od, S. Thiemjarus, and C. Polprasert, "Speaker tracking module for indoor robot navigation," in *Proceedings of the 9th International Conference on Electrical Engineering/Electronics, Computer, Telecommunications and Information Technology (ECTI-CON '12)*, 4, 1 pages, May 2012.

[23] X. Li, M. Shen, W. Wang, and H. Liu, "Real-time Sound source localization for a mobile robot based on the guided spectral-temporal position method," *International Journal of Advanced Robotic Systems*, vol. 9, article 78, 2012.

[24] B. Fransen, V. Morariu, E. Martinson et al., "Using vision, acoustics, and natural language for disambiguation," in *Proceedings of the ACM/IEEE International Conference on Human-Robot Interaction ((HRI '07)*, pp. 73–80, Arlington, Va, USA, March 2007.

[25] K. Nakadai, S. Yamamoto, H. G. Okuno, H. Nakajima, Y. Hasegawa, and H. Tsujino, "A robot referee for rock-paper-scissors sound games," in *Proceedings of the IEEE International Conference on Robotics and Automation (ICRA '08)*, pp. 3469–3474, IEEE, Pasadena, Calif, USA, May 2008.

[26] I. Nishimuta, K. Itoyama, K. Yoshii, and H. G. Okuno, "Toward a quizmaster robot for speech-based multiparty interaction," *Advanced Robotics*, vol. 29, no. 18, pp. 1205–1219, 2015.

[27] H. M. Do, W. Sheng, and M. Liu, "An open platform of auditory perception for home service robots," in *PRoceedings of the IEEE/RSJ International Conference on Intelligent Robots and Systems (IROS '15)*, pp. 6161–6166, IEEE, Hamburg, Germany, September-October 2015.

[28] A. G. Brooks and C. Breazeal, "Working with robots and objects: revisiting deictic reference for achieving spatial common ground," in *Proceedings of the 1st ACM SIGCHI/SIGART Conference on Human-robot Interaction (HRI '06)*, pp. 297–304, Salt Lake City, UT, USA, March 2006.

[29] O. Sugiyama, T. Kanda, M. Imai, H. Ishiguro, and N. Hagita, "A model of natural deictic interaction," *Human-Robot Interaction in Social Robotics*, vol. 104, 2012.

[30] H. G. Okuno, K. Nakadai, K.-I. Hidai, H. Mizoguchi, and H. Kitano, "Human–robot non-verbal interaction empowered by real-time auditory and visual multiple-talker tracking," *Advanced Robotics*, vol. 17, no. 2, pp. 115–130, 2003.

[31] A. Atkin, "Peirce's theory of signs," in *The Stanford Encyclopedia of Philosophy*, E. N. Zalta, Ed., 2013.

[32] D. Kaplan, "Dthat," in *Syntax and Semantics*, P. Cole, Ed., vol. 9, pp. 221–243, Academic Press, New York, NY, USA, 1978.

[33] L. Pineda and G. Garza, "A model for multimodal reference resolution," *Computational Linguistics*, vol. 26, no. 2, pp. 139–193, 2000.

[34] R. Mitkov, *Anaphora Resolution*, vol. 134, Longman, London, UK, 2002.

[35] J. R. Tetreault, "A corpus-based evaluation of centering and pronoun resolution," *Computational Linguistics*, vol. 27, no. 4, pp. 507–520, 2001.

[36] V. Ng and C. Cardie, "Improving machine learning approaches to coreference resolution," in *Proceedings of the 40th Annual Meeting on Association for Computational Linguistics*, pp. 104–111, Association for Computational Linguistics, 2002.

[37] R. A. Brooks, "How to build complete creatures rather than isolated cognitive simulators," in *Architectures for Intelligence*, K. VanLehn, Ed., pp. 225–239, 1991.

[38] R. P. Bonasso, R. J. Firby, E. Gat, D. Kortenkamp, D. P. Miller, and M. G. Slack, "Experiences with an architecture for intelligent, reactive agents," *Journal of Experimental and Theoretical Artificial Intelligence*, vol. 9, no. 2-3, pp. 237–256, 1997.

[39] L. A. Pineda, I. Meza, H. H. Avilés et al., "IOCA: an interaction-oriented cognitive architecture," *Research in Computer Science*, vol. 54, pp. 273–284, 2011.

[40] I. Meza, C. Rascon, and L. A. Pineda, "Practical speech recognition for contextualized service robots," in *Advances in Soft Computing and Its Applications: 12th Mexican International Conference on Artificial Intelligence, MICAI 2013, Mexico City, Mexico, November 24–30, 2013, Proceedings, Part II*, vol. 8266 of *Lecture Notes in Computer Science*, pp. 423–434, Springer, Berlin, Germany, 2013.

[41] L. Pineda and G. Golme, "Grupo Golem: robocup@home," in *Proceedings of the RoboCup*, 8 pages, RoboCup Federation, Eindhoven, The Netherlands, June 2013.

[42] C. Rascon, I. Meza, G. Fuentes, L. Salinas, and L. A. Pineda, "Integration of the multi-doa estimation functionality to human-robot interaction," *International Journal of Advanced Robotic Systems*, vol. 12, no. 8, 2015.

[43] C. Rascon, G. Fuentes, and I. Meza, "Lightweight multi-DOA tracking of mobile speech sources," *EURASIP Journal on Audio, Speech, and Music Processing*, vol. 2015, no. 1, pp. 1–16, 2015.

[44] P. Davis, JACK Connecting a World of Audio, http://jackaudio.org.

[45] T. van der Zant and T. Wisspeintner, "RoboCup X: a proposal for a new league where RoboCup goes real world," in *RoboCup 2005: Robot Soccer World Cup IX*, A. Bredenfeld, A. Jacoff, I. Noda, and Y. Takahashi, Eds., vol. 4020 of *Lecture Notes in Computer Science*, pp. 166–172, 2006.

[46] T. Wisspeintner, T. van der Zant, L. Iocchi, and S. Schiffer, "RoboCup@Home: scientific competition and benchmarking for domestic service robots," *Interaction Studies*, vol. 10, no. 3, pp. 392–426, 2009.

4

Learning by Teaching with Humanoid Robot: A New Powerful Experimental Tool to Improve Children's Learning Ability

Frank Jamet,[1,2,3] **Olivier Masson**,[1,2] **Baptiste Jacquet**,[1,2] **Jean-Louis Stilgenbauer**,[1,2,4] **and Jean Baratgin**[1,2,5]

[1] *P-A-R-I-S Association, 25 rue Henri Barbusse, 75005 Paris, France*
[2] *Laboratoire CHArt, Université Paris VIII, EPHE, 4-14 rue Ferrus, 75014 Paris, France*
[3] *Université de Cergy-Pontoise, 33 boulevard du Port, 95011 Cergy-Pontoise Cedex, France*
[4] *Facultés Libres de Philosophie et de Psychologie (IPC), 70 av. Denfert-Rochereau, 75014 Paris, France*
[5] *Institut Jean Nicod (IJN), École Normale Supérieure (ENS), 29 rue d'Ulm, 75005 Paris, France*

Correspondence should be addressed to Jean Baratgin; jean.baratgin@univ-paris8.fr

Academic Editor: Gordon R. Pennock

Browsing the literature shows that an increasing number of authors choose to use the *learning by teaching* approach in the field of educational robotics. The goal of this paper is, on the one hand, to produce a review of articles describing the effects of this approach on learning and, on the other hand, to review the literature in order to explore the characteristics at the core of this approach. We will only focus on the work using a humanoid robot. The areas of learning studied are writing, reading, vocabulary, and reasoning, but also there are some metacognitive abilities like task commitment and mental state attribution. Their targets are from very young children to preadolescents. We can already notice some studies on pupils with special educational needs. In all of these domains, the results show a nonnegligible effect of *learning by teaching* both on learning and on metacognitive abilities. If the concept of *learning by teaching* is clear, a careful investigation of the different studies shows that experimental paradigms do not use the same basic characteristics. For some, it is the robot's weakness, the care that must be given to it, which is the main requirement for the approach, while for others it is the unbalanced distribution of knowledge which is at the heart of it. The *learning by teaching* approach we will study has two components: the robot and the child tutor. The characteristics of the robot and what is asked of the child to accomplish his or her task of the tutor will be analyzed.

1. Introduction

"It is by teaching that men learn." With this sentence, the Greek philosopher Seneca (4BC-65AC) lays the foundations, a bit less than 2000 years ago, of a new teaching method consisting in learning through the act of teaching. This method is now commonly referred to as *learning by teaching*. Its main interest is that it allows the tutor (the teacher) to conceptually increase his or her level of knowledge [1, 2]. For some years now, we see an increasing number of studies in which a human teaches a humanoid robot [3–11]. Teachings were on a great variety of areas such as reading [4], writing [6, 7, 10], language [5], reasoning [11, 12], but also metacognitive abilities like commitment to a task and mental states attribution [13]. These studies concentrate on young children (3 to 6 years old) [5, 11, 12], children (7 to 8 years old) [4, 6], preteenagers (13 to 14 years old) [7], and adults [8, 14]. They also focus both on typical students and on those with special educational needs: children with visuoconstructive deficits, with poor handwriting, and even poorer self-confidence or with attention disorders [7], with autism spectrum disorders [15], and with potential symptoms of a developmental disability [16]. In this review we will focus on studies with children.

2. A Pupil Named NAO

The use of humanoid robot [for a review of possible robots in education see [17]] with children has numerous advantages

Figure 1: NAO and the fruits used in [11].

[18]. For example, humanoid robot facilitates the consideration of the robot as a social person by children [19–21] and seems to help children concentrate [22]. Furthermore, children are motived and have a positive attitude when seeing robots in their classroom [23–25].

In the majority of these studies on *learning by teaching*, the robot used was NAO (Figure 1): a bipedal, humanoid robot of 58 cm, with 25 degrees of freedom, cameras, and vocal capacities (synthesis and recognition), which can be programmed to carry out autonomously a set of tasks. This robot has many specific advantages that can be beneficial for experimental work with children, especially in the field of education [31]. By taking up the classification of the types of use of robots in the robotics studies applied by [32], NAO has five advantages when using it for *learning by teaching*.

(1) Height. This characteristic is mainly advanced to justify the choice of NAO to establish proximity, by its resemblance to a child [33].

(2) Movement precision. NAO has 25 degrees of freedom 25 degrees of freedom allowing it to perform movements close to human movements (walk, gestures, tilt its head, and appearing to look at something). NAO is able to manipulate small objects [5, 11], to make deictic gestures useful for learning [4, 11], and to mimic gestures or signs [34].

(3) Verbal interaction abilities. NAO has a speech synthesis module, which allows him to utter sentences aloud and to speak to the child orally [35]. In addition, the rate and tone of his voice are both easily parameterizable, which is particularly useful in *learning by teaching*, for example, in learning how to read.

(4) Nonverbal and emotional interaction abilities. Although devoid of facial expressions, NAO can express emotions (anger, fear, joy, sadness, and surprise) as much as a robot with this ability, such as iCat [33]. The way NAO can simulate emotions is through body movements and LEDs around the eyes (blinking and changing color). Hence the absence of facial expression of NAO does not constitute a barrier to the proximity felt by the child.

(5) Accessibility in terms of use and programming. NAO can be programmed by robotic and computer science neophytes thanks to the software Choregraphe (Aldebaran, Softbank Robotics), presenting a visual interface by which it is possible to program relatively complex behaviors in a simple way.

The aims of this paper are (1) to list the effects on learning when a human teaches a humanoid robot; (2) to analyze the different experimental characteristics of the humanoid robot; (3) to carefully analyze what is asked of the human; and finally (4) to show how the *learning by teaching* approach is a new paradigm for developmental and cognitive psychology.

3. Effects of the *Learning by Teaching* Approach on Learning

In this first part, we will be interested in the concept of *learning by teaching* by only focusing on the results of experimental studies in educational robotics which relies on the use of humanoid robots with children.

Five great areas of essential learning are set out: reading, writing, language, reasoning, and metacognitive abilities. They are summarized in Table 1.

3.1. Reading. To our knowledge, only the study of Yadollahi et al. [4] uses the approach of *learning by teaching* with a humanoid robot to develop reading abilities. In their setup described below, the authors [4] are interested in the effect of the robot's deictic gestures during reading on the reading ability of the child. In this study, the deictic gestures consist in the finger moving during reading.

The setup is the following.

The child is seated at a table. In front of them is a book. NAO is on the table. The child and the robot can both read the book. The task of the robot is to read a part of the book and then, indicating that it is getting tired, the child continues reading. If the child finds a mistake, they signal it to the robot by pressing a red button. If they need to hear it again, they press a yellow button, and they congratulate it by pressing a green button. Depending on the feedback of the child, the robot reacts verbally, physically, and emotionally (eyes changing color).

Two groups of children between 6 and 7 years old are defined, based on their reading ability (high reading ability versus low reading ability). Good readers will work on a 36 pages long text, and the others on a 24 pages long text. Both groups are split between two conditions: in the first condition, the robot reads aloud with its finger (pointing), and in the second condition it does not follow with its finger (nonpointing). Two sessions of 30 minutes over two days are offered. If a good reader teaches NAO how to read, NAO reads the first 10 pages then indicates getting tired, and the child continues until page 16. The following day, the child finds NAO telling them that it is eager to know the rest of the story. The task continues following the same principles. NAO reads pages 16 to 26 and then indicates getting tired. The child finishes reading. For less advanced readers, the protocol is identical, except for the number of pages: NAO reads 9 pages

Table 1: Main studies of *Learning by Teaching* according to the characteristics of the "pupil robot" and the "child teacher".

Research	Robots Type	Robot's characteristics	Children's characteristic	Collaboration' phenomenon
Reading				
[4]	Nao	Autonomous	6 to 7 years old (Deictic gesture, pointing)	Protégé effect
Writing *Handwriting*				
[6]	Nao V4	Autonomous	6 to 8 years old 7 to 8 years old	Induce agency anthropomorphizing. Cognitively engage
[7]	Nao V4, V5	Autonomous	5 to 8 years old with SEN*	Induce agency anthropomorphizing. Cognitively engage
[26]	Nao	Autonomous	7 to 9 years old	Corrective feedback as a peer assessment approach Alone with NAO
[27]	Nao	Autonomous	5 to 6 years old	Protégé effect Corrective feedback
[10]	Nao	Autonomous	7 to 9 years old age	
Language				
[5]	Nao	Tele-operated	3 to 6 years old	Care-receiving robot
[28]	Pepper	Total Physical Response	4 to 5 years old	Care-receiving robot
Reasoning				
[11, 12]	Nao V5	Tele-operated, designed to give the impression to be ignorant	5 to 6 years old	Alone with NAO NAO is silly Teacher status
Metacognitive abilities *Feasibility of collaborative learning* *Learning time*				
[15, 16]	Ifbot	Tele-operated, Limited number of expressions and its arms and body are immobile.	8 to 11 years old with SEN**	Unable to answer correctly Learning capability
Children's engagement Attribution of mental states				
[29]	Epi	Designed to give the impression of being a child while still being decidedly robotic.	5 to 9 years old	
Handwriting Children's engagement				
[30]	NaoV4, V5	Tele-operated	5 years old 6,05 to 8 years old with SEN*	Protégé effect

*Children with special educational needs: visuoconstructive deficits and poor handwriting and even poorer self-confidence.
**Children with special educational needs: autism spectrum disorder and with potential symptoms of developmental disability.

and indicates getting tired, and the child continues reading 3 pages. NAO's reading speed can be adjusted. It is set to 60% of normal speed for weaker readers and 80% for good readers.

Three types of mistakes are studied: mistakes of type I: reading "elephant" instead of "penguins," mistakes of type II: reading "start" instead of "stop," and mistakes of type III: pronunciation mistakes or syntax issues. These mistakes correspond to Goodman's Miscue corpus [36, 37]. The NAO robot makes 12 mistakes during each session.

Results show (1) that there is no effect of the condition (pointing *versus* nonpointing) on the quality of the correction; (2) that pointing actually improves the detection of mistakes of type I ("elephant" instead of "penguins"); and (3) that children of both levels of reading ability benefit from pointing when they must recognize mistakes that do not correspond to the context picture (type I). For readers with a low reading ability, pointing does not improve text comprehension.

3.2. Writing. Learning how to write is a complex task as it is a lengthy process and requires repeating the same gesture. Discouragement can quickly disturb the acquisition of the correct gesture. The challenge faced is thus consequent. Despite these difficulties, this area has numerous studies conducted with the approach of *learning by teaching* with the NAO robot [4, 6, 9]

In its design, the experimental setup must allow the child to draw many times and to be motivating in order for it to be efficient. To this we can add the additional difficulty: giving NAO the ability to write, even though it only has three fingers, and as a result, drawing cursive letters is impossible. As a consequence, the questions that need to be answered are the following: How to make a robot that only has three fingers be able to write? Does the setup allow the robot and the child to draw enough? Is it not tiresome? Does the setup allow the child to evaluate the robot? Are the relationships between children and the robot not influenced by the existence or absence of the robot's progress? Is the child able to adapt his or her teaching? Does the progress of NAO give the child the impression that they are a better teacher?

To address the issue of NAO's three fingers, the authors [6] synchronize a tablet. Thus, NAO is able to simulate with its finger the drawing of a letter on the tablet. It contains two writing zones: one on the left, spanning the 2/3rd of the tablet for NAO, and the one for the child, on the right, smaller. The child can draw and erase letters.

The task of the child of 6-8 years old is to teach NAO how to draw letters (lowercase and uppercase). NAO draws one of these letters and asks the child to correct it. NAO improves. The effects of feedback [38] and regular evaluation are determining factors [39]. Mistakes made by NAO when drawing letters are not the result of chance; they are observed during the development of the learning task: proportions, breaks, and the alignment [40, 41].

Does the setup allow the robot and the child to draw enough? Is it not tiresome? Results show that over the two sessions (respectively, 65 and 160 minutes long), the interaction is effective and efficient. The robot and the child both produce many drawings of letters: 96 letters, among which 49 in response to demonstrations with 6-7-year-old children, and 335 traces among which 152 in response to interventions of 7-8-year-old children.

In the teaching act, evaluating is an important concept. Was the setup efficient in this aspect? Did the progress of NAO influence the relationship with the child? To answer this question, the authors [10] randomly assign 25 7-9-year-old children in two conditions: one in which NAO makes progress: "NAO learns," and the other in which it does not: "NAO does not learn." NAO keeps all of its social qualities. Each condition contains four 15 minutes long sessions. The authors add a pretest and a posttest in which they measure the quality of the child's writing (the ability to recognize letters and the ability to reproduce them) and ask the child about the progress of NAO along with the quality of their relationships with NAO. Results show that, in the condition "NAO learns," children after the third session evaluate more favorably the performance of NAO than in the condition "NAO does not learn." The same difference is observable in the fourth session. The relationships with the robot (attachment) are not influenced by the robot's performance.

Is the child able to adapt his or her teaching? Does the progress of NAO give the child the impression that they are a better teacher? The authors [26] use the same experimental setup as in previous studies but add to the posttest an evaluation of children's perceived self-efficacy towards tutoring. The robot is called Michael. The experimenter explains to the children (from 7 to 9 years old) that they will meet Michael four times. It needs help to learn how to write. The experimenter explains the role of Michael as a pupil and the role of the children as teachers. NAO is then placed in the room alone with the child. NAO then politely introduces itself and tells the child about its difficulties with learning how to write by showing them his grades. Children are randomly assigned to the two conditions "NAO learns" and "NAO does not learn." Each condition contains four 15 minutes long sessions, one per week.

Does the robot's space arrangement (face-to-face or side-by-side) affect the child's focus of attention and perception of the robot's performance in a situation of *learning by teaching* of writer? The authors [27] use the same experimental setup as in previous studies [4, 6, 9] on the cowriting. The children are distributed under two conditions: the face-to-face condition with the NAO and the condition where NAO is next to the child.

Results show that children (1) do pay attention both to the grades and to the robot's progress; (2) do notice the robot's progress starting in the third session only in the condition "NAO learns"; (3) do give a better evaluation during the last session in the condition "NAO learns" compared to the condition "NAO does not learn"; (4) enjoy teaching more and more with the robot's progress; (5) do feel like better teachers after each interaction; (6) in the condition "NAO learns" do improve their performance between the pretest and the posttest; (7) are not influenced by the robot's performance in their attachment towards it; (8) only found the robot to be intelligent after the last session in the condition "NAO learns"; and (9) if the face-to-face versus side-by-side position does not affect the quality of the learning, the quality of the feedback changes. Children provide more positive feedback in a side-by-side position.

3.3. Language. The works of Tanaka et al. [5] are interesting, on the one hand because it deals with learning in the domain of language, and on the other hand because it enriches the conceptual framework of *learning by teaching* with *care-receiving-robots* (CRR). The framework of CRR was introduced for the first time in 2007 by [42]. The authors [42] show that in this framework, children present many caretaking and caregiving behaviors. They increase the motivation of the child in their teaching task building on the fact that the robot is fragile, and that it needs to be cared for. Subsequently, the authors [28] added to CRR the framework of Total Physical Response (TPR) developed by [43] in the 1960s. The TPR approach is a specific second language teaching method, which consists of associating a sentence with physical movement; for example, to learn the sentence "driving a car," the student simulates driving by maneuvering a fictional steering wheel with both hands. This teaching method applies to both children and adults. It allows a pleasant learning and stress-free learning.

3.3.1. Learning English Action Verbs. The goal of the experiment of [5] is to evaluate the incidence of *learning by teaching* on the learning of English action verbs in Japanese children between 3 and 6 years old. The experiment is divided into four phases: (1) pretest; (2) interaction with the NAO robot in two parts; (3) free time; and (4) posttest.

The goal of the pretest is to select, with the child, six action verbs from a set of graphic cards representing an action like drinking, sweeping, brushing, and playing. The task of the child is to associate the word card to the graphic card. If the child succeeds, the experimenter says: "it is correct." We keep playing until the child makes four mistakes.

The interaction with the NAO robot happens in two steps. In a first step, it is the experimenter who teaches the child, and then the child teaches NAO. The "lesson" happens as follows: the experimenter puts in front of NAO four items corresponding to four action verbs: a glass to drink, a hand brush to sweep, a brush to brush, and a toy to play. NAO, teleoperated, welcomes the child with his name and happily introduces itself by shaking their hand.

The experimenter has four cards at their disposal, among which two have the mention "with CRR" written on them. The experimenter takes a card randomly and asks the child: "Show me how to < verb >." If the child cannot answer, the experimenter points to an item while pronouncing the verb and continues. If the experimenter takes a CRR card, they ask NAO to answer. NAO nods and says: "Please take . . ." while opening its hand. The experimenter takes the item and gives it to NAO. NAO mimics the action but makes mistakes. The experimenter tells it: "No, it's incorrect." NAO then asks: "Please, teach me." The experimenter takes the robot's hand and teaches it step by step. The teleoperator record the learning sequence. Once the learning is done, NAO acts the action while pronouncing the verb. The experimenter tells it: "Yes, it's correct." For the three other action verbs, the experimenter proceeds in the same way. During the second part, the protocol is identical but it is the child's turn to teach the robot.

During the free time, the child is invited to play with NAO. The experimenter observes the child for 10 minutes. Finally, the experimenter begins a first posttest. The content is the same as in the pretest. A second posttest will be done 3 to 5 weeks later.

Results show that (1) performances in learning the action verbs are significantly improved when the child teaches the robot instead of when the experimenter teaches the child; (2) three to five weeks later, performances have increased even further; (3) the child uses different forms of teaching: a verbal teaching (the child states the action in English); a gestural teaching (the child moves NAO to do the action); a direct teaching (the child helps and back NAO step by step). The two first forms of teaching are the most commonly used (verbal and gestural), and (4) the fact of having brothers, sisters, or a dog has a significant influence on learning performance. The question of the incidence of gender should be investigated in future studies.

3.3.2. Learning of Nonnative Language. To allow learning of the nonnative language, at home, [28] proposes two methodologies in using the humanoid robot Pepper. This robot is 1.21m high and 42 cm wide and 48 cm thick. It weighs 28 kg and can move at a maximum speed of 3 km/h. A touch screen is placed on his chest. The English learning course can take place in two ways: (1) with a teacher in real time at the view on a screen. In this situation, Pepper is a student like any other. It is programmed to follow the teacher's gaze. The results show that he favors the "circle time," that is to say the moment when students gather around the teacher to listen to the instruction and (2) with a video that children can see on the chest of the robot Pepper. The originality of the learning situation is that it crosses the *learning by teaching* and the TPR. The child learns English using the TPR. He must also teach the Pepper robot using *learning by teaching* and TPR. These two learning modalities (*learning by teaching* and TPR) are added.

3.4. Reasoning. To our knowledge, studies of this kind in educative robotics are rare. Only one study [11] shows an interest in the question of class inclusion, a fundamental notion in the cognitive development of children. Studying class inclusion allows us to understand the links between "all" and "parts." "All" is the sum of all parts. It is Piaget [44] who is the first to study the development of this notion in his famous flower task. He shows a bunch of flowers composed of four asters and two tulips. He asks the child "Are there more flowers or more asters?" It is necessary to wait until the age of 8 years old to have children answering correctly "there are more flowers." Before the age of 7, the answer given is "There are more asters." Piaget et al. [45] explain the incorrect answer of the children by the absence of reversibility in the child's thinking. If the class flowers = asters + tulips, then tulips + asters = flowers. Politzer [46] suggests a new pragmatic linguistic relevance-theoretic explanation of children's answer "more asters." The task is a difficult one [45] because it requires the child to be mastering the logic of classes but also judgments of relations (more than, less than, same as). Furthermore, the question is ambiguous [46] in the sense that two interpretations are possible and relevant. Indeed, if the child interprets the notion of flowers as being the "all," then there are indeed more flowers than asters, but if the child interprets the term flowers as what is not an aster then the answer "there are more asters" is actually correct. From the linguistic point of view, these two interpretations are correct. From the point of view of the pragmatics of language, comparisons generally happen on classes at the same level (in our case, between hyponyms apple and pear or asters, and tulips). The answer of the 7-year-old children illustrates this: "there are more apples." For psychologists, this is an important question: do children master or not the notion of class inclusion at the age of 6-7? Age when, for Piaget et al. [44], the majority of subjects gives the answer "more asters?" Indeed, children of 5-6 years old are perfectly capable of understanding who is intelligent and who is not, who is qualified and who is not [47].

The originality of the work of Masson et al. [11, 12] is twofold: the authors first investigate the evolution of this central question in 1921 and then suggest an experimental paradigm to remove the pragmatic ambiguity within the question itself: after having indicated that the robot is called NAO, the experimenter explains to children their task: teaching new things to NAO, who knows nothing. The experimenter explicitly indicates to the children that they know much more than NAO. Here, children do not know beforehand what they will teach NAO. The experimenter tells the children that they will have to be careful as NAO's questions may seem silly to them. The children will have to answer spontaneously without thinking too much about the question itself or the robot's intention. The experimenter insists that if children do not agree with a suggestion from the robot, they are welcome to correct it. It is important to emphasize that children will be left alone with the robot. The experimental design here is a semiteleoperated situation where the experimenter is in another room without the child being able to see him or her. NAO is programmed to ask in an initial stage questions that should reinforce the children's belief in them and in the robot's great ignorance. Hence the child is in a position of a teacher, a person who knows, in opposition to NAO who is the "young" pupil.

The authors have integrated into the robot a program to execute the inclusion task. The important aspect is here. It is NAO during its questioning that will drive the child to teach it the link between the class and the subclass. It concerns the class fruit: 5 apples and 3 pears. Through a computer, the robot is teleoperated without the child's knowledge by one of the authors hidden behind a wall from the child's sight. The second experimenter brings the child from their classroom to the room where the experience takes place. On the way, the experimenter explains to the child the didactic contract. The child enters the room and sits at a table on which NAO is placed in a sitting position. The child and NAO are in front of each other, with the robot at the same level as the child so that it does not feel dominating. The experimenter that accompanied the child then leaves the room. The child is alone with the robot. NAO introduces itself, asking for the child's name and if they want to be its teacher. NAO will use the child's name to address them. NAO tells them that it knows nothing and that it might ask really silly questions. NAO points at the different items on the table and asks what they are. The child answers. NAO makes sure that it has understood by pointing at the different items again and repeating their name, but it makes mistakes. The child repeats the teaching phase until NAO is able to identify the class (fruits) and the two subclasses (apples and pears). In order to perfect its knowledge, NAO tries to know if "there are more fruits than apples?" The child answers. NAO when reformulating the answer makes a mistake: if the child tells it: "there are more fruits," NAO answers: "Oh ok so there are more apples." If the child says "there are more apples," NAO continues by saying: "Oh ok so there are more fruits." The child corrects the robot's mistake. The dialogue between the child and NAO cannot be reduced to a system of preestablished question-answers, but we are in a true conversation between two conversation partners which explains why the robot is teleoperated.

This experiment has been carried out with 40 children in preschool aged between 5 and 6. They were randomly assigned to two conditions: one condition human-human and one condition NAO-human. The authors notice that one child out of five answers correctly in the human-human condition, and more than six children out of ten do in the condition NAO-human. The difference is significant. This result shows that (1) children master the notion of class inclusion earlier than Piaget et al. [45] indicated and (2) the pragmatic factors identified by Politzer [46] constitute a real obstacle in the original experiment.

A qualitative analysis shows that children are actually immersed in a conversation rather than in a list of questions and answers. If at the beginning of the conversation children are rather calm and patient, accepting that NAO makes mistakes in the identification of the different fruits ("this is an apple, this is a pear"), it should not take too long. Indeed, children will usually take the initiative and show NAO that an object is an apple and that another is a pear by pointing at them. Although they remain polite, the tone of their voice is firm and decisive: "Listen little robot, this is an apple and this is a pear." That way they put an end to any ambiguity. They implicitly tell NAO that it must pay attention and try to understand! Similar expressions are found when NAO has some trouble with the concept of fruit. After the robot fails a couple of times to answer the question: "are there more apples or more fruits?," the children move closer to NAO and tell it: "Listen little robot, it's all fruits." In all the conversations, the children remain polite and show respect to the robot.

3.5. Metacognitive Abilities. In this section, we will investigate the question of the effect of *learning by teaching* through three points of view: task commitment, motivation and self-confidence, and mental states attribution. These studies are interested in both typical children and those with special educational needs.

3.5.1. Task Commitment. Task commitment is an essential question for teachers [5, 30, 48]. It is one of the factors on which the efficiency of learning relies. The *learning by teaching* approach results in a well-known "protégé effect" [13]. In the teacher-pupil relationship, the teacher is responsible for the pupil. This responsibility manifests itself through a stronger time investment of the teacher in the tasks. This phenomenon is observed in children (10-11 years old) and in preteenagers (13-14 years old) spending more time on one task when they must teach a teachable agent than if they focused on learning for themselves [13]. Between a teachable agent and a social robot, children are more interested in the social robot [29]. The spatial position of the robot in the interaction also has an influence. It is a powerful way of communicating nonverbal information [49, 50]. When face-to-face, we learn better [51].

Commitment is an even more delicate matter with children with special educational needs. Indeed, these pupils often show specific troubles associated with difficulties in focusing their attention or with an actual attention disorder. A preliminary question arises: Is the *learning by teaching* approach relevant? Does it allow to decrease the specific learning difficulties? Can the time dedicated to learning be increased for these children?

Is the *learning by teaching* approach relevant? In a first paper [15], authors wonder about the possibility for children with autism spectrum disorders to work collaboratively such as in a task of *learning by teaching*. They suggest an experimental setting in which two people (an 11-year-old child with an autism spectrum disorder, and a teacher) are with the conversational robot Ifbot. This robot has a limited number of expressions and does not move neither its arms nor its body. Ifbot is teleoperated and is programmed to make mistakes. The session takes place in two stages, with at first a time of familiarization with the robot (10 minutes), in which the child, the teacher and the robot play a game. Then the session continues with a time of collaborative work (40 minutes). The teacher asks alternatively to the child and to the robot to read a page of a comic book on the History of Japan aloud. Each time the robot makes a mistake, it says: "This question is difficult for me" or "Please teach me the right answer." It will then answer: "It's amazing" or "Yes, the answer was right. Very good" and might ask for clarifications: "What is it?"

Results show that (1) when Ifbot makes a mistake, the child says: "That is not correct." The child explains the robot its mistake quite willingly. To do this, the child comes closer to the robot and (2) if the robot makes too many consecutive mistakes, the child will tend to be a bit more reluctant in correcting it.

In another experiment, these same authors [15] use the same experimental design described above, but this time with two children with an autism spectrum disorder, and a teacher. The collaborative work consists in building, using Lego blocks, a story. During the conception of the story, the robot asks: "What is it?" and "Please explain that?" Each time the child answers the robot, the machine answers: "It is very good."

The results show that children are engaged in the task and that they explain the story to the robot. Thus, children with an autism spectrum disorder communicate with the robot. They are able to teach it and can explain the meaning of correct answers. We can conclude that the *learning by teaching* approach with a humanoid robot is perfectly relevant to pupils with specific educational needs. It is now up to us to see if it could be possible to increase the time these pupils dedicate to learning.

Results on learning of how to write show that these pupils with special educational needs are able to remain committed to a task during the entire length of the sessions: over two weeks, each child participated to a three hours long session [7]. In a case study [7], the authors show that a participant sent, months later, a letter to NAO to ask how far it had progressed in its learning. These pupils were able to evaluate the performance of NAO not only positively but also negatively. Thus, they show their ability as teachers, and this even while being very young (5 years old and 5 months). If these results are promising, it should not be dismissed that an adult (experimenter or therapist) was also with the child during the interaction

To answer the question of the increase of the time dedicated to learning in children with potential symptoms of a developmental disability, authors [16] once again use Ifbot. Learning sessions were distributed across three weeks, but their actual duration is not reported. Children were distributed between three groups: a *learning by teaching* group (G1), a group where the child is alone (G2), and a group where children learn together. Effective learning time during sessions is measured. They show that (1) learning time in G1 is higher than in the two other groups (G2 and G3). In particular, there was a significant difference between group G1 (*learning by teaching*) and group G2 (child learning alone); (2) the duration of children teaching is higher in group G1 (*learning by teaching*) than in group G3 (children learning together); and (3) children consider the collaborative experience of teaching the robot quite enjoyable.

3.5.2. Motivation and Self-Confidence. The *learning by teaching* paradigm is known to increase motivation and self-confidence, and this independently from educational contents [48]. For pupils with specific educational needs, learning can only be done with a strong motivation and great self-confidence. It is by using the concept of the "Protégé Effect" [13] that authors [30] have designed an experimental setup allowing to motivate pupils with specific educational needs with long-term writing learning, despite the fact that they encounter difficulties in this area.

The experimental setup is as follows: NAO and the child both have a tactile tablet with a customized program. This program responds to the specific needs of the child (his or her drawing errors). The child draws and then moves the tablet towards NAO. NAO then simulates drawing on the tablet. The child gets the tablet back and corrects the drawing and presses a button to give it to NAO. NAO then simulates a new drawing, and so on. In order to build up the conversation with the child, NAO asks: "What do you think of it?" Depending on the abilities of the child, a basic scenario can be used for contextualization (NAO needs to know how to write because it tries to communicated with another NAO). To evaluate the robot, two buttons are offered: a red one, with a thumb pointing down for negative evaluations, and a green one with a thumb pointing up for positive evaluations.

Between three and four sessions of 60 minutes have been suggested, one per week to (1) a child with a strong inhibition and a weak motivation for writing; (2) a child with visuoconstructive deficits and focusing difficulties under the care of an occupational therapist; and (3) a group of height children with difficulties to lean cursive writing of different magnitudes. Some are under the care of occupational therapists, others of neurologists.

The results indicated that (1) the number of corrections increases with the sessions even though the drawings that must be realized get more and more complex for the child with writing difficulties and strong inhibition. The child spends more time correcting the robot and even creates an affective relationship with the robot (cries during the last session). The child even sent letters a month after the last session to inquire about the progress of NAO. The parents have reported noticing an evolution in the quality of their child's writing skill; (2) the child with visuoconstructive defects and attentional difficulties remained engaged in the task more than 40 minutes each of the 4 sessions. The child's progress is massive (13 out of 26 during the first session, 26 out of 29 in the last one); and (3) all children remained in the activity during all sessions. They gave more positive evaluations to NAO (99) than they gave negative ones (33).

3.5.3. Mental States Attribution. The incidence of *learning by teaching* on mental states attribution has also been explored. Studies have shown that children 10-11 years old and also preteenagers (13-14 years old) consider a teachable agent as a social entity through the attribution of mental states and responsibility [13]. With younger children (5 to 9 years old), Lindberg et al. [29] observe that mental states attribution to a social robot is similar to a teachable agent. The authors notice wide between-subject variations regarding the recognition of the robot's intelligence. In contrast, the teachable agent received a higher score in anthropomorphism than the social robot [29].

4. Characteristics of the *Learning by Teaching* Paradigm

After briefly recalling the historical context of the *learning by teaching* approach, we will extract from the literature the specificities attributed to the humanoid robot, then we will do the same regarding what is asked of the tutor (the child).

4.1. Brief Historical Context. After Seneca, the idea that it is by teaching that we learn will be used again in the XVIIIth century, in England, with the "*monitorial-system.*" Facing a lack of teachers, some children or adults are asked, after taking into consideration their abilities, to teach their peer. A similar concept can be found in France with schools of mutual teaching at the beginning of the XIXth century [52], and later in the Freinet pedagogy, or of Montessori. Starting in the eighties, Martin, professor in language didactics, used the approach of "Lernen durch Lehren" to develop the teaching of the French language to German students. This educational method will then widely spread in France [53–56] then, during recent years, in educational robotics.

4.2. Features of Robot as Tutee. Three main characteristics can be highlighted. The humanoid robot is programmed to make mistakes [15, 16, 56]. The mistakes are never random. They always correspond to specific mistakes observed in the development of the ability, studied for reading [5], for writing [4], and for reasoning [11, 12]. These mistakes decrease during learning and they increase the tutor's motivation. The humanoid robot must show learning abilities by improving its skills, displaying as a consequence its intelligence [47]. In all the experiments, the humanoid robot displays a strong social dimension. From our point of view, Tanaka et al. [5] bring a real added value by using the possibility of recording in real time the new behavior that was just learned. Furthermore, it allows for the integration in the robot of different kinds of teaching (verbal, gestural, etc.) and thus to answer more accurately the pupils' various needs. Masson et al. [11, 12] give the humanoid robot a status of ignorant in order to strengthen the position of the child as being the teacher. Other authors use a fictive grading of the robot's performance to show its progress [26] and in doing so display the efficiency of the teaching offered by the child.

4.3. Features of the Child as Tutor. The specificities of the tutor, or rather what is asked of the child, are relatively less developed in the literature. In the studies on learning of drawing letters, the experimenter asks the child to learn to draw letters in lowercase and in uppercase. The experimenter tells the child that Michael (name of the robot) needs help to learn how to write. The didactic contract is more implicit than it is explicit. Tanaka et al. [5] add to the approach of *learning by teaching* the framework of CRR. This concept is interesting for it strengthens the position of teacher for the child. In contrast, the child does not receive explicit instructions to teach.

Masson, et al. [11, 12] bring another clear added value by elaborating a clear and explicit didactic contract between the tutee (the robot) and the tutor (the child). The methodology used can be generalized to any form of learning. This contract clarifies some points: the goal targeted, the role of the child, the difference of knowledge between tutor and tutee is quantified, the position of teaching is prepared (the child is alone with the robot and can express themselves without any judgmental attitude).

The goal targeted: the experimenter only gives the child the instruction that they must teach NAO a lot of things.

The function is defined: "The child is NAO's teacher." NAO asks the child if they want to be its teacher.

The difference between the knowledge of the child and the knowledge of NAO is at its maximum: "You know a lot of things, and NAO knows nothing." The use of the binary quantification is easily understandable for the child. The experimenter adds a value judgment: "NAO is really silly." The difference is quantified through examples given by the experimenter to the child. This strengthens the role of teacher for the child: a teacher, by definition, knows. From the child's point of view, the situation suggested becomes believable.

The teaching position is also prepared. The experimenter indicates to the child that they will need to answer surprising questions; some might not even make sense. The child should not be surprised to be facing this kind of questions, because otherwise when facing a nonsensical question, the child might be driven to find a meaning that is not really there. Indeed, asking a question that does not make sense violates Grice's cooperation principle [57]. In the class inclusion task, the child is driven to answer another question than the one that has been initially asked. "Are there more flowers or more asters" implies a comparison between the number of flowers and the number of asters. If we consider that this question is surprising because there are necessarily more in "all" than in one "part," the child might infer that the real question was to compare numerically the subclass "asters" to the only other subclass "the other flowers," or "the flower-tulips." This question then gains meaning. Furthermore, it is at that age that the child learns how to count and learns to compare quantities. Children evaluate each other on these skills. How many times can we hear things such as: "You do not even know how to count." This, in turn, triggers in response the counting rhyme. Adults themselves to valorize the child can ask this kind of comparison. The fact that the child is alone in front of the robot, that no adult is present in the room, reduces pragmatic interferences and gives this teaching context all of its authenticity.

The fact that the child is alone with the robot strengthens their status of teacher, liberating them from constraints. The child cannot be judged, and their responses are more natural, more spontaneous.

5. Conclusion

To summarize, we will remember of this work the following points.

The *learning by teaching* approach with a humanoid robot is truly relevant. As we have shown in our review, it allows children to make real progress in various areas of learning such as reading, writing, language, and reasoning but also allows them to improve their tasks commitment and their focus. This approach is just as efficient in very young children (3 years old) as in older children. Its efficiency is also noticeable in children with special educational needs: children with visuoconstructive deficits, with poor handwriting or even poorer self-confidence, or with attention disorders.

If the characteristics of the tutee (the social robot) are well defined in the literature, it is our belief that a more clearly defined description of the didactic contract with the tutor is required. From our point of view, the targeted goal must be explicitly told to the tutor. The child must know (1) that they are asked to teach NAO; (2) that they have the status of the teacher, and in order to make it believable, it should be quantified; (3) the difference of knowledge between the child and NAO; (4) that the position of the teacher should be anticipated so that they will not be surprised by questions that do not make sense; and (5) that a child alone with the robot considerably decreases the weight of pragmatic implicatures and allows the psychologist to record more objective measures.

For psychologists, the *learning by teaching* approach with a human-humanoid robot interaction is a new experimental paradigm extremely promising. As we have shown, we can identify the genesis of the building of concepts by reducing effects of the pragmatics of language [58]. It is our belief that this new experimental paradigm is very important since it opens the way to a reexamination of the acquisitions of all the great concepts (number conservation, quantity, matter, weight, and volume) that participate in human intelligence (concepts that can remain problematic in teenagers [59]). New research is also incoming with NAO in the study of cognitive bias and their influence on decision-making [60, 61].

Conflicts of Interest

The authors declare that there are no conflicts of interest regarding the publication of this paper.

References

[1] L. Fiorella and R. E. Mayer, "The relative benefits of learning by teaching and teaching expectancy," *Contemporary Educational Psychology*, vol. 38, no. 4, pp. 281–288, 2013.

[2] W. J. Mchugh, "Team Learning in Skills Subjects in Intermediate Grades," *Journal of Education*, vol. 142, no. 2, pp. 22–51, 2018.

[3] G. Biswas, K. Lee, L. K. Belynne et al., "Incorporating self regulated learning techniques into learning by teaching environments," in *Proceedings of the 26th Annual Meeting of the Cognitive Science Society*, pp. 120–125, 2004.

[4] E. Yadollahi, W. Johal, A. Paiva, and P. Dillenbourg, "When deictic gestures in a robot can harm child-robot collaboration," in *Proceedings of the the 17th ACM Conference*, pp. 195–206, New York, NY, USA, June 2018.

[5] F. Tanaka and S. Matsuzoe, "Children Teach a Care-Receiving Robot to Promote Their Learning: Field Experiments in a Classroom for Vocabulary Learning," *Journal of Human-Robot Interaction*, pp. 78–95, 2012.

[6] D. Hood, S. Lemaignan, and P. Dillenbourg, "When Children Teach a Robot to Write: An Autonomous Teachable Humanoid Which Uses Simulated Handwriting," in *Proceedings of the 10th Annual ACM/IEEE International Conference on Human-Robot Interaction, HRI 2015*, pp. 83–90, New York, NY, USA, March 2015.

[7] S. Lemaignan, A. Jacq, D. Hood, F. Garcia, A. Paiva, and P. Dillenbourg, "Learning by Teaching a Robot: The Case of Handwriting," *IEEE Robotics and Automation Magazine*, vol. 23, no. 2, pp. 56–66, 2016.

[8] E. S. Kim, D. Leyzberg, K. M. Tsui, and B. Scassellati, "How people talk when teaching a robot," in *Proceedings of the 4th ACM/IEEE International Conference on Human-Robot Interaction, HRI'09*, pp. 23–30, USA, March 2009.

[9] W. Johal, J. Kennedy, V. Charisi, H. W. Park, G. Castellano, and P. Dillenbourg, "Robots for Learning - R4L," in *Proceedings of the Companion of the 2018 ACM/IEEE International Conference*, pp. 397-398, Chicago, IL, USA, March 2018.

[10] S. Chandra, R. Paradeda, H. Yin, P. Dillenbourg, R. Prada, and A. Paiva, "Affect of Robot's Competencies on Children's

Perception," in *Proceedings of the inProceedings of the 16th Conference on Autonomous Agents and MultiAgent Systems (AAMAS '17*, pp. 1490–1492, Richland, CA, USA, 2017.

[11] O. Masson, J. Baratgin, F. Jamet, F. Ruggieri, and D. Filatova, "Use a robot to serve experimental psychology: Some examples of methods with children and adults," in *Proceedings of the 2016 International Conference on Information and Digital Technologies, IDT 2016*, pp. 190–197, Rzeszo, Poland, July 2016.

[12] O. Masson, J. Baratgin, and F. Jamet, "NAO robot, a social clues transmitter: what impacts? The example with endowment effect," in *Proceedings of the 30th International Conference on Industrial, Engineering, Other Applications of Applied Intelligent Systems (IEA/AIE)*, pp. 559–568, Université d'Artois, Arras, France, 2017.

[13] C. C. Chase, D. B. Chin, M. A. Oppezzo, and D. L. Schwartz, "Teachable agents and the protégé effect: Increasing the effort towards learning," *Journal of Science Education and Technology*, vol. 18, no. 4, pp. 334–352, 2009.

[14] T. Chaffey, H. Kim, E. Nobrega, N. Lubold, and H. Pon-Barry, "Dyadic Stance in Natural Language Communication with a Teachable Robot," in *Proceedings of the Companion of the 2018 ACM/IEEE International Conference*, pp. 85-86, Chicago, IL, USA, March 2018.

[15] F. Jimenez, T. Yoshikawa, T. Furuhashi, M. Kanoh, and T. Nakamura, "Feasibility of Collaborative Learning between Robots and Children with Autism Spectrum Disorders," in *Proceedings of the International Workshop on Intervention of Children with Autism Spectrum Disorders using a Humanoid Robot*, 2015.

[16] F. Jimenez, T. Yoshikawa, T. Furuhashi, M. Kanoh, and T. Nakamura, "Collaborative learning between robots and children with potential symptoms of a developmental disability," in *Proceedings of the 2017 IEEE Symposium Series on Computational Intelligence (SSCI)*, pp. 1–5, November 2017.

[17] O. Mubin, C. J. Stevens, S. Shahid, A. A. Mahmud, and J. Dong, "A review of the applicability of robots in education," *Technology for Education and Learning*, vol. 1, no. 1, 2013.

[18] A. K. Pandey and R. Gelin, "Humanoid robots in education: A Short Review," in *Humanoid Robotics: A Reference*, A. Goswami and P. Vadakkepat, Eds., Springer Science, 2017.

[19] M. Alemi, A. Meghdari, and M. Ghazisaedy, "The Impact of Social Robotics on L2 Learners' Anxiety and Attitude in English Vocabulary Acquisition," *International Journal of Social Robotics*, vol. 7, no. 4, pp. 523–535, 2015.

[20] A. N. Meltzoff, R. Brooks, A. P. Shon, and R. P. N. Rao, ""Social" robots are psychological agents for infants: A test of gaze following," *Neural Networks*, vol. 23, no. 8-9, pp. 966–972, 2010.

[21] C.-W. Chang, J.-H. Lee, P.-Y. Chao, C.-Y. Wang, and G.-D. Chen, "Exploring the possibility of using humanoid robots as instructional tools for teaching a second language in primary school," *Journal of Educational Technology & Society*, vol. 3, no. 2, pp. 13–24, 2010.

[22] J. Han, M. Jo, V. Jones, and J. Jo, "Comparative Study on the Educational Use of Home Robots for Children," *Journal of Information Processing Systems*, vol. 4, no. 4, pp. 159–168, 2008.

[23] J. Choi, J. Lee, and J. Han, "Comparison of Cultural Acceptability for Educational Robots between Europe and Korea," *Journal of Information Processing Systems*, vol. 4, no. 3, pp. 97–102, 2008.

[24] . Han, "The Cross-cultural Acceptance of Tutoring Robots with Augmented Reality Services," *International Journal of Digital Content: Technology and its Applications*, vol. 3, no. 2, 2009.

[25] Y. C. Lin, T. C. Liu, M. Chang, and S. P. Yeh, "Exploring children's perceptions of the robots," *Lecture notes in computer science*, vol. 5670, pp. 512–517, 2009.

[26] S. Chandra, R. Paradeda, H. Yin, P. Dillenbourg, R. Prada, and A. Paiva, "Do Children Perceive Whether a Robotic Peer is Learning or Not?" in *Proceedings of the the 2018 ACM/IEEE International Conference*, pp. 41–49, Chicago, IL, USA, March 2018.

[27] W. Johal, A. Jacq, A. Paiva, and P. Dillenbourg, "Child-robot spatial arrangement in a learning by teaching activity," in *Proceedings of the 25th IEEE International Symposium on Robot and Human Interactive Communication, RO-MAN 2016*, pp. 533–538, August 2016.

[28] F. Tanaka, K. Isshiki, F. Takahashi, M. Uekusa, R. Sei, and K. Hayashi, "Pepper learns together with children: Development of an educational application," in *Proceedings of the 15th IEEE RAS International Conference on Humanoid Robots, Humanoids 2015*, pp. 270–275, Seoul, Republic of Korea, November 2015.

[29] M. Lindberg, K. Månsson, B. Johansson, A. Gulz, and C. Balkenius, "Does a robot tutee increase children's engagement in a learning-by-teaching situation?" in *Intelligent Virtual Agents*, J. Beskow, C. Peters, G. Castellano, C. O'Sullivan, I. Leite, and S. Kopp, Eds., Lecture Notes in Computer Science, pp. 243–246, 2017.

[30] A. Jacq, S. Lemaignan, F. Garcia, P. Dillenbourg, and A. Paiva, "Building successful long child-robot interactions in a learning context," in *Proceedings of the 11th Annual ACM/IEEE International Conference on Human-Robot Interaction, HRI 2016*, pp. 239–246, March 2016.

[31] T. Belpaeme, J. Kennedy, A. Ramachandran, B. Scassellati, and F. Tanaka, "Social robots for education: A review," *Science Robotics*, vol. 3, no. 21, p. eaat5954, 2018.

[32] F. Erich, M. Hirokawa, and K. Suzuki, "A Systematic Literature Review of Experiments in Socially Assistive Robotics using Humanoid Robots," 2017, https://arxiv.org/abs/1711.05379.

[33] E. Mwangi, M. Diaz, E. Barakova, A. Catala, and M. Rauterberg, "Can children take advantage of nao gaze-based hints during gameplay?" in *Proceedings of the 5th International Conference on Human Agent Interaction, HAI 2017*, pp. 421–424, October 2017.

[34] N. Akalin, P. Uluer, H. Kose, and G. Ince, "Humanoid robots communication with participants using sign language: An interaction based sign language game," in *Proceedings of the 2013 IEEE Workshop on Advanced Robotics and Its Social Impacts, ARSO 2013*, pp. 181–186, Japan, November 2013.

[35] E. J. G. Van Der Drift, R.-J. Beun, R. Looije, O. A. B. Henkemans, and M. A. Neerincx, "A remote social robot to motivate and support diabetic children in keeping a diary," in *Proceedings of the 9th Annual ACM/IEEE International Conference on Human-Robot Interaction, HRI 2014*, pp. 463–470, March 2014.

[36] K. S. Goodman, "Analysis of Oral Reading Miscues: Applied Psycholinguistics," *Reading Research Quarterly*, vol. 5, no. 1, p. 9, 1969.

[37] K. S. Goodman, "Miscue analysis: Applications to reading instruction," *ERIC Clearinghouse on Reading and Communication Skills*, 1973.

[38] J. Hattie and H. Timperley, "The power of feedback," *Review of Educational Research*, vol. 77, no. 1, pp. 81–112, 2007.

[39] J. Bitchener, "Evidence in support of written corrective feedback," *Journal of Second Language Writing*, vol. 17, no. 2, pp. 102–118, 2008.

[40] S. Graham, K. R. Harris, L. Mason, B. Fink-Chorzempa, S. Moran, and B. Saddler, "How do primary grade teachers teach handwriting? A national survey," *Reading and Writing*, vol. 21, no. 1-2, pp. 49–69, 2008.

[41] S. Graham, N. Weintraub, and V. Berninger, "Which manuscript letters do primary grade children write legibly?" *Journal of Educational Psychology*, vol. 93, no. 3, pp. 488–497, 2001.

[42] F. Tanaka, A. Cicourel, and J. R. Movellan, "Socialization between toddlers and robots at an early childhood education center," *Proceedings of the National Acadamy of Sciences of the United States of America*, vol. 104, no. 46, pp. 17954–17958, 2007.

[43] J. J. Asher, "The Total Physical Response Approach to Second Language Learning," *The Modern Language Journal*, vol. 53, no. 1, pp. 3–17, 1969.

[44] J. Piaget, "La notion de partie chez lenfant , [The notion of part in children]," *Journal de psychologie normale et pathologique*, vol. 18, no. 6, pp. 359–480, 1921.

[45] J. Piaget and A. Szeminska, *La genèse du nombre chez l'enfant [The origin of number in children]*, Delachaux et Niestlée, Neuchâtel, Switzerland, 1941.

[46] G. Politzer, "The class inclusion question: a case study in applying pragmatics to the experimental study of cognition," *SpringerPlus*, vol. 5, no. 1, 2016.

[47] J. D. Lane, H. M. Wellman, and S. A. Gelman, "Informants' traits weigh heavily in young children's trust in testimony and in their epistemic inferences," *Child Development*, vol. 84, no. 4, pp. 1253–1268, 2013.

[48] C. A. Rohrbeck, M. D. Ginsburg-Block, J. W. Fantuzzo, and T. R. Miller, "Peer-assisted learning interventions with elementary school students: A meta-analytic review," *Journal of Educational Psychology*, vol. 95, no. 2, pp. 240–257, 2003.

[49] L. Takayama and C. Pantofaru, "Influences on proxemic behaviors in human-robot interaction," in *Proceedings of the 2009 IEEE/RSJ International Conference on Intelligent Robots and Systems, IROS 2009*, pp. 5495–5502, USA, October 2009.

[50] A. Kristoffersson, K. Severinson Eklundh, and A. Loutfi, "Measuring the Quality of Interaction in Mobile Robotic Telepresence: A Pilot's Perspective," *International Journal of Social Robotics*, vol. 5, no. 1, pp. 89–101, 2013.

[51] D. Leyzberg, S. Spaulding, and B. Scassellati, "Personalizing robot tutors to individuals' learning differences," in *Proceedings of the 9th Annual ACM/IEEE International Conference on Human-Robot Interaction, HRI 2014*, pp. 423–430, Germany, March 2014.

[52] P. Lesage, "La pédagogie dans les écoles mutuelles au XIXe siècle," *Revue française de pédagogie*, vol. 31, no. 1, pp. 62–70, 1975.

[53] J.-P. Martin, "Quand les élèves font la classe" [When children teach]," *Franтais dans le monde*, no. 224, pp. 51–55, 1989.

[54] A. Renkl, *Lernen durch Lehren*, Deutscher Universitätsverlag, Wiesbaden, 1997.

[55] J.-P. Martin, "Treibhäuser der Zukunft, Wie in Deutschland Schulen gelingen [How German schools succeed]," *eine Dokumentation von Reinhard Kahl und der Deutschen Kinderund Jugendstiftung*, 2004.

[56] L. Pareto and R. A. Tutee, *Robotics in Education*, Springer International Publishing, 2017.

[57] B. Jacquet, J. Baratgin, and F. Jamet, "The Gricean Maxims of Quantity and of Relation in the Turing Test," in *Proceedings of the 2018 11th International Conference on Human System Interaction (HSI)*, pp. 332–338, Gdansk, July 2018.

[58] B. Jacquet, O. Masson, F. Jamet, and J. Baratgin, "On the Lack of Pragmatic Processing in Artificial Conversational Agents," in *Human Systems Engineering and Design*, vol. 876 of *Advances in Intelligent Systems and Computing*, pp. 394–399, Springer International Publishing, Cham, 2019.

[59] F. Jamet, J. Baratgin, and D. Filatova, "Global warming and sea level rise: the intellect development study of preadolescents and adolescents from 11 to 15 years old," *Studia Pedagogiczne*, vol. 24, pp. 361–380, 2014.

[60] O. Masson, J. Baratgin, and F. Jamet, "NAO robot and the "endowment effect"," in *Proceedings of the IEEE International Workshop on Advanced Robotics and its Social Impacts, ARSO 2015*, Lyon, France, July 2015.

[61] O. Masson, J. Baratgin, and F. Jamet, "NAO robot as experimenter: Social cues emitter and neutralizer to bring new results in experimental psychology," in *Proceedings of the 2017 International Conference on Information and Digital Technologies, IDT 2017*, pp. 256–264, Slovakia, July 2017.

5

A Review on Compliant Joint Mechanisms for Lower Limb Exoskeletons

Miguel A. Gálvez-Zúñiga and Alejandro Aceves-López

Tecnologico de Monterrey, CEM, 52926 Atizapan, Mexico City, Mexico

Correspondence should be addressed to Miguel A. Gálvez-Zúñiga; a01160915@itesm.mx

Academic Editor: Liwei Shi

Lower limb exoskeletons are experiencing a rapid development that may suggest a prompt introduction to the market. These devices have an inherent close interaction with the human body; therefore, it is necessary to ensure user's safety and comfort. The first exoskeletal designs used to represent the human joints as simple revolute joints. This approximation introduces an axial misalignment issue, which generates uncontrollable internal forces. A mathematical description of the said misalignments is provided to better understand the concept and its consequences. This review will only focus on mechanisms aiming to comply with its user.

1. Introduction

Lower limb exoskeletons (LLE) are a type of humanoid service robot designed to improve its user's physical performance, commonly during locomotion. As the name implies, the support is given to the lower limbs through the ankle, knee, and/or hip joints by means of actuated wearable mechanisms.

According to its application, LLE can be classified as augmentation, rehabilitation, or assistive exoskeletons. The first category aims to enhance characteristics of the human movement such as strength [1], endurance [2], speed [3], or metabolic cost [4–6]. This kind of exoskeleton is of great interest to the military as established by the development of systems such as the Berkeley Lower Extremity Exoskeleton (BLEEX) [7], Human Universal Load Carrier (HULC) [8], Hercule [9] and XOS2 [10].

The rehabilitation category focuses on restoring motion function and improving gait patterns of patients affected by spinal cord injuries [11], strokes [12], foot drop [13], or even poliovirus [14]. The exoskeletons belonging to this category specifically help to provide the therapy needed to (partially) restore normal gait. This can be achieved, for example, by controlling the trajectory or torque of each user's joint independently [15]. This category can be illustrated by Lokomat [16], Active Leg Exoskeleton (ALEX) [17], ReoAmbulator [18], and Lower Extremity Powered Exoskeleton (LOPES) [19].

Lastly, assistive exoskeletons' objective is to compensate physical disabilities caused by any sort of injury, trauma, or weakness. Taking this into account, only the impaired joints should be artificially supported by the robot, keeping the others unaffected. Robots like ReWalk [20], Hal-5 [21], and Ekso [22] belong to this category.

In the past ten years, this type of technology has been the subject of an increasing number of publications (Figure 1). According to Scopus [23], just in 2014 there were more works associated with lower limb exoskeleton than the accumulated publications between 1995 and 2010 (120 versus 110 published items).

The rapid development is a result of advances in actuators, sensors, materials, batteries, and computer processors. These innovations had led Ferris [24] to state that, by 2024, people will be walking down the street, in the malls, and to their homes wearing robotic exoskeletons in a portable, svelte, and fashionable way. However, it is yet a long way until we can seize the full potential of exoskeletons.

There are still many things not completely understood about the physiology and biomechanics of human movement [25]. This unawareness makes it impossible, for example, to

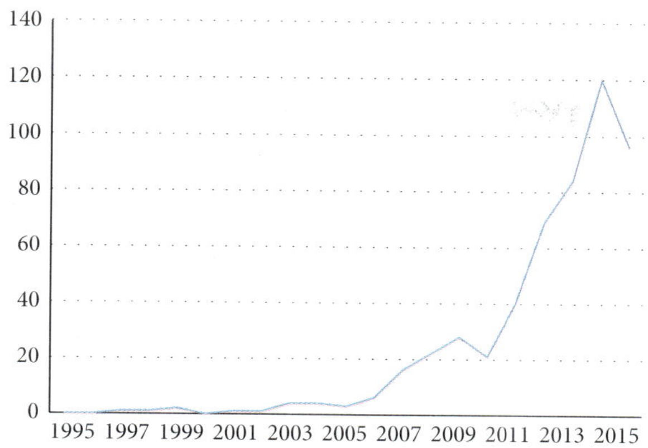

FIGURE 1: Publications per year reported in citation indexing service Scopus [23] when searching for the term lower limb exoskeleton.

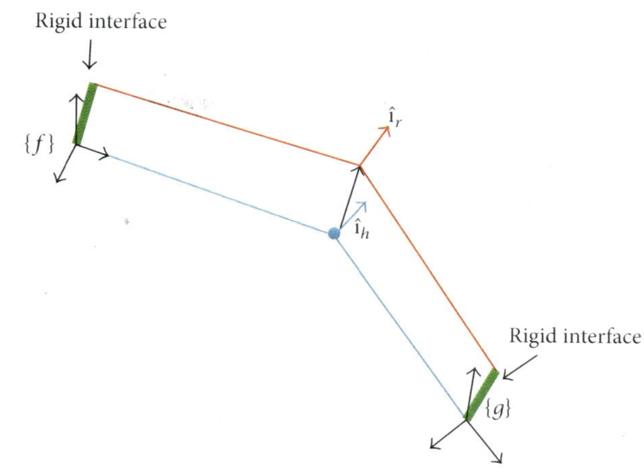

FIGURE 2: Representation of the simplified close kinematic chain formed by the exoskeleton (human in blue, robotic in red).

predict how exoskeletons will affect locomotion when there is little knowledge of the cognitive mechanism involved in the process [26]. Therefore, to fulfill Ferris' vision, researchers must overcome several challenges in all areas of expertise associated with exoskeletons: control, instrumentation, computer science, electronics, and mechanics.

Irrespective of the different application fields, exoskeletons share a main objective for a proper functioning: transfer power to the user's limb. Additionally, due to the close interaction between robot and user, the robots must assure safety of its wearer without compromising their own efficiency. Both concepts relate directly to the mechanical aspect of exoskeletons and will be the main focus for this paper.

Mechanically, the robot must provide support so that the added weight, of either the loads or the exoskeleton itself, does not affect the user. It also must adapt to the wearer's anatomy so it is beneficial for a wide range of the population. All of this while maintaining a small size to ensure its portability.

2. The Axial Misalignment Problem

There are two main concerns when designing the physical part of an exoskeleton. The first one is the added mechanical impedance to the legs which, in absence of a specialized control scheme that masks the inertia, hinders the user's agility and increases the metabolic energy consumption [27]. The other is the mismatch between the human and robotic joints which cause kinematic incompatibility. We focus on the latter as it compromises the user's safety and comfort.

According to Chasles' theorem, the motion of any rigid body can always be expressed as a translation along a line followed by a rotation about the same line. In screw theory, said line can be obtained by a twist which is a six-dimensional vector that represents all the velocities of a body. This twist, as any other vector, is formed by magnitude and a unit vector. The first takes the value of the angular velocity, while the latter is a combination of the pitch (which is the ratio of linear velocity to its angular counterpart) and a Plücker line. The said line describes the position and orientation of an infinite line that can be interpreted as the *instantaneous axis of rotation* (IAR), an extension of the Instantaneous Center of Rotation in planar movement, which is also the line described by the theorem.

Additionally, in any open kinematic chain, no matter the number of links, there is a twist (T_c) that describes the movement of the final link of the chain relative to its first. It is equal to the sum of the twists (T_x) of every joint that conforms the chain

$$T_c = T_1 + T_2 + \cdots + T_n. \quad (1)$$

The case is quite different for close kinematic chains; adding every twist in the chain would imply calculating the velocity of a body relative to itself, which is zero:

$$T_1 + T_2 + \cdots + T_n = 0. \quad (2)$$

Individually, exoskeletons and human limbs are both open chains, each one having a characteristic twist, ${}^fT_g^{(R)}$ and ${}^fT_g^{(H)}$, respectively, that describes the instantaneous velocities of body $\{g\}$ relative to the reference body $\{f\}$.

The exoskeletal application inherently requires connecting both chains together, thus forming one closed kinematic chain. Figure 2 is a representation of the said kinematic chain that aims to illustrate the appearance of hyperstatic forces.

If we consider that the interfaces, as well as the elements of the human and robotic kinematic chains, are rigid, the twists ${}^fT_g^{(R)}$ and ${}^fT_g^{(H)}$ must be equal. Failing in doing so will result in the inability of the system to move. Hence, applying external forces when the twists are different will not initiate motion but instead they will propagate through the system as hyperstatic forces affecting the human limb. In practice, this phenomenon is mitigated by the use of flexible elements in the interfaces, yet it is never completely eliminated.

A formal description can be obtained by analyzing the system through screw theory. According to (2), the sum of the twists should be equal to zero

$${}^fT_g^{(H)} + {}^gT_f^{(R)} = 0. \quad (3)$$

If we apply the twist's symmetric property and considering that $\hat{\imath}_x$ represents the unit vector of twist ${}^{f}T_{g}^{(x)}$, we can rewrite (3) as follows:

$$\begin{gathered} {}^{f}T_{g}^{(H)} = -{}^{g}T_{f}^{(R)} = {}^{f}T_{g}^{(R)}, \\ \left|{}^{f}T_{g}^{(H)}\right| \hat{\imath}_h = \left|{}^{f}T_{g}^{(R)}\right| \hat{\imath}_r. \end{gathered} \tag{4}$$

The equation presents two scenarios; in the first, both magnitudes are zero, despite the values of the unit vectors, which can be interpreted as no motion in either chain. The second scenario requires that the unit vectors and the magnitudes of both chains are equal meaning that the chains share the same angular velocity, pitch, and IAR.

It is important to remember that ${}^{f}T_{g}^{(H)}$ and ${}^{f}T_{g}^{(R)}$ are the twists that describe the open human and robotic chains, meaning that theoretically the IARs need to be equal for the closed chain to move. Nevertheless, in exoskeletons, movement can be obtained despite discrepancies between the IARs. This reveals that there is a velocity of a body in relation to itself, which can be expressed as the existence of a third twist:

$${}^{f}T_{g}^{(H)} + {}^{g}T_{f}^{(R)} + {}^{f}T_{f'}^{(R)} = 0. \tag{5}$$

The twist ${}^{f}T_{f'}^{(R)}$ represents the velocity of deformations present in the system. Due to its hyperstatic nature, it is impossible to determine the precise geneses of this twist. The soft tissues surrounding the interfaces (i.e., skin) and in the effective human joint (i.e., meniscus in the knee) are capable of producing the new twist by deforming themselves [28]. Furthermore, the flexible elements are used in the connection between human and exoskeleton, which also contribute to ${}^{f}T_{f'}^{(R)}$. Though deformations in flexible connections do not occur explicitly in the human limb, they can chafe or damage the tissues in contact [29] during prolonged usage.

The manifestation of deformations implies that there are forces generated compressing the human limb rather than helping it move. These forces, known as uncontrollable forces [30], are responsible for compromising both safety and comfort.

It could seem that the solution for this problem is to simply design mechanism so that the IARs are aligned, yet mapping $\hat{\imath}_h$ is not an easy task. It is important to notice that most human articulations act on several planes of the body (sagittal, transverse, and coronal) at the same time as well as having small translations due to bones sliding. The actual location and direction of $\hat{\imath}_h$ depend on the geometry of the articulating surfaces, payload, and health of tissues such as cartilages and ligaments [31].

Clinically, a noninvasive way to estimate the IAR pathway is by taking successive radiographies of the joint of interest, yet it has significant errors [32] as the radiography only obtains a point in a plane rather than a line in space. Predicting the position of $\hat{\imath}_h$ requires a lot of information, especially the description of bones' contact surfaces. These surfaces are very complex and cannot be determined exactly through noninvasive methods. Additionally, biomechanical models of joints, such as [33], usually rely on several calibration parameters to adjust for the variability of dimensions among people. Thus, failing to set these parameters to the correct value for each person will produce an error estimating the position and orientation of $\hat{\imath}_h$.

Taking into account the excessive variance of morphology among subjects and the intrinsic complexity of the human joints, it is impossible to determine a general design that maximizes the performance while minimizing the mechanical interference [26] for every wearer.

Therefore, it is a common practice to simplify the human joint kinematics when designing the robotic joints. Using simple robotic elements such as revolute or spherical joints to model human anatomy only considers the dominant rotations and neglects the small dose of concealed translations [34] making it impossible to correctly comply with the user [35]. Yet the great advantage is the mechanical simplicity obtained, which translates into simpler designs, actuation, control schemes, and (consequently) cheaper prototypes.

3. State of Art

During gait, the whole body is engaged, from toes to arms. Nonetheless, the ankle, knee, and hip have been identified to be more relevant than any other joint, as they perform the major motions and torques. This led pioneer researchers to neglect the motion of the bones in the foot and to treat them like one continuous link, as well as considering the upper body as a concentrated mass, concentrating their efforts in these three joints.

Another common simplification is to only consider the effects of the joints along the sagittal plane (which divides the right side of the body from the left), as these also represent the majority of motion and torque compared to the effects of the joints in the two remaining planes.

Researchers have acknowledged the axial misalignment situation and have been proposing new mechanisms that redefine the paradigms of exoskeletal design. They no longer design structures that facilitate the engineering that is kinematics or control but design them with a human-centered perspective looking for a better adaptation to the user's morphology.

From a physiological perspective, each of these joints has a specific geometry, workspace, function, and characteristic number of elements involved (bones, muscles, and ligaments). As a consequence, we will group the reviewed mechanisms according to the joint they are supporting.

3.1. Ankle. The ankle involves the relative movement among four bones in three joints (Figure 3(a)). The tibia and fibula converge in the tibiofibular joint, which produces almost no motion but serves as a bracket to align the talus. The tibiofibular joint in contact with the talus is known as the talocrural joint; it is responsible for plantar/dorsiflexion (movement on the sagittal plane). And lastly, the joint between talus and calcaneus, called the subtalar joint, is responsible for the remaining movements [36].

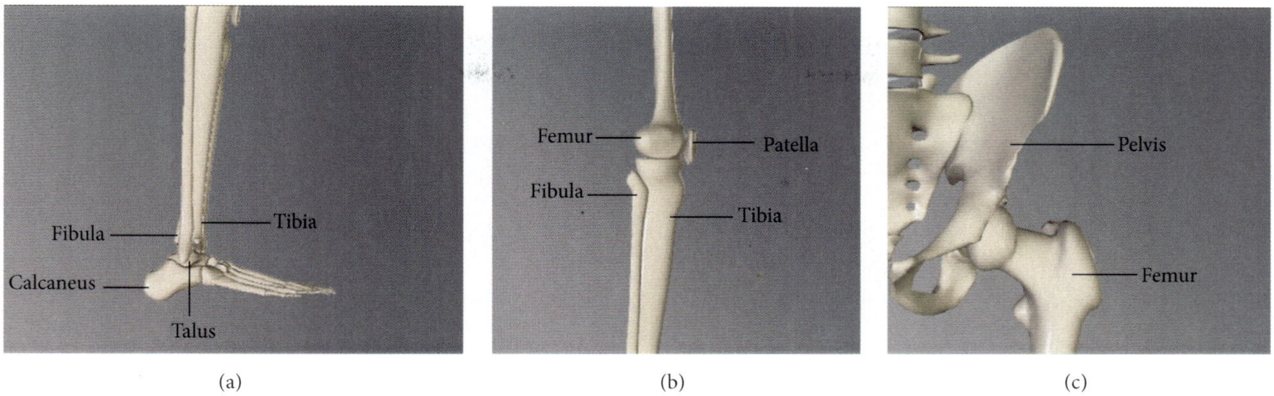

FIGURE 3: Anatomy of the human joints: (a) ankle, (b) knee, and (c) hips. Powered by BioDigital [66].

The ankle joint registers the greatest torque during gait compared to the hip and knee [37]; this phenomenon is associated with the ankle's distal position along the limb which creates greater lever arms.

As explained before, the focus is on plantar/dorsiflexion as these movements represent the majority of motion during gait. As a consequence, previous designs tend to be just a revolute joint [5, 38–42]. Nonetheless, the torque generated at the human ankle is so large that inconsistent rotation center of the ankle mechanism may be a potential risk against the wearer's safety [43]. Furthermore, balance may be compromised when limiting the motion of the ankle as it depends on its proper functioning [44].

Agrawal et al. [45] proposed an exoskeleton that has an active plantar/dorsiflexion joint combined with a passive eversion/inversion joint. This prototype was one of the first to use an alternative to the oversimplified approach. The results show a good performance with a simple PD controller regardless of being an underactuated mechanism. On the downside, its dimensions and weight (2.5 kg) limit its portability and its applicability, as well as considerably increasing the inertia.

Fan and Yin [46] made a 3-RPS (Rotational-Prismatic-Spherical) parallel mechanism. The design is capable of moving along all of the ankle DOF's while providing a complying couple with the user. The results proved that the ankle joint meets the kinematical and dynamical requirement for rehabilitation, yet its application for a portable exoskeleton is also limited due to its large size.

At the university of Bristol, Carberry et al. [47] used a 2 DOF 2-RP parallel mechanism that allows the same movements as the one of Agrewal. The proposed design is lighter (weighting only .75 kg) and easier at donning and doffing. In addition, the exoskeleton addresses intersubject variability by means of a parametric design, thus making it a more suitable solution for assistive exoskeletons.

Hong et al. [43] proposed a passive spherical five-bar linkage mechanism parallel to an active RSU-chain (Rotational-Spherical-Universal). The design intends to closely locate the axis of rotation of the ankle exoskeleton to that of the wearer's ankle. The results show that the design is capable of reaching the whole spectrum of positions as the human ankle, though they are in early design stages as no prototype has been developed yet.

3.2. Knee. The knee is considered a condyloid joint [48] formed by the femur and tibia (Figure 3(b)). It allows flexion/extension and internal/external rotations; nevertheless, internal/external rotations are severely constrained when the knee is loaded under body weight or fully extended [49]. During flexion/extension, tibia rolls and slides on femur causing the instantaneous axis of rotation to be displaced up to 30 mm [33].

Typically, the knee is simply modeled as a revolute joint, as seen in [40, 41, 50–52]. This approximation simplifies greatly the mechanical and kinematical considerations during exoskeletal design at the expense of ergonomics and mechanical compatibility with the human anatomy. According to [53], this oversimplification produces parasitic forces in the user's knee that may lead to discomfort or injury over prolonged time of use.

To avoid the said problem and provide a better user experience, several researchers are proposing new mechanisms that comply with the users' anatomy. Some of the first designs to do so are those proposed by Wang et al. [53]. They proposed four compliant designs using combinations of simple kinematic element as an alternative of the previous approximation: (1) pin and slider, (2) cam and slider, (3) pin and pinned slider, and (4) cam with pinned slider. They used an anatomically correct commercial model of the knee (A82 Functional Knee Joint Model, 3B Scientific GmbH, Germany) to have a more approximate measure of the internal forces without recurring to invasive instrumentation setups. According to the results obtained, the implementation of an extra joint as a compliant element reduces the magnitude of the forces generated at the knee almost to a tenth.

Parallel to Wang et al.'s work, Ergin and Patoglu [54] developed a self-aligning knee exoskeleton by means of a more complex mechanism. The design was based on a 3-RRP planar parallel mechanism which included 3 DOF, two passive translations in the plane and one active rotation along the perpendicular axis. The reported space where the ICR could exist was a 120 mm diameter circle in the sagittal plane and infinite number of revolutions along the z-axis. However, the dimensions of the prototype are too big, compromising the portability of a complete LLE that implements this design.

A couple of years later, Celebi et al. [49] developed a new self-aligning knee exoskeleton, ASSIST-ON Knee. This design was based on the Schmidt coupling [55], this coupling allows for axial misalignments between its input and output links without losing power in the transmission. The mechanism allowed the same DOF as its predecessor but a smaller diameter (48 mm), which indicates a more compact prototype.

Another proposal for a compliant design was given by Kuan et al. under the name of Adaptive Coupling Joint (ACJ) [56]. The ACJ is based on a 2 DOF five-bar linkage mechanism. It allows an external torque source to be applied in the mechanism while permitting the displacement of the axis of rotation through a set trajectory. According to the authors, the ACJ is being manufactured and prepared for further bench testing.

The presented designs use passive DOFs to be driven by the human limb and couple accordingly; however, this tactic sacrifices the weight support capability of the exoskeleton. This negligence may imply a reduced area of opportunity for further applications. Considering this, and trying to reduce pain for people with knee osteoarthritis, Lee and Wang [57] built a prototype that was capable of providing weight support during the first 20° of flexion and a compliant coupling for the rest of the motion. Unlike previous designs, the approach of this prototype was through deformable links rather than rigid links. The results show that this design was capable of supporting up to 20% of human weight during walking on level floor.

3.3. Hip. The hip joint is a multi-axial ball-and-socket synovial joint where the femur head acts as the ball and the pelvis as the socket (Figure 3(c)). The hip joint moves in different planes that pass through the joint center, behaving as a kind of spherical joint, allowing three DOF's. All these DOF's are important in order to allow stable locomotion, even in a straight line [58]. Considerably, the hip is the least studied joint in terms of proposed designs for compliant exoskeletons.

Singla et al. [59] utilized the ball-and-socket model of the hip and implemented a spherical 3-RRR mechanism. Similar to Hong's ankle mechanism discussed previously, the mechanism tries to locate its own axis of rotation as close as possible to the one of the hip. It is important to notice that this design has a fixed center of rotation with respect to the waist and thigh attachments. Results yield an acceptable mechanism kinematically but dynamic performance was not considered.

Pan et al. [60] proposed a more compliant exoskeleton than Singla et al.'s. The design consisted of a 5-DOF 3-PUU parallel mechanism. Once coupled with the user, two of those DOFs are passive and are used to align with the user's hip, and the remaining are active and are used to control its position. The kinematic analysis suggests that the configuration is reliable; however, there is no prototype yet.

Yu and Liang [61] proposed, similar to Pan et al.'s design, a 6-DOF 3-UPS parallel mechanism. This exoskeleton used 3 DOFs to align the user's hip. The design was optimized using the authors' proposed Manipulability Inclusive Principle, which compares the manipulability ellipsoid [62] of both kinematic chains to ensure the robotic kinematic chain can shadow all human's movements. Using this principle should be proof of its assisting ability and quicker reaction.

3.4. Combined. To our knowledge, there is only one lower limb exoskeleton with more than one active compliant joint: the design of Yang et al. [63] known as Bionic Lower Extremity Rehabilitation Exoskeleton (BLERE). At the knee, they used a pinned slider alongside a cam as proposed in [53], while at the hip, they used a curved slider and two revolute joints allowing free movement in all of the 3 DOFs available. Only the flexion/extension motion is powered in both mechanisms. The prototype proved the compliance of the design yet its big size may be still a problem for massive implementation.

4. Methodologies

All of the works mentioned represent an advancement in the solution of the misalignment problem; nevertheless, each one is focused on a specific case of study. To attend the general problem, a couple of researchers have proposed design methodologies to facilitate the coupling between an exoskeleton and the human user.

The first one was proposed by Jarrassé and Morel [30] after recognizing the hyperstaticity generated when connecting the human and robotic kinematic chains. Its main objective is to prevent the manifestation of uncontrollable forces rather than align the chains. The method, given an exoskeleton designed to replicate a limb kinematic model, provides three equations to design passive mechanisms that act as interfaces between the kinematic chains. These equations derive from the *Global Isostaticity Condition* (GIC), which establishes that no forces are to be transmitted to the human chain through the interfaces and no velocities are allowed in the robot chain when the user is static. This condition is first stated in terms of the spaces of twists and wrenches and then it is expressed in terms of the connectivity of the mechanisms involved (robot, human, and interfaces) so that at the end we can calculate the number of passive DOFs necessary in the interfaces to fulfill the GIC.

The methodology was applied in two experiments with the same exoskeleton, one with a manikin and the other with human users. The results in the first experiment reduced the appearance of uncontrollable forces by an average of 96% while those of the second have a less impressive yet significant reduction of about 30%. Nevertheless, the reduction of uncontrollable forces came at the cost of also reducing the assistive forces meaning a lower efficiency for the robot.

This new approach to exoskeleton design focuses on the interfaces leaving the original robot structure untouched. The direct implication is that any exoskeleton already available could be benefitted by new interfaces without altering its current design, which should not affect their control scheme. On the other hand, its greatest restriction is that the method can only be applied in robots that mimic the human limb anatomy and that have all of their joints powered.

Later on, the *Kinetostatic Compatibility Method* was developed by Cempini et al. [64]. Different from the hyperstaticity approach from the aforementioned method, this one focuses on avoiding the misalignment between chains. The objective is to design *self-aligning mechanisms* (SAM), much more similar to the mechanisms described in the previous section, so that these fulfill four requirements: kinematic closure, adaptability, controllability of the human joint position, and assistance effectivity. Regardless of the specific application of the exoskeletons, the requirements are the same.

Cempini's strategy is to add the necessary degrees of freedom to the kinematic models of the human limbs so that the minimal set of variables that define the pose is covered; that is, when modeling a hip as a spherical joint, the added DOFs are the three remaining translations to cover the six parameters that define 3D space. These extra DOFs are called deviations and are used as an independent parameter that helps define the homogeneous transformations matrices for the human chain.

The expanded model then helps to set the restrictions necessary to achieve all the requirements leaving three elements that can be set by the designer: the desired workspace, the human articulation variability, and the SAM chain itself.

While this method presents a more general solution in terms of the possible robotic kinematic chain to be used, there is no experimental result that would let us quantify its success. It is also important to emphasize that the choice of a suitable kinematic model is out of the scope for this method as it is extremely difficult to generalize.

Lastly, it is highly recommended to revise both of these works to further understand the concepts and the formalization of the individual objectives into mathematical equations.

5. Discussion

The presented lower limb exoskeletons are capable of accepting large misalignments between axes, which allows them to be used by different sized people without the need to manually adjust them. Due to the lack of adequate "in vivo" measurements of human morphology, these exoskeletons represent a solution for the misalignment problem as well as an improvement on the first proposals; however, some disadvantages still need to be addressed.

A great disadvantage of these self-aligning mechanisms is that they all increase design complexity and mechanical impedance compared with exoskeletons based on simplified human articulations. The increased mass is especially critical for machines interacting with human users, since it amplifies the risk of injuries in case of failure [65]. Consequently, it is essential to find the best tradeoff between a theoretically misalignment-free design and a simple mechanism.

Also, most of the designs discussed use passive DOFs to be driven by the motion of the user's leg, thus letting the exoskeleton be self-aligning. These underactuated systems provide torque assistance in the desired joint but lack weight support. This means that the forces generated by the exoskeleton weight, as well as the ones generated by its actuators, will not be translated to ground but to the user. Though it has not yet been proved, the lack of weight support for prolonged time may cause damage or pain in human joint tissues, thus limiting applications for assistive and augmentation purposes where long-time usage is fundamental.

Another perspective is given by Zanotto et al.'s investigation [65]. They compare the effects of the axial misalignment to the inertial effects. Human-robot interactions are sensed by two six-axis sensors mounted at the interfaces while torques exerted by the actuators are measured by means of single-axis sensors. The results suggest that the inertial effects are considerably larger than the ones produced by misalignment, as well as pointing out that the traditional approach may be sufficient to guarantee the wearer's comfort in spite of substantial misalignments. While this study may contradict the purpose of specifically designing mechanism to comply with the axial misalignment, it is important to emphasize that the research does not consider the internal forces that act on the limb; therefore, the conclusions given are purely from the robot point of view.

Finally, it is fundamental to consider that there must exist balance between exoskeleton compliance and inertial effects, since adopting larger, more complex mechanisms can result in an impediment for development and massive distribution of this type of robots.

6. Conclusion

The exoskeleton development has had remarkable progress regarding correct compliance between robot and user. The new human-centered design approach has led to several designs that have successfully attached the robot without any major alignment problems, as well as methodologies that help designers to create robots capable of complying with the human. Yet further studies are needed to ensure the mechanisms' capability of assistance as well as weight support capability of the exoskeleton. Physical prototypes are needed to better understand the capabilities and limitations of each design.

Moreover, there is still a lot of work to be done to achieve Ferris' futuristic vision of exoskeletons. The next step towards achieving ideal mechanical design is to ensure that no forces (caused by weight, inertia, or hyperstaticity) are introduced at any point of the lower limb. We actuate the systems in such a way that the efficiency of the robot and the safety of the user are not compromised and, finally, design complete lower limb exoskeletons to assist every joint during gait.

Competing Interests

The authors declare that they have no competing interests.

References

[1] A. Zoss, H. Kazerooni, and A. Chu, "On the mechanical design of the Berkeley Lower Extremity Exoskeleton (BLEEX)," in *Proceedings of the IEEE IRS/RSJ International Conference*

on Intelligent Robots and Systems (IROS '05), pp. 3465–3472, Edmonton, Canada, August 2005.

[2] J. E. Pratt, B. T. Krupp, C. J. Morse, and S. H. Collins, "The RoboKnee: an exoskeleton for enhancing strength and endurance during walking," in *Proceedings of the IEEE International Conference on Robotics and Automation (ICRA '04)*, vol. 3, pp. 2430–2435, April 2004.

[3] Y. Hasegawa and K. Ogura, "First report on passive exoskeleton for easy running: PEXER IV," in *Proceedings of the International Symposium on Micro-NanoMechatronics and Human Science (MHS '13)*, pp. 1–6, IEEE, Nagoya, Japan, November 2013.

[4] A. M. Grabowski and H. M. Herr, "Leg exoskeleton reduces the metabolic cost of human hopping," *Journal of Applied Physiology*, vol. 107, no. 3, pp. 670–678, 2009.

[5] P. Malcolm, W. Derave, S. Galle, and D. De Clercq, "A simple exoskeleton that assists plantarflexion can reduce the metabolic cost of human walking," *PLoS ONE*, vol. 8, no. 2, Article ID e56137, pp. 1–7, 2013.

[6] M. B. Wiggin, G. S. Sawicki, and S. H. Collins, "An exoskeleton using controlled energy storage and release to aid ankle propulsion," in *Proceedings of the IEEE International Conference on Rehabilitation Robotics (ICORR '11)*, pp. 1–5, Zurich, Switzerland, June 2011.

[7] J. Ghan and H. Kazerooni, "System identification for the Berkeley Lower Extremity Exoskeleton (BLEEX)," in *Proceedings of the IEEE International Conference on Robotics and Automation (ICRA '06)*, pp. 3477–3484, IEEE, Orlando, Fla, USA, May 2006.

[8] Medical Applications, "Protonex powers Lockheed Martin HULC exoskeleton," *Fuel Cells Bulletin*, vol. 2010, no. 2, pp. 4–5, 2010.

[9] RB3D, "Hercule," 2014, http://www.rb3d.com/en/exo/.

[10] Army-Technology, Raytheon XOS 2 Exoskeleton, Second-Generation Robotics Suit, United States of America, http://www.army-technology.com/projects/raytheon-xos-2-exoskeleton-us/.

[11] W. Banchadit, A. Temram, T. Sukwan, P. Owatchaiyapong, and J. Suthakorn, "Design and implementation of a new motorized-mechanical exoskeleton based on CGA Patternized Control," in *Proceedings of the IEEE International Conference on Robotics and Biomimetics (ROBIO '12)*, pp. 1668–1673, Guangzhou, China, December 2012.

[12] P. Beyl, *Design and Control of a Knee Exoskeleton Powered by Pleated Pneumatic Artificial Muscles for Robot-Assisted Gait Rehabilitation*, Brussels University, 2010.

[13] D. P. Ferris, J. M. Czerniecki, and B. Hannaford, "An ankle-foot orthosis powered by artificial pneumatic muscles," *Journal of Applied Biomechanics*, vol. 21, no. 2, pp. 189–197, 2006.

[14] B. G. Do Nascimento, C. B. S. Vimieiro, D. A. P. Nagem, and M. Pinotti, "Hip orthosis powered by pneumatic artificial muscle: voluntary activation in absence of myoelectrical signal," *Artificial Organs*, vol. 32, no. 4, pp. 317–322, 2008.

[15] M. F. Levin, J. A. Kleim, and S. L. Wolf, "What do motor 'recovery' and 'compensation' mean in patients following stroke?" *Neurorehabilitation and Neural Repair*, vol. 23, no. 4, pp. 313–319, 2009.

[16] G. Colombo, M. Joerg, R. Schreier, and V. Dietz, "Treadmill training of paraplegic patients using a robotic orthosis," *Journal of Rehabilitation Research and Development*, vol. 37, no. 6, pp. 693–700, 2000.

[17] S. K. Banala, S. H. Kim, S. K. Agrawal, and J. P. Scholz, "Robot assisted gait training with active leg exoskeleton (ALEX)," in *Proceedings of the 2nd Biennial IEEE/RAS-EMBS International Conference on Biomedical Robotics and Biomechatronics (BioRob '08)*, pp. 653–658, Scottsdale, Ariz, USA, October 2008.

[18] Motorika, *ReoAmbulator*, Motorika, Mount Laurel, NJ, USA, 2014, http://www.motorika.com/?categoryId=90004.

[19] J. F. Veneman, R. Kruidhof, E. E. G. Hekman, R. Ekkelenkamp, E. H. F. Van Asseldonk, and H. Van Der Kooij, "Design and evaluation of the LOPES exoskeleton robot for interactive gait rehabilitation," *IEEE Transactions on Neural Systems and Rehabilitation Engineering*, vol. 15, no. 1, pp. 379–386, 2007.

[20] A. Esquenazi, M. Talaty, A. Packel, and M. Saulino, "The rewalk powered exoskeleton to restore ambulatory function to individuals with thoracic-level motor-complete spinal cord injury," *American Journal of Physical Medicine and Rehabilitation*, vol. 91, no. 11, pp. 911–921, 2012.

[21] Y. Sankai, "Leading edge of cybernics: robot suit HAL," in *Proceedings of the SICE-ICASE International Joint Conference*, pp. P-1–P-2, Busan, South Korea, October 2006.

[22] L. A. Rapp, V. Huang, S. Firpi, L. Tovar, J. M. Zanca, and J. Baeza-Dager, "The age of the bionic man has arrived: the use of Ekso exoskeleton in acute paraplegia," *The Journal of Spinal Cord Medicine*, vol. 36, no. 5, p. 507, 2013.

[23] Elsevier, "Scopus," 2015, http://www.scopus.com/.

[24] D. P. Ferris, "The exoskeletons are here," *Journal of NeuroEngineering and Rehabilitation*, vol. 6, article 17, 2009.

[25] D. P. Ferris, G. S. Sawicki, and M. A. Daley, "A physiologist's perspective on robotic exoskeletons for human locomotion," *International Journal of Humanoid Robotics*, vol. 4, no. 3, pp. 507–528, 2007.

[26] M. Cenciarini and A. M. Dollar, "Biomechanical considerations in the design of lower limb exoskeletons," in *Proceedings of the IEEE International Conference on Rehabilitation Robotics (ICORR '11)*, pp. 1–6, Zurich, Switzerland, June 2011.

[27] G. Aguirre-Ollinger, J. E. Colgate, M. A. Peshkin, and A. Goswami, "Inertia compensation control of a one-degree-of-freedom exoskeleton for lower-limb assistance: initial experiments," *IEEE Transactions on Neural Systems and Rehabilitation Engineering*, vol. 20, no. 1, pp. 68–77, 2012.

[28] A. Schiele, "An explicit model to predict and interpret constraint force creation in pHRI with exoskeletons," in *Proceedings of the IEEE International Conference on Robotics and Automation (ICRA '08)*, pp. 1324–1330, Pasadena, Calif, USA, May 2008.

[29] A. Nilsson, K. S. Vreede, V. Häglund, H. Kawamoto, Y. Sankai, and J. Borg, "Gait training early after stroke with a new exoskeleton—the hybrid assistive limb: a study of safety and feasibility," *Journal of NeuroEngineering and Rehabilitation*, vol. 11, no. 1, article 92, 2014.

[30] N. Jarrassé and G. Morel, "Connecting a human limb to an exoskeleton," *IEEE Transactions on Robotics*, vol. 28, no. 3, pp. 697–709, 2012.

[31] J. D. Bronzino, *The Biomedical Engineering Handbook*, vol. 1, CRC Press, Boca Raton, Fla, USA, 2nd edition, 2000.

[32] M. J. Simmonds and T. B. Hoshizaki, "Instantaneous center of rotation shifts in symptomatic anterior cruciate ligament deficient knee joints," *Journal of Biomechanics*, vol. 23, no. 7, p. 739, 1990.

[33] K.-M. Lee and J. Guo, "Kinematic and dynamic analysis of an anatomically based knee joint," *Journal of Biomechanics*, vol. 43, no. 7, pp. 1231–1236, 2010.

[34] M. K. Vukobratovic, "When were active exoskeletons actually born?" *International Journal of Humanoid Robotics*, vol. 4, no. 3, pp. 459–486, 2007.

[35] S. H. Scott and D. A. Winter, "Biomechanical model of the human foot: kinematics and kinetics during the stance phase of walking," *Journal of Biomechanics*, vol. 26, no. 9, pp. 1091–1104, 1993.

[36] D. S. B. Williams and J. Taunton, "Physical therapies in sport and exercise," in *Physical Therapies in Sport and Exercise*, G. S. Kolt and L. Snyder-Mackler, Eds., Churchill Levingstone, New York, NY, USA, 2nd edition, 2007.

[37] S. H. Collins, M. B. Wiggin, and G. S. Sawicki, "Reducing the energy cost of human walking using an unpowered exoskeleton," *Nature*, vol. 522, no. 7555, pp. 212–215, 2015.

[38] N. Costa and D. G. Caldwell, "Control of a biomimetic 'soft-actuated' 10DoF lower body exoskeleton," in *Proceedings of the 1st IEEE/RAS-EMBS International Conference on Biomedical Robotics and Biomechatronics (BioRob '06)*, pp. 495–501, Pisa, Italy, February 2006.

[39] G. S. Sawicki, K. E. Gordon, and D. P. Ferris, "Powered lower limb orthoses: applications in motor adaptation and rehabilitation," in *Proceedings of the IEEE 9th International Conference on Rehabilitation Robotics (ICORR '05)*, pp. 206–211, Chicago, Ill, USA, July 2005.

[40] C. Teng, Z. Wong, W. Teh, and Y. Z. Chong, "Design and development of inexpensive pneumatically-powered assisted knee-ankle-foot orthosis for gait rehabilitation-preliminary finding," in *Proceedings of the International Conference on Biomedical Engineering (ICoBE '12)*, pp. 28–32, IEEE, Penang, Malaysia, February 2012.

[41] W. Van Dijk, H. Van der Kooij, and E. Hekman, "A passive exoskeleton with artificial tendons," in *Proceedings of the IEEE 12th International Conference on Rehabilitation Robotics*, Zurich, Sweden, June-July 2011.

[42] C. J. Walsh, K. Pasch, and H. Herr, "An autonomous, under-actuated exoskeleton for load-carrying augmentation," in *Proceedings of the IEEE/RSJ International Conference on Intelligent Robots and Systems (IROS '06)*, pp. 1410–1415, Beijing, China, October 2006.

[43] M. B. Hong, Y. J. Shin, and J.-H. Wang, "Novel three-DOF ankle mechanism for lower-limb exoskeleton: kinematic analysis and design of passive-type ankle module," in *Proceedings of the IEEE/RSJ International Conference on Intelligent Robots and Systems (IROS '14)*, pp. 504–509, Chicago, Ill, USA, September 2014.

[44] J. Han, J. Anson, G. Waddington, R. Adams, and Y. Liu, "The role of ankle proprioception for balance control in relation to sports performance and injury," *BioMed Research International*, vol. 2015, Article ID 842804, 8 pages, 2015.

[45] A. Agrawal, S. K. Banala, S. K. Agrawal, and S. A. Binder-Macleod, "Design of a two degree-of-freedom ankle-foot orthosis for robotic rehabilitation," in *Proceedings of the IEEE 9th International Conference on Rehabilitation Robotics (ICORR '05)*, pp. 41–44, Chicago, Ill, USA, June 2005.

[46] Y. Fan and Y. Yin, "Mechanism design and motion control of a parallel ankle joint for rehabilitation robotic exoskeleton," in *Proceedings of the IEEE International Conference on Robotics and Biomimetics (ROBIO '09)*, pp. 2527–2532, IEEE, Guilin, China, December 2009.

[47] J. Carberry, G. Hinchly, J. Buckerfield et al., "Parametric design of an active ankle foot orthosis with passive compliance," in *Proceedings of the 24th International Symposium on Computer-Based Medical Systems (CBMS '11)*, pp. 1–6, Bristol, UK, June 2011.

[48] J. Hamill and K. M. Knutzen, *Biomechanical Basis of Human Movement*, Lippincott Williams & Wilkins, 3rd edition, 2009.

[49] B. Celebi, M. Yalcin, and V. Patoglu, "AssistOn-Knee: a self-aligning knee exoskeleton," in *Proceedings of the 26th IEEE/RSJ International Conference on Intelligent Robots and Systems (IROS '13)*, pp. 996–1002, Tokyo, Japan, November 2013.

[50] Y. Zhu, J. Cui, and J. Zhao, "Biomimetic design and biomechanical simulation of a 15-DOF lower extremity exoskeleton," in *Proceedings of the IEEE International Conference on Robotics and Biomimetics (ROBIO '13)*, pp. 1119–1124, Shenzhen, China, December 2013.

[51] K. Shamaei, M. Cenciarini, A. A. Adams, K. N. Gregorczyk, J. M. Schiffman, and A. M. Dollar, "Design and evaluation of a quasi-passive knee exoskeleton for investigation of motor adaptation in lower extremity joints," *IEEE Transactions on Biomedical Engineering*, vol. 61, no. 6, pp. 1809–1821, 2014.

[52] C. Chen, D. Zheng, A. Peng, C. Wang, and X. Wu, "Flexible design of a wearable lower limb exoskeleton robot," in *Proceedings of the IEEE International Conference on Robotics and Biomimetics (ROBIO '13)*, pp. 209–214, IEEE, Shenzhen, China, December 2013.

[53] D.-H. Wang, J. Guo, K.-M. Lee, C.-J. Yang, and H. Yu, "An adaptive knee joint exoskeleton based on biological geometries," in *Proceedings of the IEEE International Conference on Robotics and Automation (ICRA '11)*, pp. 1386–1391, Shanghai, China, May 2011.

[54] M. A. Ergin and V. Patoglu, "A self-adjusting knee exoskeleton for robot-assisted treatment of knee injuries," in *Proceedings of the IEEE/RSJ International Conference on Intelligent Robots and Systems*, pp. 4917–4922, IEEE, San Francisco, Calif, USA, September 2011.

[55] R. Schmidt, "Torque transmitting system," US3242694, 1966.

[56] J.-Y. Kuan, K. A. Pasch, and H. M. Herr, "Design of a knee joint mechanism that adapts to individual physiology," in *Proceedings of the 36th Annual International Conference of the IEEE Engineering in Medicine and Biology Society (EMBC '14)*, pp. 2061–2064, Chicago, Ill, USA, August 2014.

[57] K.-M. Lee and D. Wang, "Design analysis of a passive weight-support lower-extremity-exoskeleton with compliant knee-joint," in *Proceedings of the IEEE International Conference on Robotics and Automation (ICRA '15)*, pp. 5572–5577, IEEE, Seattle, Wash, USA, May 2015.

[58] J. L. Pons, *Wearable Robots: Biomechatronic Exoskeletons*, John Wiley & Sons, New York, NY, USA, 2008.

[59] E. Singla, B. Dasgupta, K. Kondak, C. Fleischer, and G. Hommel, "Optimal design of an exoskeleton hip using three-degrees-of-freedom spherical mechanism," *VDI Berichte*, no. 1956, p. 73, 2006.

[60] M. Pan, D. Zhang, and Z. Gao, "Novel design of a three degrees of freedom hip exoskeleton based on biomimetic parallel structure," in *Proceedings of the IEEE International Conference on Computer Science and Automation Engineering (CSAE '11)*, pp. 601–605, Shanghai, China, June 2011.

[61] Y. Yu and W. Liang, "Design optimization for lower limb assistive mechanism based on manipulability inclusive principle," in *Proceedings of the IEEE International Conference on Robotics and Biomimetics (ROBIO '12)*, pp. 174–180, IEEE, Guangzhou, China, December 2012.

[62] T. Yoshikawa, "Manipulability of robotic mechanisms," *International Journal of Robotics Research*, vol. 4, no. 2, pp. 3–9, 1985.

[63] W. Yang, C.-J. Yang, and Q.-X. Wei, "Design of an anthropomorphic lower extremity exoskeleton with compatible joints," in

Proceedings of the IEEE International Conference on Robotics and Biomimetics (IEEE ROBIO '14), pp. 1374–1379, Bali, Indonesia, December 2014.

[64] M. Cempini, S. M. M. De Rossi, T. Lenzi, N. Vitiello, and M. C. Carrozza, "Self-alignment mechanisms for assistive wearable robots: a kinetostatic compatibility method," *IEEE Transactions on Robotics*, vol. 29, no. 1, pp. 236–250, 2013.

[65] D. Zanotto, Y. Akiyama, P. Stegall, and S. K. Agrawal, "Knee joint misalignment in exoskeletons for the lower extremities: effects on user's gait," *IEEE Transactions on Robotics*, vol. 31, no. 4, pp. 978–987, 2015.

[66] BioDigital, *The Biodigital Human*, 2015, https://www.biodigital.com/.

6

Bounded Control of an Actuated Lower-Limb Exoskeleton

Michael Oluwatosin Ajayi,[1,2] Karim Djouani,[1,2] and Yskandar Hamam[1,3]

[1]*F'SATI, Department of Electrical Engineering, Tshwane University of Technology, Staatsartillerie Road, Pretoria West, Pretoria, South Africa*
[2]*LISSI, University of Paris Est Creteil (UPEC), 94400 Vitry Sur Seine, France*
[3]*LISV, Bâtiment Boucher, Pole Scientifique et Technologique de Vélizy, 10-12 avenue de l'Europe, 78140 Vélizy, France*

Correspondence should be addressed to Michael Oluwatosin Ajayi; ajayimo@tut.ac.za

Academic Editor: Gordon R. Pennock

A bounded control strategy is employed for the rehabilitation and assistance of a patient with lower-limb disorder. Complete and partial lower-limb motor function disorders are considered. This application is centered on the knee and the ankle joint level, thereby considering a user in a sitting position. A high gain observer is used in the estimation of the angular position and angular velocities which is then applied to the estimation of the joint torques. The level of human contribution is feedback of a fraction of the estimated joint torque. This is utilised in order to meet the demands for a bounded human torque; that is, $\tau_h \leq N_{2,n} \leq N_{1,n}$. The asymptotic stability of the bounded control law without human contribution and the convergence analysis of the high gain observer is verified using Lyapunov-based analysis. Simulations are performed to verify the proposed control law. Results obtained guarantee a fair trajectory tracking of the physiotherapist trajectory.

1. Introduction

Wearable robots such as exoskeletons fall into a class of robot manipulators. However, they have to be worn by humans for the purpose of enhancing their movement. One of the interesting features of wearable robots is the fact that they enable healthy wearers to perform stressful activities easily over a lengthy period of time. This sole attribute has made their use in the domain of rehabilitation and assistance quite appreciated, with an extension to the medical, military, and agricultural field. Wearable exoskeletons are electromechanical devices driven by actuators in order to induce the movement of the embodied limbs. In the process of inducing this movement, there remains a possibility of human torque contribution via these limbs through the activities of the muscles, thus delivering part of the torque during rehabilitation.

In the near future, it is envisaged that the number of elderly people in world population will increase considerably; therefore this will further increase the burden of treating the health risks associated with aging by physiotherapists [1]. Robotic therapy solutions that give rise to the use of intelligent machines that are capable of offering solutions that promote motor recovery and better understanding of motor control [2, 3] remain the only efficient way to alleviate physiotherapists from this time and energy consuming exercise; thereby reducing the demands on these assist practitioners [4]. Certainly, the goal of robotic therapy is to assist in a way by automating the performance of repetitive motor-therapy for humans [5]. Robotic solutions could be either assistive, rehabilitative, or both, depending on the purpose of the design. However, the conceptual design with regard to performance-augmenting applications is always based on maintaining the human-exoskeleton kinematic compatibility [6]. This was addressed in [7]; nevertheless, for the purpose of simplicity, in this work, human-exoskeleton misalignment is not considered, and the human-exoskeleton joint axes are assumed to align. In addition, the development of active exoskeletons determines its future and some eventual activities associated with the rehabilitation systems in conjunction with expanding its working capabilities with humans [8].

In rehabilitative robotics, the solutions developed are aimed at optimising existing therapeutic approaches in order to improve or cure motor lost functions after a disabling encounter. Assistive robotics provides solutions that will assist and interact with individuals with reduced motor functionalities so as to help them increase their self-reliance and in sustaining a fulfilling lifestyle within their personal environment [9]. Nevertheless, there exist technical similarities between them, in that the primary subject of interest is centered on the motor performance of the patient.

Utilising robotic systems for assistance and rehabilitation, the wearer could be passive or active throughout the whole session of training. When passive, the wearer voluntary movement tends to be limited since the lower-limb rehabilitation device is only designed to assist the user to follow predefined movement patterns with the aid of certain control methods. Although this method has been proven to be an effective therapeutic measure [10, 11], the need for patients to be actively involved in robotic therapy brings about more success in the results obtained [12]. One rehabilitative method which further validates this proof is the "assist-as-needed" paradigm [13, 14]. Kong and Tomizuka in [15] also designed a rehabilitation device connected to the human via elastic components which are meant to assist the human by generating assistive force to enable the human lower limbs to follow the motion of the links. Most especially, the need to ensure the safety of the wearer is essential as highlighted in [16].

Often the contribution of the wearers may be determined using on-board sensors such as EMG, Electroencephalogram (EEG), Inertial Measurement Unit (IMU), camera, and adequate algorithms to detect and classify the intention. Electromyography (EMG) control strategy is one that promotes the user specific support for controllers by converting EMG signals into desired human torques which is then partly applied by the exoskeleton device [17, 18]. However, this method is somewhat sensitive to noise and sensor location [19].

Furthermore, it is required that since wearable exoskeletons fall into the category of robot manipulators which require control input constraints, the need for the use of a bounded control law is paramount. Controllers used for exoskeletons also need to be robust enough against uncertainties [20] of which the bounded control may be a perfect fit. Bounded control literally refers to the control of a system within certain bounded variables. This type of control method seeks to define certain limits that correspond to the physical limit of the variable in question and therefore consequently helps minimize the power requirements for actuating the exoskeleton [21]. The variables often considered are those of the amplitude of the actuators. This is to avoid the possibility of an irreversible damage to the actuator [22, 23]. An in-depth description of this control method is given in [24]. In [25, 26] the authors propose a bounded control method based on nested saturations for the rehabilitation of the knee joint and a full exoskeleton with four degrees of freedom, respectively. By "nested saturation" it is meant that the actuator is subjected to magnitude and rate saturation simultaneously [27]. This not only allows maintaining the bounded inputs of the control torque, but also provides an efficient convergence of the position and velocity of the physical system. In [28], an adaptive observer-based controller for an upper-limb exoskeleton robot is proposed and the boundedness is also proven.

Due to increased use of wearable robots, which is basically inclined towards providing support to humans to effectively perform their daily activities, the force control has continued to attract the attention of many researchers. The force sensors have been often used; however this makes the system bulky and costly. Hence, the use of encoders to obtain the information external force has been exploited in [29, 30]. In [31] an observer-based user torque estimation is also used to evaluate the patient-active rehabilitation task in order to avoid the use of a force sensor. This has been proven to be a well effective method; nevertheless, the focus was on the estimation of task-specific interaction forces, and the extent of the force that the wearer is able to exert on system-controller is not addressed, owing to the fact that the length of the lower limb and the weight of the wearer may differ.

This work is an attempt to modify and generalise the control method in [25] and propose the same for the rehabilitation of individuals with knee or ankle impairment. This study seeks to drive the knee-ankle exoskeleton to follow a predefined rehabilitation trajectory with or without human contribution for a patient in sitting position. To verify the level of contribution of the human the system may allow, a high gain observer is designed. Although high gain observers have been extensively used for the recovery of states such as in [32], in this work the recovered states are employed to verify the level of human contribution. The asymptotic stability of the closed-loop system without human contribution is verified using Lyapunov-based analysis. The convergence of the error between the states of the wearer-exoskeleton and that of the high gain observer is also analysed using Lyapunov stability theory. The contribution of this work is further detailed in Section 5.

2. Mathematical Description of the Knee-Ankle Exoskeleton

This section presents the knee-ankle dynamics and the mathematical model and its state-space form.

2.1. Knee-Ankle Exoskeleton. The knee-ankle exoskeleton is represented by a planar two-link model with revolute joints as shown in Figure 1. The structure and parameters are the same as in [33]. It is made up of two actuators, one at each joint. This model takes into account the flexion/extension of the knee joint and the plantar-flexion/dorsal-flexion of the ankle joint, assuming the motions are performed in a sagittal plane (the basic movement performed by humans such as walking, standing up, sitting down, running, and climbing stairs are examples of sagittal plane movements). This is because exoexercising tasks for lower limbs are those of sagittal plane movements. Sagittal plane model is therefore most often employed in exoskeletal design as in [34, 35] to mention a few.

It should be noted that the parameters of the subject's shank and foot links and that of the exoskeleton are built into the model parameters. The movements of the knee-ankle exoskeleton are in the range $0\,\text{rad} \leq q_1 \leq 2.35\,\text{rad}$

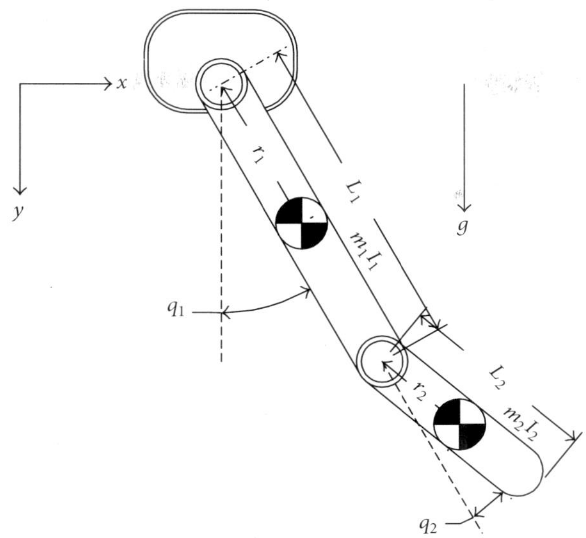

FIGURE 1: Knee-ankle orthosis.

for the knee and 0 rad $\leq q_2 \leq$ 0.87 rad for the ankle. Where 0 rad relates to the full-knee extension, 2.35 rad is the maximum flexion of the knee, and 1.57 rad corresponds to the rest position of the knee. Furthermore, 0 rad, for the ankle movements, corresponds to the rest position of the ankle, and 0.35 rad is the maximal ankle dorsal-flexion, while 0.87 rad denotes the maximal ankle plantar-flexion. The task will be to devise a control method that will enable the movement of the shank and foot about their respective joints as exoskeleton swings about joint axes concurrently with defined trajectories. This will help provide assistive measures to humans with lower-limb disorders, and consequently requires rehabilitation for improved mobility performance. A sitting position is envisaged for this task; hence the contact forces are not computed and the analysis in [36] concerning floating base systems is not considered. Figure 1 illustrates the model parameters used for this analysis. q_1 is the absolute knee joint angle while q_2 is the relative ankle joint angle (i.e., foot angle).

The need for a reduce number of degrees of freedom (DoF), that is, 2-DoF, for a knee-ankle exoskeleton is to restrict the muscle actuation to the behaviour of each joint required to participate in the actualisation of a certain control task. In addition, numerous DoF may become problematic when included in mathematical models and may eventually cause a drawback in fulfilling the engineering control tasks. Lower numbers of DoF are being assumed for the purpose of simplification [37] so as to reduce its complexity.

The dynamic model of the system may be derived by the Euler-Lagrange principle and written as [38]

$$M(q)\ddot{q} + C(q,\dot{q})\dot{q} + G(q) = \tau. \quad (1)$$

For the purpose of this study, the friction torque, disturbance torque, and the human torque are included; therefore (1) may be modified to give

$$M(q)\ddot{q} + C(q,\dot{q})\dot{q} + G(q) + F(\dot{q}) + \tau_d = \tau + \tau_h, \quad (2)$$

where $q \in \mathfrak{R}^n$ is a vector of angular positions and is assumed available by measurement, $M(q) \in \mathfrak{R}^{n\times n}$ is the inertia matrix of the links, while $\tau \in \mathfrak{R}^n$ is the control torque, which serves as the control inputs of the system, $C(q,\dot{q})\dot{q} \in \mathfrak{R}^n$ is the vector of the Coriolis and centripetal torques, $G(q) \in \mathfrak{R}^n$ is the gravitational torque, and $\tau_h \in \mathfrak{R}^n$ is the human torque, which allows the investigation of the passive and active mode of the wearer. That is, $\tau_h = 0$ implies no human effort and $\tau_h \neq 0$ implies the presence of human effort, and $\tau_d \in \mathfrak{R}^n$ is the disturbance torque, which may be due to measurement noise and is represented as $\tau_d = k_n \mathrm{sgn}(\dot{q})$. k_n is a sine signal and n is the degree of freedom of the system.

The human torque τ_h is stated as

$$\tau_h = \kappa\hat{\tau}, \quad (3)$$

where κ is the variable factor of the level of contribution of the human and $\hat{\tau}$ is the estimated torque.

For simplicity the friction torque $F(\dot{q})$ is written in the form above but may be represented as

$$F(\dot{q}) = F_v\dot{q} + F_d \,\mathrm{sign}(\dot{q}), \quad (4)$$

where $F_v = \mathrm{diag}[v_n]$ is the viscous friction torque and $F_d = \mathrm{diag}[d_n]$ is the dynamic friction torque and v_n and d_n are vector quantities associated with their respective torques. $\mathrm{sign}(\cdot)$ is a Signum function.

For clarity purposes, the human and exoskeleton system dynamics may be defined as being summed up in (2). Hence, they may be regarded as $i \in [1,2]$, which stands for the respective human and exoskeleton components; that is, $M(q) = \sum_{i=1}^{2} M_i(q)$, $C(q,\dot{q}) = C_i(q,\dot{q})$, $G(q) = G_i(q)$, $F_v = F_{v_i}$, and $F_d = F_{d_i}$. In (4), it should be noted that the dynamic friction is balanced so as to make the exoskeleton most transparent to the wearer. The gravity torque is also balanced since the exoskeleton is usually used for rehabilitation purpose.

The matrices associated with $M(q) \in \mathfrak{R}^{n\times n}$, $C(q,\dot{q})\dot{q}$, and $G(q) \in \mathfrak{R}^n$ are given below.

Mass Matrix

$$\begin{aligned} M(q)_{1,1} &= I_1 + I_2 + m_1 L_2^2 + 2m_2 L_1 r_2 \cos(q_2) \\ M(q)_{1,2} &= I_2 + m_2 L_1 r_2 \cos(q_2) \\ M(q)_{2,1} &= I_2 + m_2 L_1 r_2 \cos(q_2) \\ M(q)_{2,2} &= I_2. \end{aligned} \tag{5}$$

Coriolis/Centripetal Vector

$$\begin{aligned} C(q,\dot{q})_{1,1} &= -2m_2 L_1 r_2 \sin(q_2)\dot{q}_2 \\ C(q,\dot{q})_{1,2} &= -m_2 L_1 r_2 \sin(q_2)\dot{q}_2 \\ C(q,\dot{q})_{2,1} &= m_2 L_1 r_2 \sin(q_2)\dot{q}_1 \\ C(q,\dot{q})_{2,2} &= 0. \end{aligned} \tag{6}$$

Gravity Vector

$$\begin{aligned} G(q)_1 &= (m_1 r_1 + m_2 L_1) g \sin(q_1) \\ &\quad + m_2 g r_2 \sin(q_1 + q_2) \\ G(q)_2 &= m_2 g r_2 \sin(q_1 + q_2) \end{aligned} \tag{7}$$

2.2. State-Space Description. To describe the nonlinear state-space form of (2), the human torque τ_h is neglected. This is because it is a fraction of an estimate of the control torque. Since the matrix $M(q)$ is positive definite, it is invertible. Hence, (2) may be written as

$$\ddot{q} = M(q)^{-1}(\tau - C(q,\dot{q})\dot{q} - G(q) - F(\dot{q}) - \tau_d). \tag{8}$$

The nonlinear state-space form of the dynamic model (1) is given by

$$\dot{x} = f(x) + g(x)\tau, \tag{9}$$

where the state vectors $x : x_1 = [q_1 \;\; q_2]^T$ and $x_2 = [\dot{q}_1 \;\; \dot{q}_2]^T$. The functions $f(x)$ and $g(x)$ are assumed to be continuously differentiable a sufficient number of times.

3. High Gain Observer Structure and Estimation

In wearable robotic systems the coordinates may be precisely measured by an encoder for each joint. However, velocity measurements obtained via tachometers are easily perturbed by noise. These constraints may be overcome by employing an observer for state estimation. The high gain observer is employed for this purpose. Prior to defining the observer dynamics it is necessary to reorganise the state-space model of system in (9) as

$$\begin{aligned} \dot{x}_1 &= x_2 \\ \dot{x}_2 &= f(x) + g(x)\tau = \phi(x_1, x_2, t) \\ y &= x_1, \end{aligned} \tag{10}$$

where y is the measurable position vector.

The state observer is based on the application of high gains, for the system to estimate the angular positions and velocities. The observer in state-space is given by

$$\begin{aligned} \dot{\hat{x}}_1 &= \hat{x}_2 - \frac{1}{\mu} H_p (y - \hat{x}_1) \\ \dot{\hat{x}}_2 &= \phi(x_1, x_2, t) - \frac{1}{\mu^2} H_v (y - \hat{x}_1) \end{aligned} \tag{11}$$

with $\hat{x}_1, \hat{x}_2$ denoting the generalised positions and velocities; that is, $\hat{x}_1$ is the estimation of the joint angular position $[q_1 \;\; q_2]$ and $\hat{x}_2$ is the estimation of the joint angular velocities $[\dot{q}_1 \;\; \dot{q}_2]$ and $\hat{x} = [\hat{x}_1^T \;\; \hat{x}_2^T]^T$. μ is a scalar parameter chosen with the limit; $0 < \mu < 1$ while H_p and H_v are the two constant observer gains which have the form

$$\begin{aligned} H_p &= \operatorname{diag}[h_{pn}] \\ H_v &= \operatorname{diag}[h_{vn}]. \end{aligned} \tag{12}$$

3.1. Observer Error Dynamics. The observer error dynamics between the knee-ankle exoskeleton and the high gain observer may be given by

$$\begin{aligned} \dot{\tilde{x}}_1 &= \hat{x}_2 - \frac{1}{\mu} H_p (y - \tilde{x}_1) \\ \dot{\tilde{x}}_2 &= \phi(x_1, x_2, t) - \frac{1}{\mu^2} H_v (y - \tilde{x}_1), \end{aligned} \tag{13}$$

where $\tilde{x}_1 = x_1 - \hat{x}_1$ and $\tilde{x}_2 = x_2 - \hat{x}_2$. $\tilde{x} = [\tilde{x}_1^T \;\; \tilde{x}_2^{T}]^T$ is the error between the system states and the observer states.

For the purpose of the convergence analysis, the error is transformed into a more convenient form as

$$\begin{aligned} \tilde{e}_1 &= \tilde{x}_1 \\ \tilde{e}_2 &= \mu \tilde{x}_2 \\ \tilde{e} &= [\tilde{e}_1^T \;\; \tilde{e}_2^T]^T. \end{aligned} \tag{14}$$

Based on the new transformation the error dynamics may be rewritten as

$$\mu \dot{\tilde{e}} = H\tilde{e} + \mu^2 W \phi(x_1, x_2, t), \tag{15}$$

where $H = \begin{bmatrix} -H_p & I_{2\times 2} \\ -H_v & 0_{2\times 2} \end{bmatrix}$ and $W = \begin{bmatrix} 0_{2\times 2} \\ I_{2\times 2} \end{bmatrix}$.

3.2. Convergence Analysis. For clarity, the Lyapunov-based approach in [32] is employed to ascertain the error convergence of the orthosis or exoskeleton states and the high gain observer.

Definition 1. Define the upper bound for the terms $M(q)$, $C(q,\dot{q}), G(q), F_v\dot{q} + F_d\operatorname{sign}(\dot{q})$ and τ_d, in (2), as follows: $\|M(q)\| \le \lambda_0$, $\|C(q,\dot{q})\| \le \lambda_1\|\dot{q}\|^2$, $\|G(q)\| \le \lambda_2$, $\|F_v\dot{q} + F_d\operatorname{sign}(\dot{q})\| \le v\|\dot{q}\| + d_1$, and $\|\tau_d\| \le d_2$. $\|\cdot\|$ is any appropriate norm and λ_i are positive nonzero constants, where $0 < \lambda_i < \infty, \forall i$, and all these properties are applicable $\forall q, e \in \mathfrak{R}^2$ with $\|\tilde{e}\| < \infty$.

With regard to these definitions, the condition for which the nonlinear error dynamics convergence to zero may be found.

Proposition 2. *Consider the dynamic of the system composed of the wearer and the exoskeleton given in (2), with $\tau_h = 0$. With $\tau_h = 0$, no force is being exerted by the user.*

Proof. Consider the quadratic Lyapunov function V for the error dynamics in (15) to be

$$V = \tilde{e}^T P\tilde{e}, \tag{16}$$

where P is the solution to the Lyapunov equation $H^T P + PH = -I$ of a linear system. P is a positive definite matrix and it is independent of μ.

The derivative of V is then given by

$$\dot{V} = \dot{\tilde{e}}^T P\tilde{e} + \tilde{e}^T P\dot{\tilde{e}}. \tag{17}$$

Substituting (15) in (17) gives

$$\begin{aligned}&\left(\frac{1}{\mu}H\tilde{e} + \mu W\phi\right)^T P\tilde{e} + \tilde{e}^T P\left(\frac{1}{\mu}H\tilde{e} + \mu W\phi\right)\\ &= \frac{1}{\mu}\tilde{e}^T\left(H^T P + PH\right)\tilde{e} + 2\mu\phi^T W^T P\tilde{e} \\ &= -\frac{1}{\mu}\|\tilde{e}\|^2 + 2\mu\phi^T W^T P\tilde{e},\end{aligned} \tag{18}$$

where ϕ is a vector function of the states x_1 and x_2.

Based on (18), the inequality below may be obtained:

$$\dot{V} \le -\frac{1}{\mu}\|\tilde{e}\|^2 + 2\mu\left\|PW\phi\left(x_1, x_2, t\right)\right\|\|\tilde{e}\|. \tag{19}$$

Define the upper bounded term and a set for the least upper bound as

$$\begin{aligned}\psi(t) &= \left\|PW\phi\left(x_1, x_2, t\right)\right\|,\\ c_T &= \sup_{t\in[0,T]}\psi(t).\end{aligned} \tag{20}$$

Since $\phi(x_1, x_2, t)$ may be defined as the acceleration of the orthotic or exoskeletal device (see (10)), it is therefore upper bounded based on Definition 1 and shows that $\psi(t)$ is upper bounded and c_T exists. P and W are constant matrices.

Assumption 3. Assume that $c_T = \sup_{t\in[0,T]}\psi(t)$ exists for any finite value $T > 0$.

From (19), it follows that

$$\begin{aligned}\dot{V} &\le -\frac{1}{\mu}\|\tilde{e}\|^2 + \frac{1}{\mu}K_e\left(\mu, c_T\right)\|\tilde{e}\|, \quad \forall t \in [0, T]\\ &= \frac{1}{\mu}\left(K_e\left(\mu, c_T\right) - \|\tilde{e}\|\right)\|\tilde{e}\|,\end{aligned} \tag{21}$$

where $K_e(\mu, c_T) = 2\mu^2 c_T$. K_e is a value determined by μ and c_T. This value remains unchanged when the observer of the orthosis is designed. From (21), it may be deduced that if $\|\tilde{e}\| > K_e(\mu, c_T)$, then $\dot{V} < 0, \forall t \in [0, T]$. Also, based on (20), it may be seen that different cases could arise.

Assumption 4. Assume that $c_T < \infty$.

Then it means that $\|\tilde{e}(t_0)\| \le K_e(\mu, c_T)$. Therefore, it may be deduced that $\|\tilde{e}\| \le K_e(\mu, c_T)$, $\forall t \in [0, T]$. The reason is that once $\|\tilde{e}(t_1)\| > K_e(\mu, c_T)$, then the derivative $\dot{V} < 0$, and therefore the norm of the error $\|\tilde{e}\|$ diminishes. By choosing small enough values for μ, the errors between the states of the system and the observer state may be kept small for any finite T.

Assumption 5. Also assume that $\psi(t) \le \eta$.

And here η represents the bounded range of $[0, \infty)$. Hence, $c = \sup_{t\in[0,\infty)}\psi(t) < \infty$. It may be also seen that if $\|\tilde{e}\| \le K_e(\mu, c)$, then $\|\tilde{e}(t)\| \le K_e(\mu, c), \forall t \in \eta$. Also $\|\tilde{e}(t_0)\| > K_e(\mu, c)$ causes the derivative $\dot{V} < 0$ which in turn diminishes $\|\tilde{e}(t)\|$. Therefore, there exists a finite value $t_\mu > 0$ such that $\|\tilde{e}(t)\| \le K_e(\mu, c)\forall t > t_\mu$.

From the transformed error coordinate in (14), it may be deduced that, for $\mu < 1, \|\tilde{x}\| \le (1/\mu)\|\tilde{e}\|$. Hence if $\|\tilde{e}\| \le K_e(\mu, c_T)$, then $\|\tilde{x}\| \le K_x(\mu, c_T)$, where $K_x(\mu, c_T) = 2\mu c_T$. For any finite value of T, $K_x(\mu, c_T)$ may be made small by making μ small enough. The obtained error convergence proves that the state of the orthosis/exoskeleton may be observed by the high gain observer in (11). Therefore the torque may be estimated based on the recovered states. □

4. Control Law

The formulation of the control law is based on considering the safety of the wearer. It must therefore be sufficiently robust for the stability of the human-orthosis system whether the user is passive or active. The proposed control law is based on [25]. However the generalisation of the control law is considered in this study and it is modified as

$$\begin{aligned}\tau = {} & \mathrm{sat}_{N_1,n}\left[K_d\dot{\varepsilon} + \mathrm{sat}_{N_2,n}\left(K_p\varepsilon\right)\right] + M\left(q\right)\ddot{q}_d\\ &+ C\left(q, \dot{q}\right)\dot{q}_d + G\left(q\right) + F_v\dot{q} + F_d\,\mathrm{sign}\left(\dot{q}\right),\end{aligned} \tag{22}$$

where $\varepsilon \in \mathfrak{R}^n = q_d - q$ is a vector of the angular position error. $\dot{\varepsilon} \in \mathfrak{R}^n = \dot{q}_d - \dot{q}$ is the vector of the angular velocity error. $\ddot{\varepsilon} \in \mathfrak{R}^n = \ddot{q}_d - \ddot{q}$ is the vector of the angular acceleration error. $q_d \in \mathfrak{R}^n, \dot{q}_d \in \mathfrak{R}^n$, and $\ddot{q}_d \in \mathfrak{R}^n$ signify the desired joint angular position, angular velocity, and angular acceleration, respectively. $K_d = \mathrm{diag}[k_{dn}]$ is the derivative gain while $K_p = \mathrm{diag}[k_{pn}]$ is the proportional gain. $n = 1, 2$.

The definition of the saturation function is given below.

4.1. Saturation Function. The saturation function $\mathrm{sat}_{N_i,n}(x_n)$ is defined as

$$\mathrm{sat}_{N_{i,n}} = \min\left(N_{i,n}, \max\left(-N_{i,n}, x_n\right)\right), \quad \forall x_n \in \mathfrak{R}^n, \tag{23}$$

where $\pm N_{i,n}$ represent the saturation bounds. This bounded interval (point) is chosen such that $N_{1,n} > 2N_{2,n}$. The saturation of the control law allows the actuator of the exoskeleton to avoid exceeding its limit, thereby maintaining the linearity of the actuator and avoiding the hysteresis cycles which may result in irreversible damage. This guarantees the stability of the closed-loop trajectories. High value control law demands high power and may expose the user to risk.

4.2. Closed-Loop System. The dynamics of the closed-loop system are achieved by substituting (4) into (2) and then substituting the eventual equation into (22). The resulting equation is given by

$$M(q)\ddot{\varepsilon} = -\mathrm{sat}_{N_1,n}\left[K_d\dot{\varepsilon} + \mathrm{sat}_{N_2,n}\left(K_p\varepsilon\right)\right] - C(q,\dot{q})\dot{\varepsilon} - \tau_h. \tag{24}$$

Note that τ_d for the purpose of the proof is not included in the closed-loop system equation.

4.3. Stability Analysis. Here, the closed-loop stability of the entire system (orthotic/exoskeletal device and controller) is verified. A completely passive mode (i.e., $\tau_h = 0$) for the user is considered. Passive mode refers to a user exerting no muscular torque that may interfere with the orthosis/exoskeleton trajectory, as specified by the physiotherapist. The stability for the system's equilibrium point is proven. For the purpose of this proof consider $N_{i,n} \in \mathfrak{R}^n$.

Proposition 6. *Consider the wearer-exoskeleton dynamics defined by (2), with $\tau_h = 0$. With $\tau_h = 0$, no force is being exerted by the user.*

Assumption 7. Assume that the derivatives of the desired angle ($\dot{q}_d$, $\ddot{q}_d$ and $q_d^{(3)}$) are bounded and known.

Then the equilibrium point $x_e = [\varepsilon^T \quad \dot{\varepsilon}^T]^T = 0 \in \mathfrak{R}^{2n}$ of the closed-loop system (24) is asymptotically stable.

Proof. Let $K_p|\varepsilon| > N_{2,n}$ and $K_d|\dot{\varepsilon}| > N_{1,n} - N_{2,n} > N_{2,n}$, defining the positive definite Lyapunov function V_1 to be

$$V_1 = \frac{1}{2}\dot{\varepsilon}^T M(q)\dot{\varepsilon}. \tag{25}$$

The derivative of $\dot{V}_1$ is given by

$$\dot{V}_1 = \dot{\varepsilon}^T M(q)\ddot{\varepsilon} + \frac{1}{2}\dot{\varepsilon}^T \dot{M}(q)\dot{\varepsilon}. \tag{26}$$

Substituting (24) into (26) gives $\dot{V}_1$ as

$$\dot{V}_1 = -\dot{\varepsilon}^T\left(\mathrm{sat}_{N_1,n}\left[K_d\dot{\varepsilon} + \mathrm{sat}_{N_2,n}\left(K_p\varepsilon\right)\right] - C(q,\dot{q})\dot{\varepsilon} - \tau_h\right) + \frac{1}{2}\dot{\varepsilon}^T\dot{M}(q)\dot{\varepsilon}. \tag{27}$$

If $\tau_h = 0$, the derivative V_1 is

$$\dot{V}_1 = -\dot{\varepsilon}^T\mathrm{sat}_{N_1,n}\left[K_d\dot{\varepsilon} + \mathrm{sat}_{N_2,n}\left(K_p\varepsilon\right)\right] - \dot{\varepsilon}^T C(q,\dot{q})\dot{\varepsilon} + \frac{1}{2}\dot{\varepsilon}^T\dot{M}(q)\dot{\varepsilon}. \tag{28}$$

By factorisation, (28) may be rewritten as

$$\dot{V}_1 = -\dot{\varepsilon}^T\mathrm{sat}_{N_1,n}\left[K_d\dot{\varepsilon} + \mathrm{sat}_{N_2,n}\left(K_p\varepsilon\right)\right] + \frac{1}{2}\dot{\varepsilon}^T\left(\dot{M}(q) - 2C(q,\dot{q})\right)\dot{\varepsilon}. \tag{29}$$

The matrices $\dot{M}(q)$ and $C(q,\dot{q})$ are dependant, as the following relation is satisfied by the second term in (29); therefore, $\dot{\varepsilon}^T S\dot{\varepsilon} = (1/2)\dot{\varepsilon}^T(\dot{M}(q) - 2C(q,\dot{q}))\dot{\varepsilon} = 0$, where S is skew symmetric. Then (29) becomes

$$\dot{V}_1 = -\dot{\varepsilon}^T\mathrm{sat}_{N_1,n}\left[K_d\dot{\varepsilon} + \mathrm{sat}_{N_2,n}\left(K_p\varepsilon\right)\right]. \tag{30}$$

If $|K_d\dot{\varepsilon}| > N_{2,n}$, by observation $\dot{\varepsilon}$ and $K_d\dot{\varepsilon} + \mathrm{sat}_{N_2,n}(K_p\varepsilon)$ have the same sign and certainly $|K_d\dot{\varepsilon} + \mathrm{sat}_{N_2,n}(K_p\varepsilon)| > 0$. With regard to this, $\dot{V}_1$ may be reduced to

$$\dot{V}_1 = -\lambda\left|\dot{\varepsilon}^T\right|N_{1,n} \leq 0. \tag{31}$$

It may therefore be assumed that as $|\dot{\varepsilon}^T|$ decreases $\dot{V}_1$ decreases also. Note that λ is any positive value to derive the inequality for checking the negativeness of $\dot{V}_1$. However, based on this, no conclusion may be reached with regard to its asymptotic stability. To establish this, a second Lyapunov function is chosen considering a velocity reference signal and the filtered tracking error.

Definition 8. Define the velocity reference signal as $\dot{q}_r = \dot{q}_d - \Gamma\varepsilon$ and the filtered error as $s = \dot{q}_r - \dot{q} = \dot{\varepsilon} - \Gamma\varepsilon$, $\Gamma = \mathrm{diag}[\gamma_i] > 0$. Hence, $\dot{s} = \ddot{q}_r - \ddot{q} = \ddot{\varepsilon} - \Gamma\dot{\varepsilon}$.

Based on Definition 8, the following equation holds:

$$\begin{aligned} M(q)[\dot{s} + k_1 s] + C(q,\dot{q})s &= M(q)[\ddot{q}_r - k_1 s] + C(q,\dot{q})\dot{q}_r \\ &\quad - [M(q)\ddot{q} + C(q,\dot{q})\dot{q}], \end{aligned} \tag{32}$$

where $M(q)k_1 \geq 0$ and commutative. $k_1 = \mathrm{diag}[\alpha] \geq 0$. Substituting (2) into (32) yields

$$\begin{aligned} M(q)[\dot{s} + k_1 s] + C(q,\dot{q})s &= M(q)[\ddot{q}_r - k_1 s] + C(q,\dot{q})\dot{q}_r \\ &\quad - [\tau + \tau_h - G(q) - F_v\dot{q} - F_d\,\mathrm{sign}(\dot{q})]. \end{aligned} \tag{33}$$

Assumption 9. From (22), consider an equivalence controller of the form

$$\begin{aligned} \tau = {} & \mathrm{sat}_{N_1,n}\left[K_d\dot{s} + \mathrm{sat}_{N_2,n}\left(K_p s\right)\right] + M(q)[\ddot{q}_r - k_2 s] \\ & + C(q,\dot{q})\dot{q}_r + G(q) + F_v\dot{q} + F_d\,\mathrm{sign}(\dot{q}), \end{aligned} \tag{34}$$

where k_2 is of the same form as k_1.

Introducing (34) into (33) results in the closed-loop system with regard to (2).

$$M(q)\dot{s} = -Ks - \mathrm{sat}_{N_1,n}\left[K_d\dot{s} + \mathrm{sat}_{N_2,n}\left(K_p s\right)\right] - C(q,\dot{q})s, \tag{35}$$

where $K = M(q)(2k_1 - k_2)$.

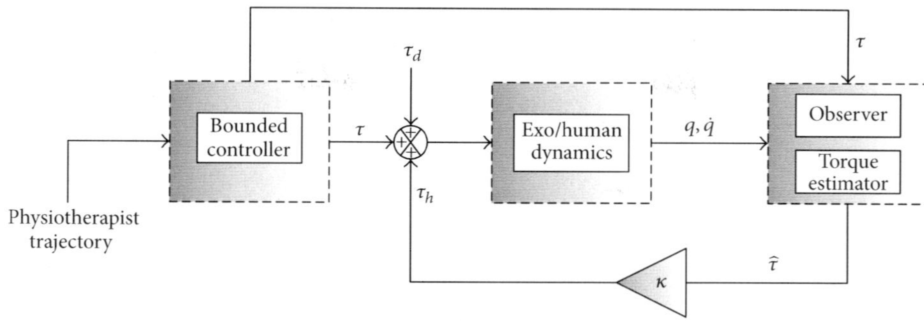

FIGURE 2: Block diagram of the observer-based controller.

Consider the Lyapunov function

$$V_2 = \frac{1}{2} s^T M(q) s. \quad (36)$$

The derivative of V_2 may be given as

$$\dot{V}_2 = s^T M(q) \dot{s} + \frac{1}{2} s^T \dot{M}(q) s. \quad (37)$$

Substituting (35) into (37) gives

$$\dot{V}_2 = -s^T K s - s^T \text{sat}_{N_1,n} \left[K_d \dot{s} + \text{sat}_{N_2,n} \left(K_p s \right) \right] + \frac{1}{2} s^T \left(\dot{M}(q) - 2C(q, \dot{q}) \right) s. \quad (38)$$

$(1/2)s^T(\dot{M}(q) - 2C(q, \dot{q}))s = 0$ based on the skew symmetric properties as previously stated. The derivative of V_2 therefore becomes

$$\dot{V}_2 = -s^T K s - s^T \text{sat}_{N_1,n} \left[K_d \dot{s} + \text{sat}_{N_2,n} \left(K_p s \right) \right]. \quad (39)$$

Using the same principle as in (30), $\dot{V}_2$ can be reduced to

$$\dot{V}_2 = -s^T K s - \lambda \left| s^T \right| N_1, \quad n \leq 0, \quad (40)$$

where λ is the same as previously defined.

This then leads to $-\dot{V}_2(0) \leq \dot{V}_2(t)$. Consequently, the boundedness of $\dot{s}$, s, and $\dot{e}$ subsequently leads to $e \to 0$ as $t \to \infty$. Based on the above choice of different Lyapunov function (V_1 and V_2), the need for the Mastrov theorem for conclusion of asymptotic stability is eliminated.

5. Problem Formulation

The objective of this paper is to formulate a strategy by which the human force (torque) sensitivity during the design and operation phases of an orthosis can be evaluated. This is to be achieved by the following:

(1) First is setting actuator size limit for the exoskeleton by means of nested saturation.

(2) Second is reconstruction of the control torque via the recovered states using the observer. A portion is then fed back to act as the human torque, mimicking the exertion of the force of the human-in-the-loop.

(3) Verification of this principle is based on the utilisation of a robust bounded control strategy with balanced gravity torque and friction to make the orthosis as transparent as possible to the wearer.

(4) Convergence analysis and asymptotic stability are therefore proven to mathematically validate the proposed principle.

The orthosis control with human-in-the-loop requires three building blocks; the dynamics of the wearer-exoskeleton, the controller, and the observer/torque estimator (see Figure 2.) Here, the mathematical equation for the torque estimator and the reference inputs are presented.

The physiotherapist trajectories for the knee and ankle are chosen to fit within the limits of their range of motions and are given by

$$q_1 = \frac{3\pi}{2} \sin\left(\frac{2\pi}{3} t\right)$$
$$q_2 = \frac{\pi}{6} \sin\left(\frac{\pi}{2} t\right). \quad (41)$$

Furthermore, the high gain observer used for the estimation of the angular positions and velocities is analysed in Section 3. The estimation of the torque is derived from these values based on the equation

$$\hat{\tau} = M(\hat{q}) \hat{\ddot{q}} + C(\hat{q}, \hat{\dot{q}}) \hat{\dot{q}} + G(\hat{q}) + F_v \hat{\dot{q}} + F_d \,\text{sign}(\hat{\dot{q}}), \quad (42)$$

where all the parameters in the equation are same as that in (2) but represent the estimated version.

6. Simulations Results and Discussion

For the purpose of clarity, three cases have been considered. For each case, the simulations were channeled towards the validation of the control strategy hence showing its robustness. In addition the following variables used in the simulation results are described below:

(i) q_1 is the knee joint angle (position).

(ii) q_2 is the ankle joint angle (position).

(iii) $\dot{q}_1$ is the derivative (velocity) of knee joint angle.

TABLE 1: Physiological parameters *(Case 1 & Case 2)*.

Parameters	Units	Values
Shank length (L_1)	m	0.2000
Foot length (L_2)	m	0.0800
Shank mass (m_1)	kg	2.8000
Foot mass (m_2)	kg	1.1700
Shank inertia (I_1)	kg·m^2	0.0750
Foot inertia (I_2)	kg·m^2	0.0120
gravity (g)	m/s^2	9.8000
Shank damping Coeff. (b_1)	Nm/s^2	0.4000
Foot damping Coeff. (b_2)	Nm/s^2	0.6000
Knee joint angle ($q_1(0)$)	rad	0.5000
Ankle joint angle ($q_2(0)$)	rad	0.2000
Knee joint friction (v_1)	rad	2.3500
Ankle joint friction (v_2)	rad	0.8700
$k_n(\tau_d)$	rad	$5\sin(4\pi t)$

(iv) $\dot{q}_2$ is the derivative (velocity) of ankle joint angle.

(v) q_{d1} is the desired knee joint angle (position).

(vi) q_{d2} is the desired ankle joint angle (position).

(vii) $\dot{q}_{d1}$ is the derivative (velocity) of desired knee joint angle.

(viii) $\dot{q}_{d2}$ is the derivative (velocity) of desired ankle joint angle.

(ix) $\hat{q}_1$ is the estimated knee joint angle (position).

(x) $\hat{q}_2$ is the estimated ankle joint angle (position).

(xi) $\dot{\hat{q}}_1$ is the derivative (velocity) of the estimated knee joint angle.

(xii) $\dot{\hat{q}}_2$ is the derivative (velocity) of the estimated ankle joint angle.

6.1. Case 1. In this case, the physiological parameters used are those of Table 1, with the exception of the d_n and v_n. Hence, arbitrary values of $d_n = 2.35$ rad and $v_n = 0.87$ rad were chosen for the purpose of simulation. For this bounded control law, it is desired that the maximal motor torque is 15 Nm for both joint torques. The saturation bounds $N_{1,1}$ and $N_{1,2}$ are chosen to be 7.6 Nm for both joints, whereas $N_{2,1}$ and $N_{2,2}$ are chosen to be 7.6/2.1 Nm for both joints as well. The controller gains for K_p are 15.2180×10^2 and 9.4500×10^2 for the knee and ankle joints and K_d are 0.2840×10^3 and 0.6720×10^3 for the respective joints. The observer gains $h_{v1} = 15.5$, $h_{v2} = 14.2$, $h_{p1} = 10.5$, and $h_{p2} = 12.5$, while μ is set to 0.01. The gains of the system-controller are manually tuned to find the best fit. This is the same for the observer gain. Due to the presence of control chattering activity induced by the saturation function, the control gains required to ensure a good tracking performance somehow increase. This justifies the high gains needed for this controller.

In *Case 1*, three scenarios were studied, where the desired trajectory is expected to be tracked by the wearer in each scenario despite the condition introduced. These scenarios are discussed below.

6.1.1. Scenario 1. In the first scenario, the wearer is completely passive (complete motor function disorder); that is, $\tau_h = 0$ (see Figures 3, 4, and 5). The controller is seen to perform the task of making the user completely track the desired angular position and velocity trajectories, at a considerable time. The torque input is within the limits of 15 Nm and hence the saturation of the actuators is avoided. The observer's ability to correctly estimate the position and velocity trajectories is also guaranteed.

6.1.2. Scenario 2. In the second scenario, two simulations that show the wearer, providing a percentage of the human torque used in resisting the orthosis control ability to track the desired angular position and velocity trajectories are presented (see Figures 6, 7, 8 and 9) for the first simulation and (see Figures 10, 11, 12, and 13) for the second simulation. In both cases, the human torque $\tau_h \neq 0$, and it is a percentage of the estimated torque given by $\tau_h = \kappa\hat{\tau}$, where $\kappa = 0.1$ and 0.2. The angular position and velocity trajectory tracking are also accomplished and the control torque inputs have a higher value. The higher the resistive torque, the higher the control torque value. This is because the controller has to produce greater torque to overcome the resistance of the wearer. The resistive human torque is achieved by inverting a fraction of the estimated torque. This may be seen in Figures 9 and 13. However, since it is within a specified limit, the torque input should not be allowed to overcome the specified limit to avoid saturation. Although the maximal torque is somewhat exceeded by a small percentage, a considerable tolerance actuator value in the physical system could compensate for this. Beyond these limits, saturation occurs. The main interest is to ensure that the wearer safety is still guaranteed; hence setting the torque limit to 15 Nm satisfies this condition.

6.1.3. Scenario 3. In the third scenario, an assistive human torque provided by the wearer to assist the orthosis control in order to track the desired angular position and velocity trajectories is verified (see Figures 14, 15, 16, and 17). The assistive human torque ($\tau_h \neq 0$) provided is 10% of the estimated torque. The angular position and velocity trajectory tracking are guaranteed as required, but the control torque inputs have a smaller value. This is due to the assistance provided by the wearer. The assistive human torque is achieved by inputting a fraction of the estimated torque directly in contrast to the resistive torque applied. This may be seen in Figure 17. It should be noted that the second and the third scenarios exemplify a partial motor function disorder.

6.2. Case 2. The value of the torque input of the system may depend on its parameters and the external loads. However, applying high value control law may lead to a high power requirement to ensure it. Wearable orthosis/exoskeleton is characteristically equipped with on-board power system and

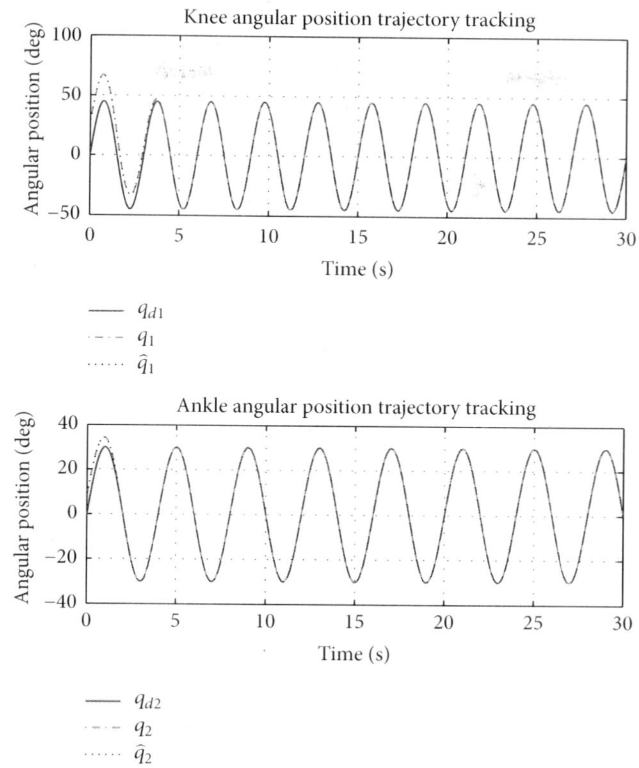

FIGURE 3: Knee-ankle position trajectory tracking. $\tau_h = 0$ *(Scenario 1 of Case 1).*

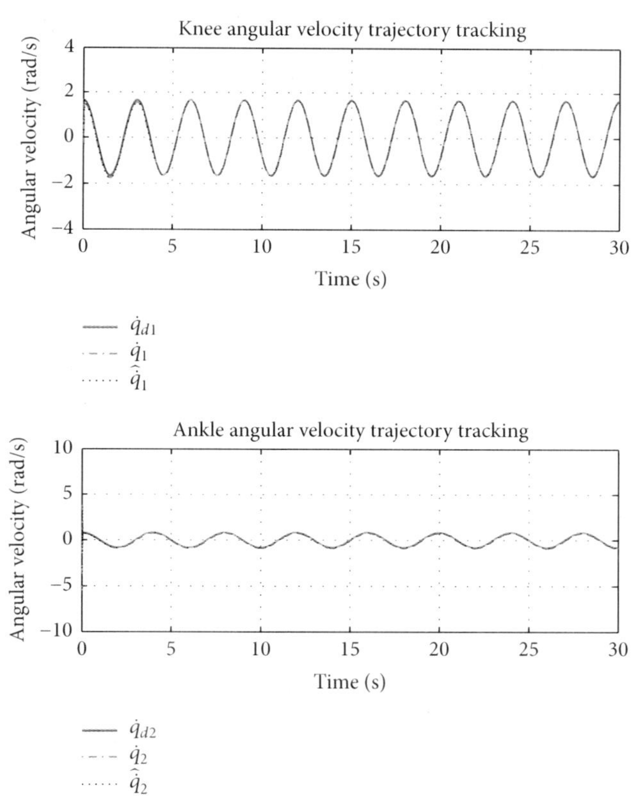

FIGURE 4: Knee-ankle velocity trajectory tracking. $\tau_h = 0$ *(Scenario 1 of Case 1).*

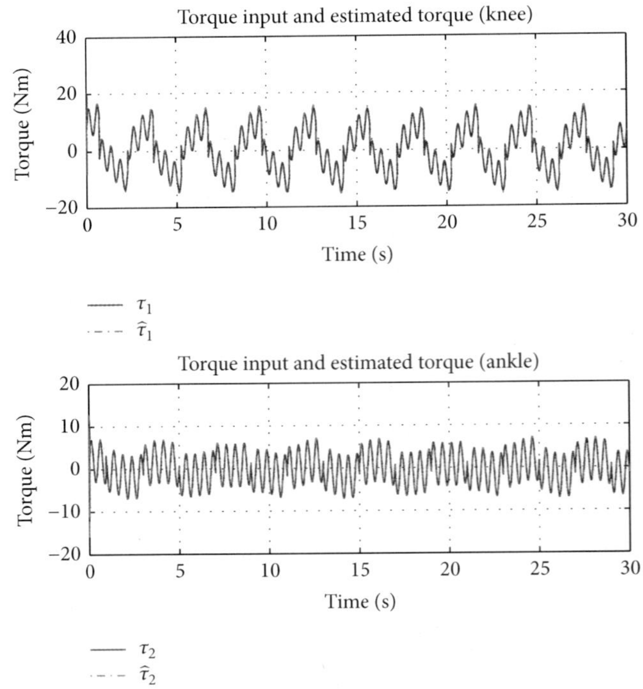

FIGURE 5: Knee-ankle torque input and estimated torque. $\tau_h = 0$. This is to exemplify a complete motor function disorder when the wearer is completely passive. The estimated torque is plotted to show that the observer can correctly estimate the exoskeleton state; hence the torque can be reconstructed *(Scenario 1 of Case 1).*

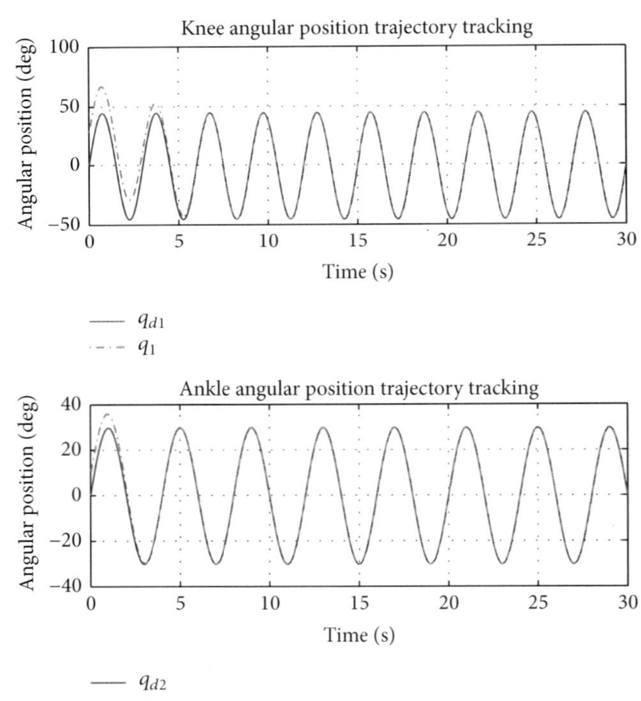

FIGURE 6: Knee-ankle position trajectory tracking. $\tau_h \neq 0$; $\kappa = 0.1$ *(Scenario 2 of Case 1).*

hence needs high power demands that may endanger the wearer. To reduce this high value control law effect the nested saturation (bounded control) is employed so that the actual actuator size limit can be set. This has been established in *Scenario 2 and 3 of Case 1.*

For the purpose of clarity, an actuator size of 30 Nm is considered. Using the same physiological parameters as in Table 1, controller gains and other parameters with the exception of the saturation bounds which are chosen as 15.1 Nm for $N_{1,1}$ and $N_{1,2}$ for both joints and

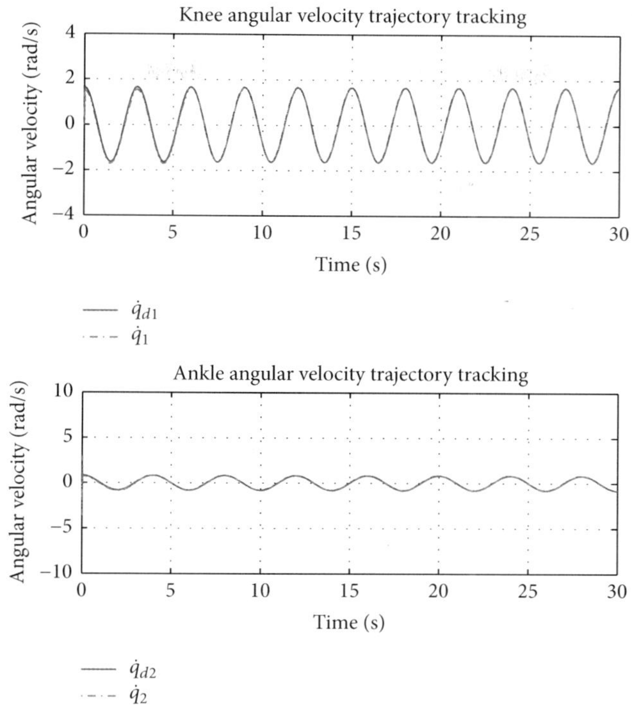

FIGURE 7: Knee-ankle velocity trajectory tracking. $\tau_h \neq 0$; $\kappa = 0.1$ *(Scenario 2 of Case 1).*

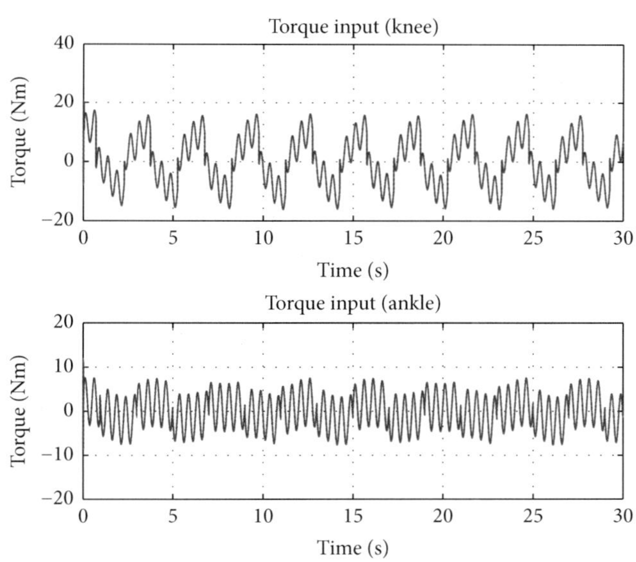

FIGURE 8: Knee-ankle torque input. $\tau_h \neq 0$; $\kappa = 0.1$. Considering the human resistive torque in Figure 9, that is, the human is exerting a force capable of resisting the orthosis control ability to track the physiotherapist trajectory, hence there is an increase in the control torque required to perform the task *(Scenario 2 of Case 1).*

also 15.1/2.1 Nm for $N_{2,1}$ and $N_{2,2}$ for both joints as well.

In this test case, a resistive human torque is applied as in *Scenario 2 of Case 1.* With the same task needed to be performed, an inverted fraction of the control torque, which is based on the state observer is fed back to the system as the human torque. Given the actuator limit as 30 Nm, the system allows a maximum of $\kappa = 0.5$ of the resistive human torque ($\tau_h \neq 0$) to be applied corresponding to 50% of the estimated torque. Beyond this limit the actuator is saturated. Figure 20 shows the control torque input required for the physiotherapist task to be performed based on the resistive human torque applied to the system. The resistive human torque applied is shown in Figure 21. Figures 18 and

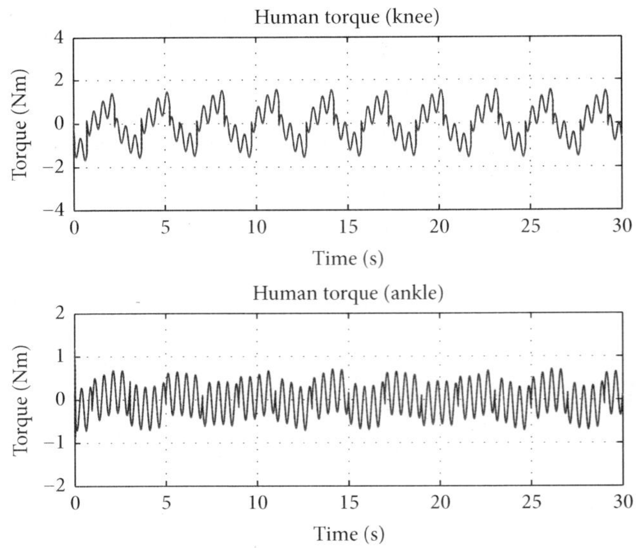

FIGURE 9: Knee-ankle human torque input (resistive torque). $\tau_h \neq 0$; $\kappa = 0.1$. The amount of human torque exerted on the orthosis is shown in this figure, which is 10% of the estimated torque *(Scenario 2 of Case 1).*

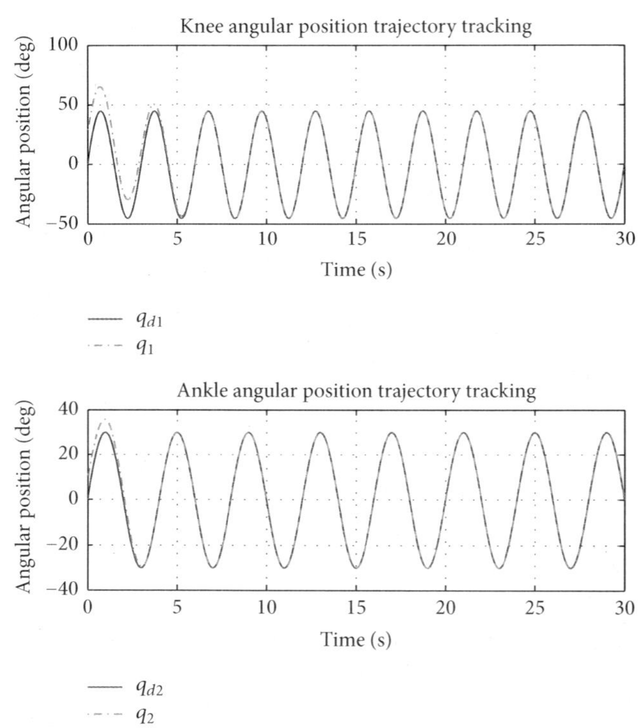

FIGURE 10: Knee-ankle position trajectory tracking. $\tau_h \neq 0$; $\kappa = 0.2$ *(Scenario 2 of Case 1).*

19 present the position and velocity trajectory tracking of the controller. This is a proof of the controller's ability to ensure the task is performed even in the presence of the resistive human force. However, the effect of the disturbance is seen in the estimated state. In addition, considering Figure 12 in comparison with Figure 20, it may be deduced that, for a system of this nature with same system parameters, the higher the actuator limit the higher the amount of resistive human torque it may allow (axis of Figure 20 is slightly increased to accommodate the increase in torque input).

6.3. Case 3. For the purpose of robustness, the bounded controller is tested using the system parameters for a lower-limb orthosis described in [39, 40]. These system parameters are given in Table 2, the values of which are calculated using 1.5 m of the human limb length and a weight of 65 kg. The

TABLE 2: Physiological parameters *(Case 3).*

Parameters	Units	Values
Shank length (L_1)	m	0.3690
Foot length (L_2)	m	0.2280
Shank mass (m_1)	kg	2.9900
Foot mass (m_2)	kg	0.9425
Shank inertia (I_1)	kg·m^2	0.0339
Foot inertia (I_2)	kg·m^2	0.0041

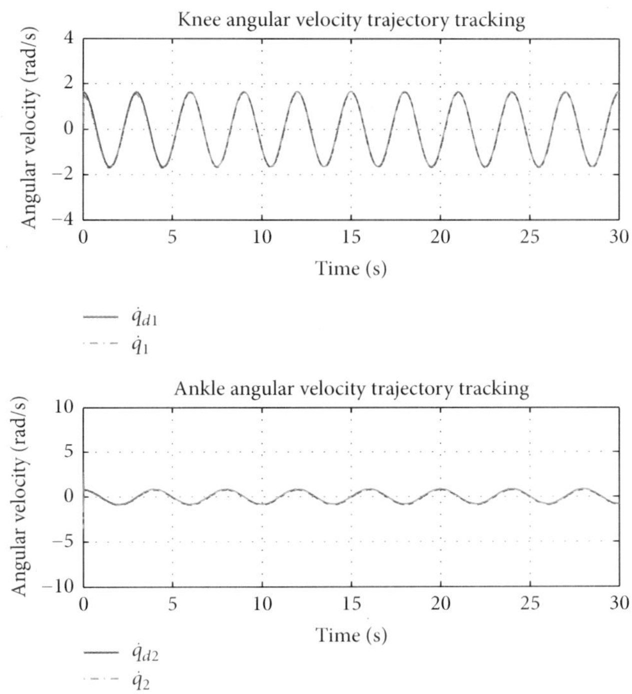

FIGURE 11: Knee-ankle velocity trajectory tracking. $\tau_h \neq 0$; $\kappa = 0.2$ *(Scenario 2 of Case 1).*

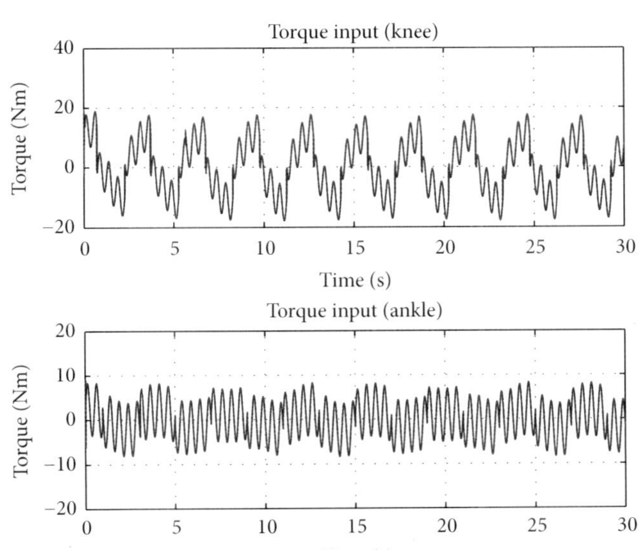

FIGURE 12: Knee-ankle torque input. $\tau_h \neq 0$; $\kappa = 0.2$. Considering the human resistive torque in Figure 13, that is, the human is exerting a force capable of resisting the orthosis control ability to track the physiotherapist trajectory, hence there is a further increase in the control torque required to perform the task as compared to Figure 8 *(Scenario 2 of Case 1).*

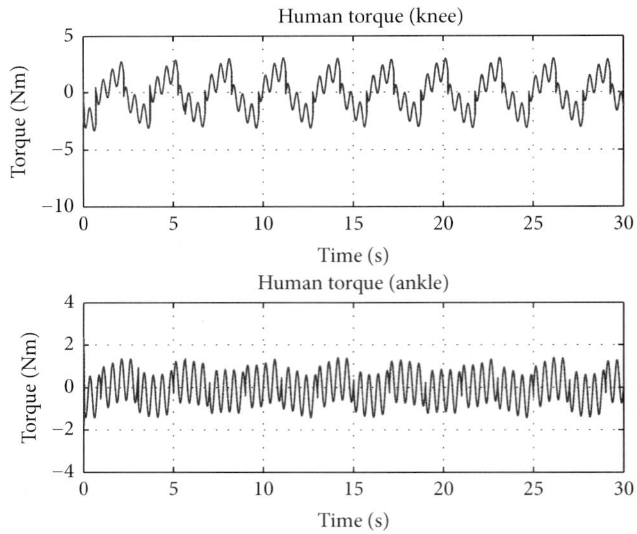

FIGURE 13: Knee-ankle human torque input (resistive torque). $\tau_h \neq 0$; $\kappa = 0.2$. The amount of human torque exerted on the orthosis is shown in this figure, which is 20% of the estimated torque. However, the torque signal is inverted for this purpose *(Scenario 2 of Case 1)*.

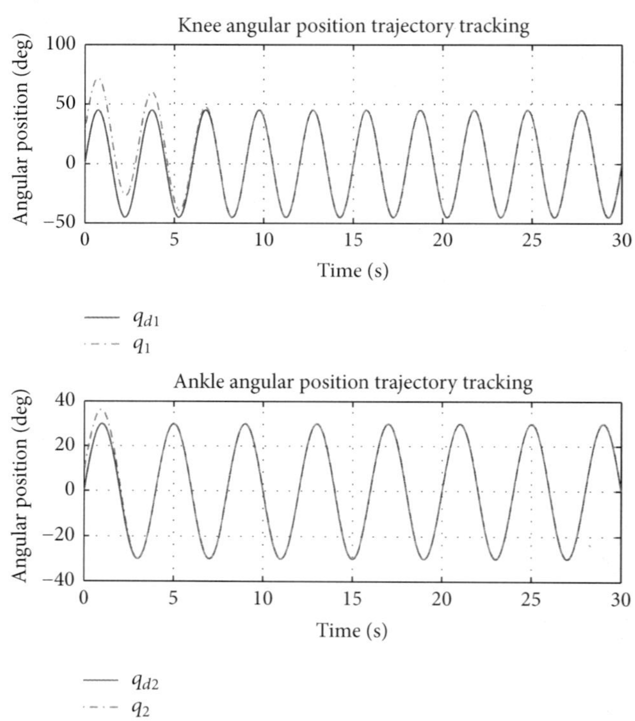

FIGURE 14: Knee-ankle position trajectory tracking. $\tau_h \neq 0$; $\kappa = 0.1$ *(Scenario 3 of Case 1)*.

moments of inertia are computed using $mL^2/12$, assuming the center of mass to be at the center of each cylindrical link (see Figure 1). Other parameters not given in Table 2 are the same as in Table 1.

In this test case, the saturation bounds are the same as in *Case 1*, and hence the maximum actuator size considered is 15 Nm. The controller gains for K_p are tuned to be 9.5435×10^2 and 4.0706×10^2 for the knee and ankle joints and for K_d are 0.3044×10^3 and 0.4716×10^2 for the respective joints. The observer gains are the same as in *Case 1*.

From Figure 24, it may be seen that the torque inputs are within the actuator size limit, and the estimated torques are correctly estimated with regard to the recovered (observer) state. Figures 22 and 23 show that the controller has the ability to ensure the wearer-exoskeleton is able to track the position and velocity of the physiotherapist trajectory in

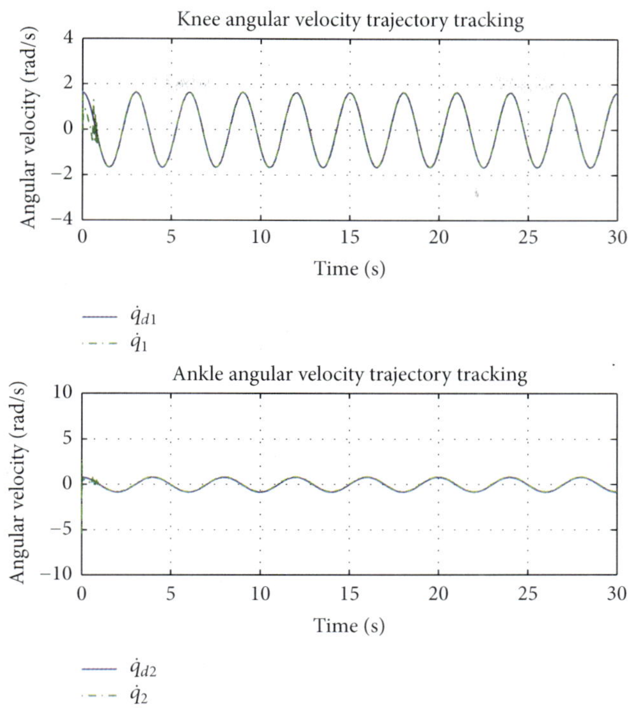

FIGURE 15: Knee-ankle velocity trajectory tracking. $\tau_h \neq 0$; $\kappa = 0.1$ *(Scenario 3 of Case 1).*

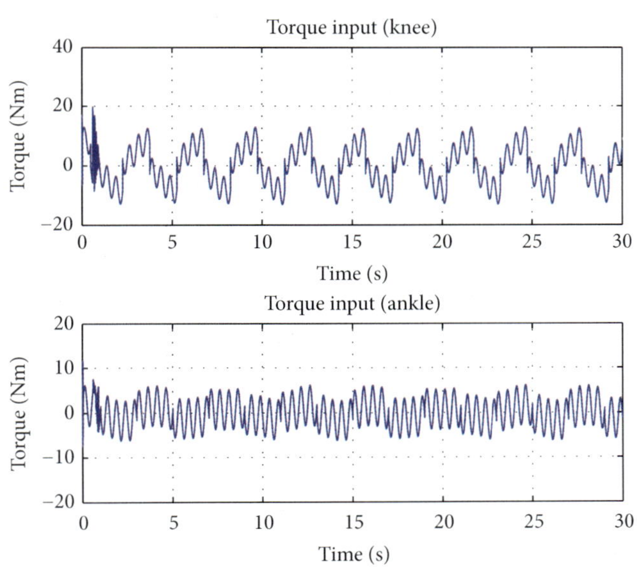

FIGURE 16: Knee-ankle torque input. $\tau_h \neq 0$; $\kappa = 0.1$. Considering the human assistive torque in Figure 17, that is, the human is exerting a force capable of assisting the orthosis control ability to track the physiotherapist trajectory, hence there is a decrease in the control torque required to perform the task *(Scenario 3 of Case 1).*

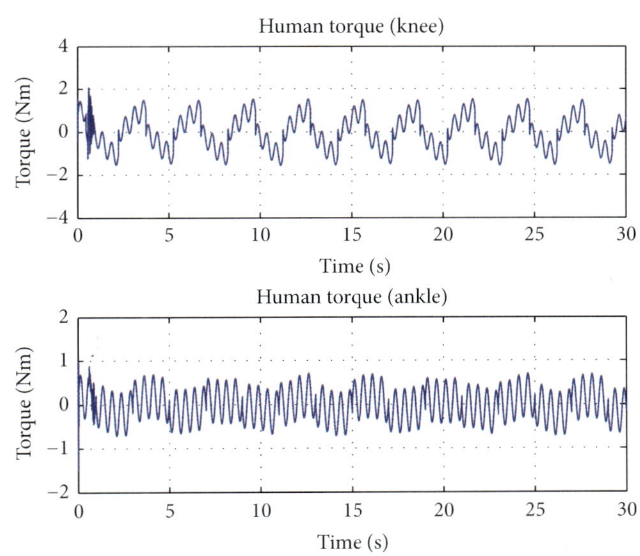

FIGURE 17: Knee-ankle human torque input (assistive torque). $\tau_h \neq 0$; $\kappa = 0.1$. The amount of human torque exerted on the orthosis is shown in this figure, which is 10% of the estimated torque *(Scenario 3 of Case 1).*

spite of the variation in system parameters. The controller robustness is therefore established. Note that the wearer is completely passive in this case and the limit is same as in *case 1.*

7. Conclusion and Future Work

Developing an exoskeleton device needs a well-designed control system so as to ensure the wearer is not further injured by the device. It is therefore necessary that the controller is robust enough to actualise its capability. In this study,

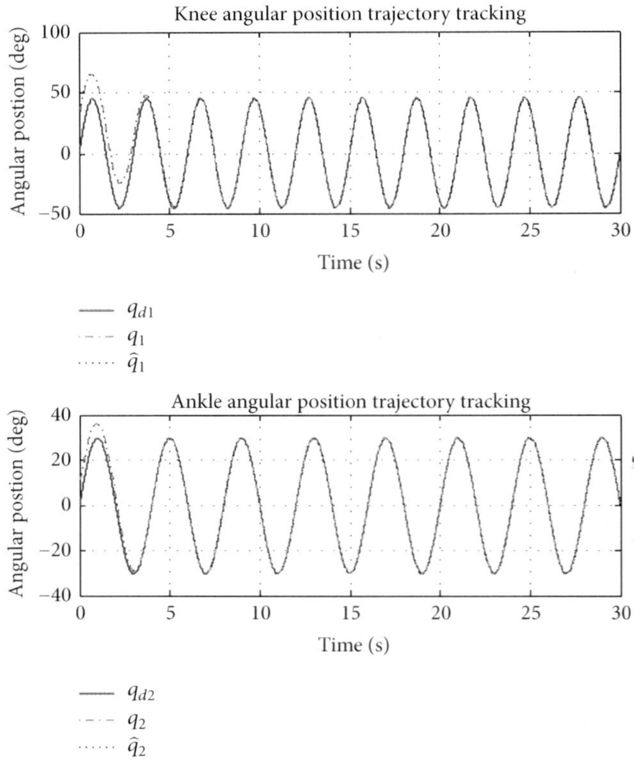

FIGURE 18: Knee-ankle position trajectory tracking. $\tau_h \neq 0$; $\kappa = 0.5$ *(Case 2)*.

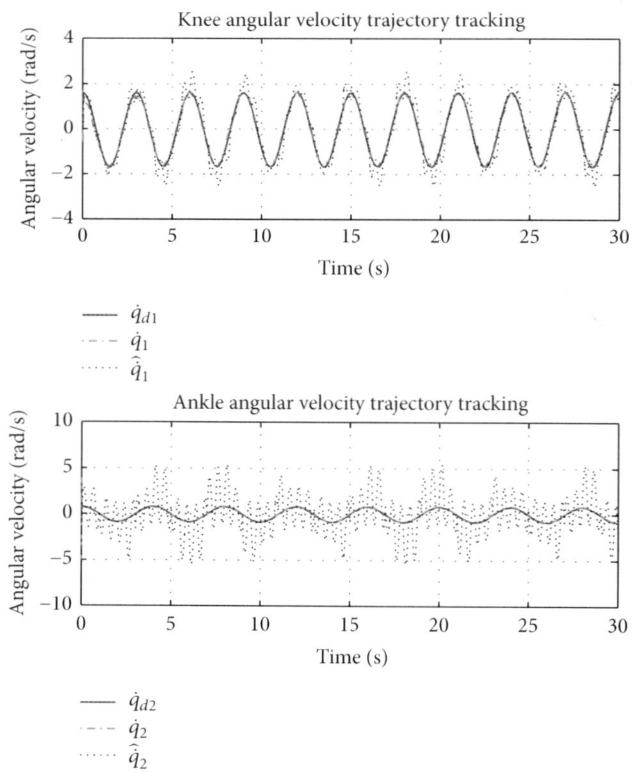

FIGURE 19: Knee-ankle velocity trajectory tracking. $\tau_h \neq 0$; $\kappa = 0.5$ *(Case 2)*.

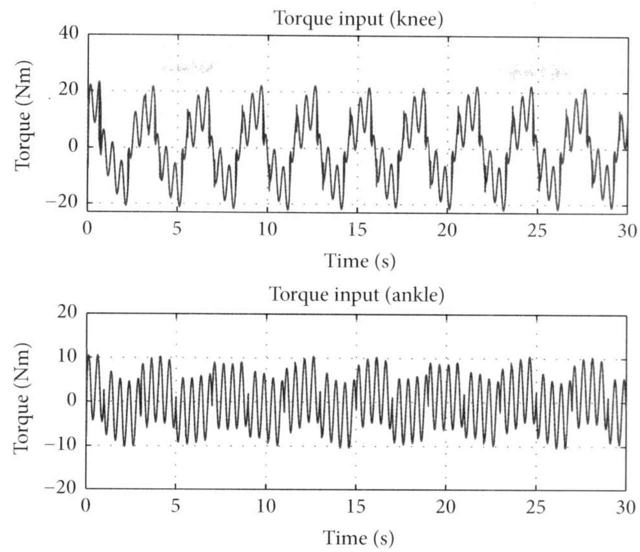

FIGURE 20: Knee-ankle torque input. $\tau_h \neq 0$; $\kappa = 0.5$. Considering the human resistive torque in Figure 17, that is, the human is exerting a force capable of assisting the orthosis control ability to track the physiotherapist trajectory, hence there is an increase in the control torque required to perform the task *(Case 2)*.

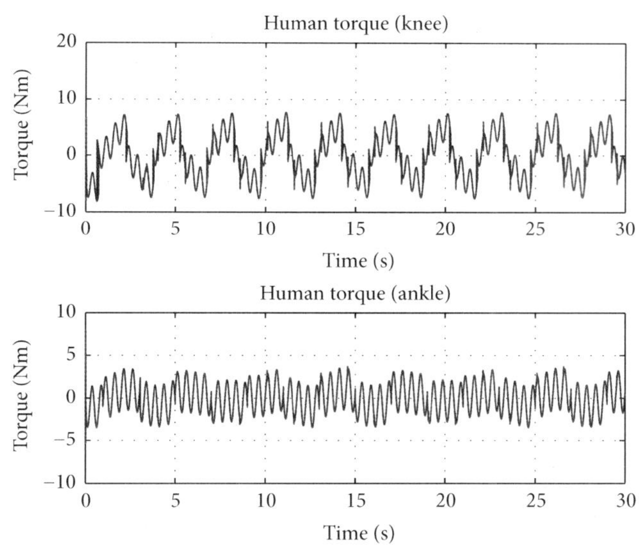

FIGURE 21: Knee-ankle human torque input (resistive torque). $\tau_h \neq 0$; $\kappa = 0.5$. The amount of human torque exerted on the orthosis is shown in this figure, which is 50% of the estimated torque *(Case 2)*.

a bounded nonlinear feedback controller in conjunction with a high gain observer is proposed for the rehabilitation of the user and the determination of the level of contribution a wearer can offer for a specific exoskeleton. The level of contribution helps demonstrate the partial motor function disorder; that is, the user may be able to exact a certain amount of human torque. This was tested via MATLAB/SIMULINK simulations for a knee-ankle orthosis model. Simulation results obtained help to establish this fact. This is established by using three test cases to analyse the exact purpose of this work. In addition to the controller robustness, its ability to avoid the saturation of the actuators has been verified. It is to be noted that the wearer is in a sitting position. Lyapunov-based analysis to prove the asymptotic stability of the controller and the convergence of the observer has also been presented.

In future works, the controller is expected to be tested in the laboratory. It is expected that parametric identification is first carried out before it is tested in real time. The input-to-state stability (ISS) will also be verified. Other future areas of research interest include robot axes misalignments, which constitute a major difficulty in the design of exoskeleton.

Conflicts of Interest

The authors declare that they have no conflicts of interest.

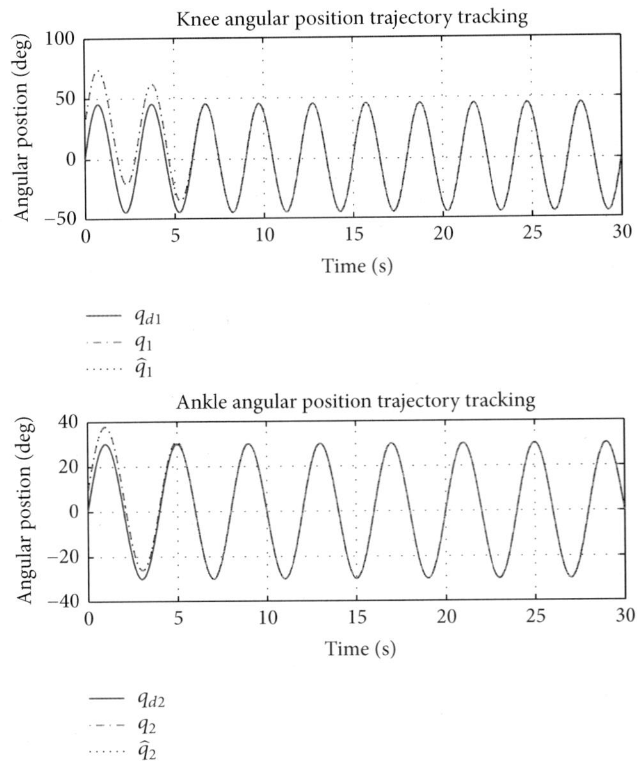

FIGURE 22: Knee-ankle position trajectory tracking. $\tau_h = 0$ *(Case 3)*.

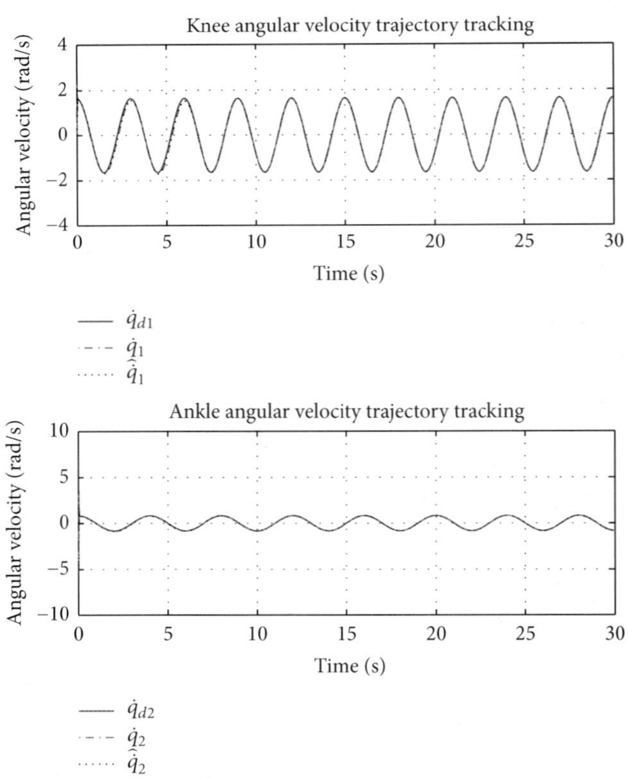

FIGURE 23: Knee-ankle velocity trajectory tracking. $\tau_h = 0$ *(Case 3)*.

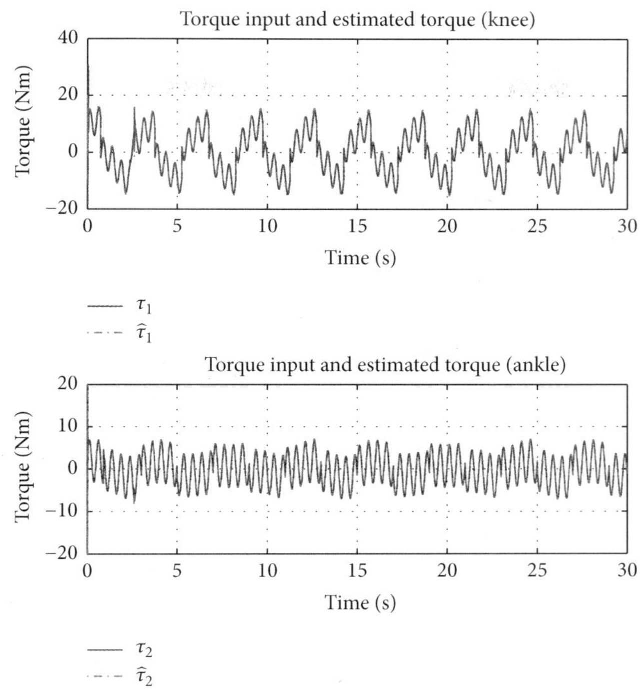

FIGURE 24: Knee-ankle torque input and estimated torque. $\tau_h = 0$. This is to exemplify a complete motor function disorder when the wearer is completely passive. The estimated torque is plotted to show that the observer can correctly estimate the exoskeleton state; hence the torque can be reconstructed *(Case 3)*.

References

[1] B. Dellon and Y. Matsuoka, "Prosthetics, exoskeletons, and rehabilitation [Grand challenges of robotics]," *IEEE Robotics and Automation Magazine*, vol. 14, no. 1, pp. 30–34, 2007.

[2] A. M. Dollar and H. Herr, "Lower extremity exoskeletons and active orthoses: challenges and state-of-the-art," *IEEE Transactions on Robotics*, vol. 24, no. 1, pp. 144–158, 2008.

[3] S. Hesse, H. Schmidt, C. Werner, and A. Bardeleben, "Upper and lower extremity robotic devices for rehabilitation and for studying motor control," *Current Opinion in Neurology*, vol. 16, no. 6, pp. 705–710, 2003.

[4] J. Hope and A. McDaid, "Development of Wearable Wrist and Forearm Exoskeleton with Shape Memory Alloy Actuators," *Journal of Intelligent & Robotic Systems*, vol. 86, no. 3-4, pp. 397–417, 2017.

[5] A. Alamdari and V. Krovi, "Design and analysis of a cable-driven articulated rehabilitation system for gait training," *Journal of Mechanisms and Robotics*, vol. 8, no. 5, Article ID 051018, 2016.

[6] K. Xu, J. Zhao, D. Qiu, and Y. Wang, "A pilot study of a continuum shoulder exoskeleton for anatomy adaptive assistances," *Journal of Mechanisms and Robotics*, vol. 6, no. 4, Article ID 041011, 2014.

[7] N. Jarrassé and G. Morel, "Connecting a human limb to an exoskeleton," *IEEE Transactions on Robotics*, vol. 28, no. 3, pp. 697–709, 2012.

[8] M. K. Vukobratovic, "When were active exoskeletons actually born?" *International Journal of Humanoid Robotics*, vol. 4, no. 3, pp. 459–486, 2007.

[9] E. Guglielmelli, M. J. Johnson, and T. Shibata, "Guest editorial: special issue on rehabilitation robotics," *IEEE Transactions on Robotics*, vol. 25, no. 3, pp. 477–480, 2009.

[10] K. P. Westlake and C. Patten, "Pilot study of Lokomat versus manual-assisted treadmill training for locomotor recovery post-stroke," *Journal of NeuroEngineering and Rehabilitation*, vol. 6, no. 1, article 18, 2009.

[11] M. H. Rahman, M. Saad, J.-P. Kenné, and P. S. Archambault, "Control of an exoskeleton robot arm with sliding mode exponential reaching law," *International Journal of Control, Automation, and Systems*, vol. 11, no. 1, pp. 92–104, 2013.

[12] N. Hogan, H. I. Krebs, B. Rohrer et al., "Motions or muscles? Some behavioral factors underlying robotic assistance of motor recovery," *Journal of Rehabilitation Research & Development*, vol. 43, no. 5, pp. 605–618, 2006.

[13] L. L. Cai, A. J. Fong, C. K. Otoshi et al., "Implications of assist-as-needed robotic step training after a complete spinal cord injury on intrinsic strategies of motor learning," *The Journal of Neuroscience*, vol. 26, no. 41, pp. 10564–10568, 2006.

[14] H. Vallery, E. H. F. Van Asseldonk, M. Buss, and H. Van Der Kooij, "Reference trajectory generation for rehabilitation robots: Complementary limb motion estimation," *IEEE Transactions on Neural Systems and Rehabilitation Engineering*, vol. 17, no. 1, pp. 23–30, 2009.

[15] K. Kong and M. Tomizuka, "Design of a rehabilitation device based on a mechanical link system," *Journal of Mechanisms and Robotics*, vol. 4, no. 3, Article ID 035001, 2012.

[16] D. G. Caldwell, N. G. Tsagarakis, S. Kousidou, N. Costa, and I. Sarakoglou, "Soft exoskeletons for upper and lower body rehabilitation - design, control and testing," *International Journal of Humanoid Robotics*, vol. 4, no. 3, pp. 549–573, 2007.

[17] K. Kiguchi, K. Iwami, M. Yasuda, K. Watanabe, and T. Fukuda, "An exoskeletal robot for human shoulder joint motion assist," *IEEE/ASME Transactions on Mechatronics*, vol. 8, no. 1, pp. 125–135, 2003.

[18] G. S. Sawicki and D. P. Ferris, "Powered ankle exoskeletons reveal the metabolic cost of plantar flexor mechanical work during walking with longer steps at constant step frequency," *Journal of Experimental Biology*, vol. 212, no. 1, pp. 21–31, 2009.

[19] C. Lintzen, Oscillator-based walking assistance.

[20] D. M. Ka, C. Hong, T. H. Toan, and J. Qiu, "Minimizing human-exoskeleton interaction force by using global fast sliding mode control," *International Journal of Control, Automation, and Systems*, vol. 14, no. 4, pp. 1064–1073, 2016.

[21] D. P. Ferris, G. S. Sawicki, and M. A. Daley, "A physiologist's perspective on robotic exoskeletons for human locomotion," *International Journal of Humanoid Robotics*, vol. 4, no. 3, pp. 507–528, 2007.

[22] J. F. Guerrero-Castellanos, N. Marchand, A. Hably, S. Lesecq, and J. Delamare, "Bounded attitude control of rigid bodies: real-time experimentation to a quadrotor mini-helicopter," *Control Engineering Practice*, vol. 19, no. 8, pp. 790–797, 2011.

[23] A. Zavala-Río and V. Santibáñez, "A natural saturating extension of the PD-with-desired-gravity-compensation control law for robot manipulators with bounded inputs," *IEEE Transactions on Robotics*, vol. 23, no. 2, pp. 386–391, 2007.

[24] Z. Flikop, "Stopping power of hot QCD plasma," *PHYSICAL REVIEW D*, 2005.

[25] H. Rifaï, W. Hassani, S. Mohammed, and Y. Amirat, "Bounded control of an actuated lower limb orthosis," in *Proceedings of the 2011 50th IEEE Conference on Decision and Control and European Control Conference, CDC-ECC 2011*, pp. 873–878, December 2011.

[26] M. O. Ajayi, K. Djouani, and Y. Hamam, "Bounded control of a full-exoskeleton device with four (4) Degree of Freedom," in *Proceedings of the IEEE International Conference on Robotics and Biomimetics, (IEEE-ROBIO '15)*, pp. 2425–2430, December 2015.

[27] A. Bateman and Z. Lin, "An analysis and design method for linear systems under nested saturation," *Systems & Control Letters*, vol. 48, no. 1, pp. 41–52, 2003.

[28] H.-B. Kang and J.-H. Wang, "Adaptive robust control of 5 DOF Upper-limb exoskeleton robot," *International Journal of Control, Automation, and Systems*, vol. 13, no. 3, pp. 733–741, 2015.

[29] Q. Van Tran, S. Kim, K. Lee, S. Kang, and J. Ryu, "Force/torque sensorless impedance control for indirect driven robot-aided gait rehabilitation system," in *Proceedings of the IEEE/ASME International Conference on Advanced Intelligent Mechatronics, AIM 2015*, pp. 652–657, July 2015.

[30] S. Oh, K. Kong, and Y. Hori, "Design and analysis of force-sensor-less power-assist control," *IEEE Transactions on Industrial Electronics*, vol. 61, no. 2, pp. 985–993, 2014.

[31] B. Ugurlu, M. Nishimura, K. Hyodo, M. Kawanishi, and T. Narikiyo, "Proof of concept for robot-aided upper limb rehabilitation using disturbance observers," *IEEE Transactions on Human-Machine Systems*, vol. 45, no. 1, pp. 110–118, 2015.

[32] G. Wang, *Observer-based feedback control methods for an underactuated robot system [Ph.D. thesis]*, Theses (School of Engineering Science)/Simon Fraser University, 2003.

[33] M. O. Ajayi, K. Djouani, and Y. Hamam, "Shank-foot trajectory control: A forward dynamics approach using computed-torque control," in *Proceedings of the 2014 14th IEEE-RAS International Conference on Humanoid Robots, Humanoids 2014*, pp. 652–657, November 2014.

[34] D. Sanz-Merodio, M. Cestari, J. C. Arevalo, X. A. Carrillo, and E. Garcia, "Generation and control of adaptive gaits in lower-limb exoskeletons for motion assistance," *Advanced Robotics*, vol. 28, no. 5, pp. 329–338, 2014.

[35] K. Radkhah, C. Maufroy, M. Maus, D. Scholz, A. Seyfarth, and O. Von Stryk, "Concept and design of the biobiped1 robot for human-like walking and running," *International Journal of Humanoid Robotics*, vol. 8, no. 3, pp. 439–458, 2011.

[36] M. Mistry, J. Buchli, and S. Schaal, "Inverse dynamics control of floating base systems using orthogonal decomposition," in *Proceedings of the 2010 IEEE International Conference on Robotics and Automation, ICRA 2010*, pp. 3406–3412, May 2010.

[37] J. L. Pons, *Wearable Robots: Biomechatronic Exoskeletons*, John Wiley & Sons, 2008.

[38] F. L. Lewis, D. M. Dawson, and C. T. Abdallah, *Robot Manipulator Control: Theory and Practice*, CRC Press, 2003.

[39] C. Souit, D. Santana Coelho, M. Szylit, F. Camargo-Junior, M. Peres Cortez Junior, and A. Forner-Cordero, "Design of a lower limb exoskeleton for experimental research on gait control," in *Proceedings of the 6th IEEE RAS/EMBS International Conference on Biomedical Robotics and Biomechatronics, BioRob 2016*, pp. 1098–1103, June 2016.

[40] D. A. Winter, *Biomechanics and Motor Control of Human Movement*, John Wiley & Sons, 2009.

7

Older Adults' Perceptions of Supporting Factors of Trust in a Robot Care Provider

Rachel E. Stuck[1] and Wendy A. Rogers[2]

[1]*Georgia Institute of Technology, 654 Cherry St., Atlanta, GA 30332, USA*
[2]*University of Illinois Urbana-Champaign, 1206 South 4th St, Champaign, IL 61820, USA*

Correspondence should be addressed to Rachel E. Stuck; stuck.rachel@gatech.edu

Academic Editor: Yunyi Jia

The older adult population is increasing worldwide, leading to an increased need for care providers. An insufficient number of professional caregivers will lead to a demand for robot care providers to mitigate this need. Trust is an essential element for older adults and robot care providers to work effectively. Trust is context dependent. Therefore, we need to understand what older adults would need to trust robot care providers, in this specific home-care context. This mixed methods study explored what older adults, who currently receive assistance from caregivers, perceive as supporting trust in robot care providers within four common home-care tasks: bathing, transferring, medication assistance, and household tasks. Older adults reported three main dimensions that support trust: professional skills, personal traits, and communication. Each of these had subthemes including those identified in prior human-robot trust literature such as ability, reliability, and safety. In addition, new dimensions perceived to impact trust emerged such as the robot's benevolence, the material of the robot, and the companionability of the robot. The results from this study demonstrate that the older adult-robot care provider context has unique dimensions related to trust that should be considered when designing robots for home-care tasks.

1. Introduction

The population of older adults is increasing worldwide at an advanced rate. A recent World Health Organization (WHO) report on aging emphasized the need to have tailored interventions for these aging populations, such as older adults with care providers [1]. As the number of older adults increases so will the demand for care providers and, with an insufficient number of humans to fill the demand, robots might help fulfill this need. Trust is essential for older adults to successfully interact with robot care providers, and we need an in-depth understanding about the factors that influence trust within this specific context of an older adult and robot care provider.

To gain an understanding of what factors might emerge, we first reviewed human-human trust research but found that it had not been specifically explored in the context of care providers. However, in general, key elements in human-human trust are the characteristics of the trustee, characteristics of the trustor [2], and the relationship [3]. For the perceived characteristics of the trustor, the ability, values, and benevolence all contributed to trust [2]. In the patient-nurse trust literature, there were several elements apart from ability, reliability, and other task specific characteristics that contributed to trust. Patients wanted to feel respected [4] and cared for, especially in situations where they felt vulnerable [5]. In addition, as trust developed, the communication between the nurse and patient became a key element in what influenced how the patient perceived the trustworthiness of the nurse [4]. These findings give insight into trust, but it is unclear if these factors will also emerge in the human-robot context.

Human-automation trust provided the basis for human-robot trust. Trust of automation is associated with use of that automation [6]. For example, if an operator trusts a system then they are more likely to use the system. This highlights the key need to understand and support trust for appropriate use of a system. Aspects that influence trust in automation include characteristics of the automation such as the quality of the system feedback [7] and the reliability of the system [8]. For example, lack of reliability and failures of the system negatively impact trust [6]. However, the relationship between failures and trust is complicated and does not have a one-to-one ratio in lowering trust [6]. Characteristics of the operator also influence trust such as the operator's self-confidence [9] and their personality traits [10].

With the increase of robots in workplace and the home, research has also focused on trust specifically in robots. A model of trust was developed by Sanders et al. [11] that identified the main components that contribute to human-robot team trust. These included the characteristics of the person, the characteristics of the robot, the environmental characteristics, and the training and design implications. The person's characteristics included aspects such as personality, self-confidence, general attitude towards robots, and knowledge of the robot [11, 12]. The qualities of the robot included elements related to the robot's performance such as the reliability and predictability [11–13]. Other qualities included the proximity of the robot to the human, type of robot, and the appearance of the robot [12, 13]. The environmental characteristics included the type of task and communication [11]. Training and design implications demonstrate how these are influenced by both the human and the robot qualities and thereby influencing trust [11, 14]. A framework that identified key components of human-robot interaction with older adults also identified many of these components including the human characteristics (e.g., psychographics, abilities), robot characteristics (e.g., appearance, capabilities), tasks constraints (e.g., proximity, accuracy requirements), and context of the interaction (e.g., living environment, safety considerations) [15].

Research has also explored older adults' attitudes towards robots in the home. Smarr et al. [16] explored healthy, independent living older adults' acceptance of robots in the home as well as what tasks they preferred a robot to do. Older adults in this study were generally accepting of a robot, but they had preferences about the type of tasks they preferred a robot to perform [16]. The older adults preferred a robot over a human for tasks related to household chores, manipulating objects, or information management [16]. In another study, older adults' acceptance of robots in the home was confirmed as they reported that they would prefer to stay at home with a robot than have to go to a facility [13]. After exposure to an assistive robot, older adults not only reported higher perceptions of usefulness but also a greater willingness to have the robot assist with various tasks [17].

One study specifically focused on trust between older adults and home robots [18]. Reliability, precision, efficiency, and safety were the top descriptors used for a trustworthy robot [18]. However, this study was limited as it did not specifically assess the role of the task, a key component of trust, and also interviewed older adults who did not necessarily have any experience with receiving care.

A first step in exploring trust in the home-care context is to understand older adults' perceptions of robot care providers and what traits the older adults want the robot care provider to have to trust it. Perceptions influence technology acceptance such as the perceived benefits or concerns of using the technology [19]. In the human-human trust literature, the trustor's perceptions of the trustee influenced the decision to trust [2, 20]. In addition, studies have shown that an individual's perception of a robot influenced how they interacted with it [21]. Insight into what factors older adults perceive as necessary for them to trust a robot will allow us to support their development of trust and diminish acceptance issues that are related to perceptions of the robot being untrustworthy.

The focus of this paper is to explore what factors older adults who currently receive care identify as supporting trust in a robot care provider. In particular, the goal was to address these specific research questions:

(1) Do the dimensions of trust identified in the literature emerge in the older adult-robot care provider relationship?

(2) Do new dimensions of trust emerge in the older adult-robot care provider relationship?

This research is part of a larger project that explored trust in human care providers as well. For details about the rest of the project see [22].

2. Method

2.1. Participants. We interviewed 24 older adults (12 in independent living and 12 in assisted living) above the age of 65 (M = 81, SD = 7.13, age range 67–96) who received 4 or more days of care a week. On average, they received around 6 days of care with the caregiver staying around 1–3 hours each visit. Participants were recruited through outreach to local independent and assisted living communities. They were primarily females. They were diverse in ethnicity and education. Overall, participants reported that they had fair health. On average, the participants cognitive function was at a scoring on the Montreal Cognitive Assessment (MOCA) that represents mild cognitive impairment [26]; note that 5 participants were unable to complete cognitive assessment due to vision or physical impairment. Participants were on average moderately confident performing daily living tasks, but there was a lot of variability between participants and tasks. When asked about technology experience, the participants reported no usage to relatively low usage. For a detailed list of descriptors see Table 1.

2.2. Procedure. Prior to inclusion in the study, participants were screened in person or via telephone to ensure that they met the eligibility criteria which included passing the Wechsler Memory Scale III [27] to ensure that they would be able to follow along with the interview. All participants gave informed consent prior to participation in the study. Participants were then given several questionnaires before the interview (Table 2).

Following the questionnaires, a semistructured interview was administered to investigate the participants' perceptions of what factors support trust. The order of the interview questions was counterbalanced between discussions of the robot care provider and the human care provider to control

Table 1: Participant characteristics.

Factor	Measure	% of participants (n)
Ethnicity	Black/African American	29% (7)
	White/Caucasian	67% (16)
	Others	4% (1)
Education	Less than high school graduate	8% (2)
	High school graduate/GED	17% (4)
	Some or in-progress college/associates degree	71% (10)
	Bachelor's degree (BA, BS)	13% (3)
	Master's degree	13% (3)
	Doctoral degree	8% (2)
General health[a]	"In general, would you say your health is..."	M = 2.38 SD = 1.06
Montreal cognitive assessment[b] (n = 19)		M = 23.05 SD = 4.01
Days assistance received each week (n = 23)*		M = 6.05 SD = 1.53
Average length of caregiver stay[c]		M = 2.00 SD = 1.05
Technology usage[d]	"In the past year, how often have you used..."	M = 0.79 SD = 0.57
Self-efficacy for daily tasks[e]*	"How confident are you in performing...?"	M = 61.26 SD = 29.13

(a) 1 = poor, 2 = fair, 3 = good, 4 = very good, 5 = excellent; (b) score ≥ 26 = normal; (c) 1 = less than an hour, 2 = 1–3 hours, 3 = 4–6 hours, 4 = 6–12 hours, and 5 = 12–24 hours; (d) 0—not used; 1 = used once, 2 = used occasionally, and 3 = used frequently; (e) 1 = not at all confident and 10 = completely confident (summed across 10 questions; range is 10–100); *one participant did not wish to answer.

Table 2: Preinterview questionnaires.

Questionnaire title	Description of measured variables	Reference
(1) Demographic/health	General descriptive information about the health and hearing/vision/motor capabilities of the participants	Czaja et al. [23]
(2) Technology experience profile	Usage and experience with various types of technologies within the last year	Barg-Walkow et al. [24]
(3) Daily living self-efficacy scale	Level of confidence on various ADL and IADL tasks	Sanford et al. [25]
(4) Assistance level with daily tasks	Level of assistance with various ADL and IADL tasks, who assists them, and how often they receive help	Locally developed
(5) Formal caregiver experience	Information about experience with caregiver either personal or through assisted living	Locally developed
(6) Montreal cognitive assessment	General cognitive ability and level of cognitive impairment	Nasreddine et al. [26]

for carry over effects (the focus of the present paper is on the discussion of the robot care provider only). To gain a deeper insight and provide context, we explored what supported trust within 4 task scenarios: bathing, medication assistance, transfer, and household tasks (see Table 3). The scenarios were developed with the help of three subject matter experts (SMEs). We chose two activities of daily living (ADLs; bathing and transferring) and two instrumental activities of daily living (IADLs). These tasks were chosen because they are some of the most frequent tasks of formal care providers for older adults in their homes [28] and bathing was added upon recommendation by the SMEs.

Bathing was discussed first as it contains the greatest amount of vulnerability. Participants were asked in general what a robot care provider would need to be like for them to trust it with that task and what would cause them to not trust the robot. Bathing was discussed in-depth with specific questions about previously identified dimensions specifically from human-human (values and benevolence), human-robot literature (precision and predictability), and dimensions discussed in both (ability, reliability, appearance, and communication). For each of the following tasks, we asked general questions about how they would want a robot care provider to be to trust it to perform the task and what would negatively impact their trust.

After completing the interview, the participants completed several follow-up questionnaires (Table 4). They were then compensated $30.00 for their time and effort and debriefed about the purpose of the study. The interviews were audio recorded and then transcribed.

TABLE 3: Scenario descriptions for each task.

Imagine you have a new formal caregiver who is going to assist you with	
Bathing	This will include them helping you remove your clothes and physically helping you bathe
Medication assistance	This means they would help remind you to take medications at the appropriate time and perhaps bring the medication bottle to you
Transferring	This will include the caregiver helping you sit up, lifting you, and moving you to the wheelchair
Household tasks	These tasks will include helping plan and prepare meals and doing some light housework such as laundry, doing the dishes, or making the bed

TABLE 4: Postinterview questionnaires.

Questionnaire title	Description of measured variables	Reference
(1) Robot self-efficacy scale	Level of confidence operating a robot	Locally developed
(2) Robot familiarity and usage	Assessed familiarity and usage of various robots	Smarr et al. [16]
(3) Trust preference checklist	Preference of human versus robot care provider for various tasks	Olson [18]

3. Results and Discussion

3.1. Overview of Analysis. We asked participants about what they perceived as supporting trust in a robot care provider across four tasks. The goal was to understand in general what factors support trust in a robot care provider, so the data presented are collapsed across tasks. However, when there were differences across tasks, we highlight those differences.

One individual coder segmented all the interviews into units of analysis. A segment was defined as a section of responses that related to one specific task. The coding scheme was developed based on prior research on trust as well as themes that emerged from the interviews [29, 30]. The coding scheme categorizes the qualitative data to understand similar attitudes amongst participants. To identify emerging themes, four interviews were randomly selected and reviewed by two coders to discuss themes that emerged that did not fit into previously identified themes.

After development of the coding scheme, three independent rounds of intercoder agreement were conducted. For each round, using MAXQDA Version 12, Cohen's Kappa was calculated to ensure that coders were in agreement at a minimum threshold of 85% based on prior research that recommends between 80 and 90% as a minimum level of agreement [31]. For each round, if agreement was not met, coders met to discuss discrepancies and used this to inform the design of the coding scheme. For the final round, a reliability of 91.61% was met. Interviews were then divided between coders and independently coded.

To distinguish the most frequent factors for each overall theme, we used a threshold of 5% as a cutoff. There is no standardization for a cutoff, but previous qualitative research has used 5% as a method of reducing attention on categories that were not prominent across participants [32]. Excel was used to calculate frequencies and descriptive statistics for the questionnaires and response frequencies.

3.2. Robot Experiences and Attitudes. In general, participants had limited to no experience with robots (M = 1.25, SD = 0.59) based on a familiarity rating of 0: not sure what this is, 1: never heard about, seen, or used this robot, 2: have only heard about or seen this robot, 3: have used or operated this robot only occasionally, 4: have used or operated this robot frequently. Only 5 participants had ever used or operated a robot only occasionally (1: domestic robots, 3: entertainment robots, and 1: research robot).

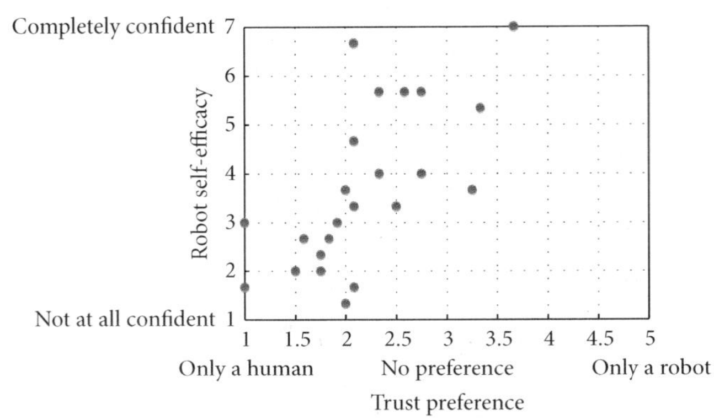

FIGURE 1: Relation of self-efficacy to trust preference.

We also assessed participants' self-efficacy (their belief in their ability to succeed) of being able to operate a robot in three situations (if there was no one around to tell them what to do, if they only had the robot manuals for reference, and if someone else showed them how to use it first). They rated on a scale of 1–7 how confident they were (1 = not at all confident, 7 = completely confident). In general participants had low to neutral self-efficacy in operating the robot (M = 3.61; SD = 1.67).

When asked if they would prefer to trust a robot or a human for a list of various home-care tasks, participants overall reported that they preferred to trust a human (M = 2.18; SD = 0.66; 1: only a human, 2: prefer a human, 3: no preference, 4: prefer a robot, and 5: only a robot).

Figure 1 illustrates the relationship between self-efficacy of operating a robot and trust preference. There was a positive

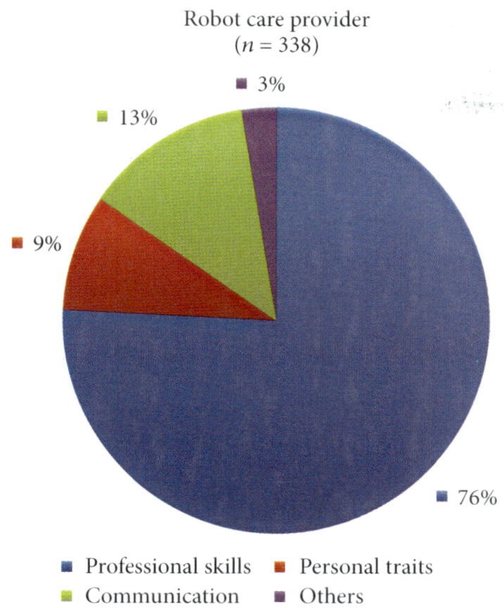

FIGURE 2: Trust factors across all tasks.

correlation (r = .65, p < 0.01) between self-efficacy and trust preference. Although this needs further exploration, perceived self-efficacy in operating a robot may make older adult users more open to trusting a robot to assist them. If this is the case, when providing robots for use by older adults, it would be important to design the robots so that they are easy to use and to train the older adults so that they feel confident in getting the robot to perform a task.

3.3. Trust Factors for Robot Care Providers. In the interviews, three main themes emerged as supporting trust in a robot care provider: professional skills, personal traits, and communication (see Figure 2). Overall, professional skills were the most frequently discussed factors that older adults perceived as supporting trust in a robot care provider. This factor was followed by communication and then personal traits. Personal traits may not have been discussed as frequently because of the participants' limited experience with robots.

Within each of these themes there were several factors from prior literature that emerged within this specific context. We will examine the details of these in the sections below.

3.3.1. Professional Skills. Within professional skills, the most frequently mentioned themes were general capability, precision, consistency of performance, safety, predictability, and gentleness (see Figure 3). A theme not previously identified in literature that emerged within this context was gentleness. Table 5 provides information on how we defined these themes, and an example quote from the interviews for the most frequently discussed themes in this category.

Within the tasks, there was some variability in the prevalence of each of these factors. For the task of bathing, 20% of participants reported wanting the robot care provider to have had prior experience with bathing, but this was not mentioned for other tasks. There were a few specific factors that seemed to be prevalent in tasks that involved touch (bathing and transfer). For example, safety and gentleness were mentioned most frequently for these two tasks. Contextual knowledge, which is defined as understanding the capabilities and sensitivities of the older adult, was also only mentioned for these two tasks. Thus, for tasks in which there is human-robot touch, there are other considerations to take into account. Particularly, older adults receiving care are commonly suffering from a chronic condition that may cause specific sensitivities that need special consideration in tasks where the robot is in contact with their body. One element that was unique for transferring was that older adults wanted to know that the robot was physically capable of lifting them, but physical capability was not mentioned as frequently for other tasks.

For medication assistance, 37.5% mentioned that the robot needed to be on time, whereas this was only mentioned once for bathing and once for household tasks. This provides another example of how, for specific tasks, there are key elements of the tasks that are more salient to the older adults than for other tasks. A key aspect of medication management is ensuring that medications are administered at the required time and this importance increases with the more medications that are being managed leading to the older adults want to be sure that the medication would be given at the appropriate time.

Previously identified themes such as general capability and reliability were also pertinent to the context of a robot care provider. Beyond these themes, there were factors that influenced the participants' perceptions related to trust such as the gentleness with which the task is performed or how safely it is performed. Some of these factors were more salient within specific tasks. When designing for specific tasks, first identify what the key components are for the user within this task and then designers should try to ensure that those expectations are met. For example, for tasks involving human-robot touch a primary goal might be that the robot is gentle and safe, but with medication assistance, the focus of design should be on ensuring the robot can deliver the medication at the appropriate time.

3.3.2. Personal Traits. For personal traits, the main factors that emerged were the material or texture of the robot, the general appearance of the robot, the compatibility of the robot, the congruence of the care provider values with the older adult's values, the benevolence and kindness of the robot, and how the robot was dressed. See Figure 4 for the frequency of factors mentioned in professional skills. See Table 6 for the list of definitions and example quotes for each of these themes.

We also asked the participants after discussing all four tasks what role benevolence played in them trusting a robot. 67% of the participants stated that benevolence would impact their trust in a robot care provider. For example, a participant responded "*that would play a lot of role. I would really trust him. If he's doing exactly what I want him to do.*"

Benevolence only emerged without specifically inquiring about it within the task of bathing which suggests that when a task is personal and vulnerable, the need to feel cared for

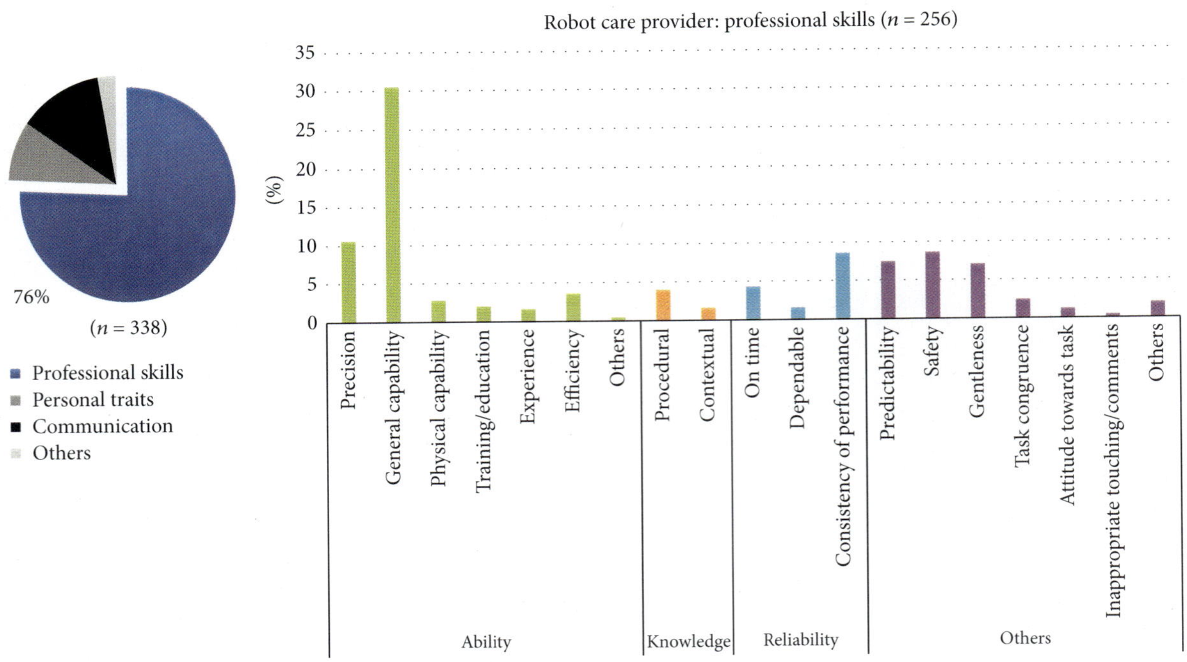

Figure 3: Professional skills.

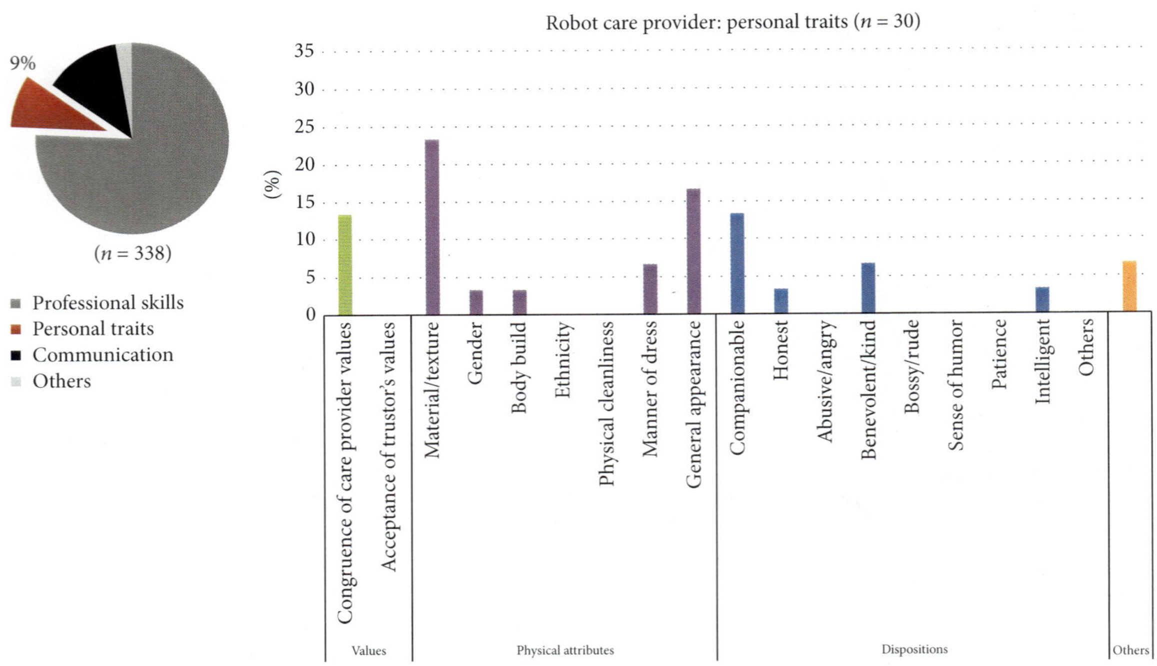

Figure 4: Personal traits.

is salient to the older adults. The only other personal trait that seemed to be impacted by task was the concern for the material and texture of the robot which was mentioned only in the tasks that involved human-robot touch like bathing and transferring. This highlights the need for comfort in tasks that involve physical interaction with the robot.

Themes related to personal traits go beyond just general appearance that was identified in human-robot literature before and demonstrate that several themes from human-human trust are also relevant to this context, such as the congruence of values and benevolence of the robot. There were also context-specific themes that emerged such as the material and texture of the robot that are a key element in human-robot tasks that involve touch. Older adults also wanted the robot to be companionable and to get along with them as well as their friends and family. All of these demonstrate that they want the robot to go beyond just performing tasks but also to provide them with a feeling of being cared for as well as something that they feel comfortable having others interact with.

TABLE 5: Professional skills: definitions and examples of factors.

Factor	Definition	Example quote
General capability	Is the robot able to perform that specific task	*"That it got the right bottles. Just if they're gonna just bring me the bottles, as long as they bring the right bottles, that's all I would require"*
Precision	Is the robot exact and accurate in their performance of the task; does it complete the task thoroughly	*"That (precision) is important to me...because I would want to feel that it is done right and I wouldn't be able to trust the robot if it's not done right"*
Consistency of performance	Is the robot consistent in their task performance	*"Well, some days a human would do it thoroughly and other days they wouldn't, so the robot would need to do it the same way every time"*
Safety	The task is performed with little to no potential to harm the older adult	*"if he dropped me or even if he hurt me while he was doing it. Now...I don't (know) if I would trust a robot"*
Predictability	The robot acts in a way that is consistent with the older adults' expectations	*"Yeah that (predictability) is really important. Much more important than that because you don't interact with a robot in the same way you do with a person, I don't think"*
Gentleness	The task is performed with little to no potential to harm the older adult	*"That they be gentle, and, because I have a lot of pain"*

TABLE 6: Personal traits: definitions and examples of factors.

Factor	Definition	Example quote
Material/texture	Is there a preference for the material or texture or temperature of the robot that would influence trust	*"Have warm hands. Definitely. I can only picture this metal concoction in my mind. I just can't conceive me going through that"*
General appearance	Is there any preference for the appearance of the care provider that would influence trust	*"You wouldn't want anything, just like you wouldn't want anything in your house ugly looking, you'd want something maybe streamlined. Like a car, whatever, like a wheelchair sometime(s) they make them- or cane people, they have colored ones. You'd want something that has a nice appearance to it"*
Companionable	Is the care provider friendly and sociable and likes people	*"That it would be friendly and be, I don't know how much personality they have... whatever is programed into him I guess. But ... I would want him to get along with baby dog if it will"*
Congruence of care provider values	Do they have the same set of values as the older adult	*"I feel like that for me to trust him, he has to have good values like I do"*
Benevolence/kind	Are they a caring person/are they doing the task because they care about the older adult	*", I feel like that for me to trust him, he has to...really show me that he wants to help me"*
Manner of dress	Is the care provider dressed in a way suitable to the older adult; what they are actually wearing	*"If it's for me only, being dressed as a female in some variety ... even if it is some pant suit"*

3.3.3. Communication. Communication has been previously identified as generally impacting trust, but not specifically how communication might be used to support trust. Within the interviews, three main communication themes emerged that older adults perceived as supporting trust: task specific communication, engaging and responsiveness in communication, and the robot able to both understand and communicate clearly with the older adult. See Figure 5 and Table 7 for more details.

These findings not only confirm that communication impacts trust, but it also deepens our knowledge of how it can impact trust by identifying exactly what type and manner of communication older adults perceive as supporting trust. For care tasks, it is not only important that the communication be used to prepare the older adult for what the robot is doing, but that the robot also demonstrates competency in being able to understand and produce appropriate responses. In addition, being able to understand the robot is key for older adults to have successful communication with the robot care provider.

3.3.4. Others. There were some other categories that did not fit into any of the major themes. However, these emerged primarily within the task of bathing. 20% of participants reported for bathing that they would need experience with the robot before they could trust it. In addition, seven participants explicitly stated that they would not trust the robot with bathing, but this was not mentioned frequently for all the tasks. There was one participant who stated for every task that they would not trust a robot.

4. Conclusion

Specialized interventions for specific groups of older adults are needed as this population continues to increase across

TABLE 7: Communication: definitions and examples of factors.

Factor	Definition	Example quote
Task specific	They explain what they are doing or about to do related to the task	*"Assuring me. . .that it could do the task that I have asked it to do, that it has done it before, give me a list of the places it has been used and how it turned out"*
Responsiveness/engagement	Answers questions; listens to what they are saying and responds appropriately	*"Answering questions and feedback as to how I feel"*
Communicates/understands clearly	Can the older adult understand them and does the robot understand the older adult	*"It would have to demonstrate that it understands its orders real well. Understands the orders and can recite them back to me or to someone who is with me"*

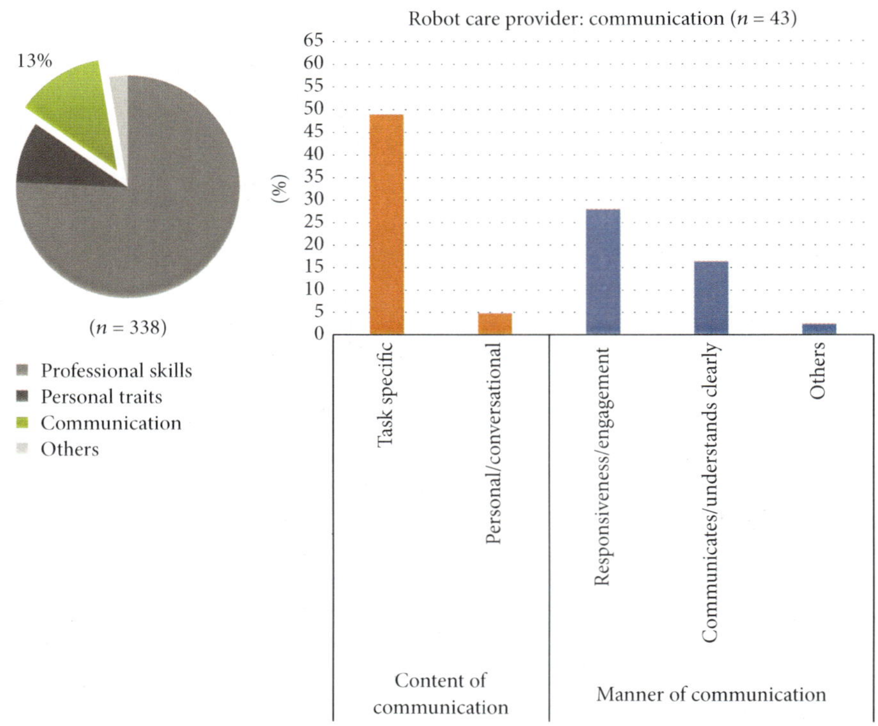

FIGURE 5: Communication.

the world. Robot care providers will be key in helping supply care for older adults and mitigating potential issues caused by decrease of human care providers. As trust is a key element in creating a successful relationship between an older adult and a robot care provider, this study explored what factors older adults perceived as supporting trust within home-care tasks.

The results from this study show that many previously identified themes from both human-human trust literature and human-robot literature emerge within this context. Themes from prior human-robot literature that were confirmed as supporting trust in this context were the capability to perform the task, precision, general appearance, reliability, predictability, and safety of the robot. Human-human themes were the knowledge, values, and benevolence of the robot.

There were also unique themes that emerged in this study including the material/texture, the gentleness, and the companionability of the robot care provider. We found support that communication is key to supporting trust. In addition, this study expands what we have previously understood about the role communication plays in supporting trust by identifying what components of communication can support trust such as the content of the communication being task specific, the responsiveness and engagement of the robot, and the ability of the robot to communicate and understand clearly

To provide better care and improve the lives of older adults, robot care providers should be designed to exemplify these factors and support trust. Although training can be used to mitigate some of these expectations, if the older adults do not perceive the robot as trustworthy then they will not accept or use the robot. We provide design recommendations in Table 8 based on our findings that might enhance the likelihood that older adults will trust a robot care provider.

For the professional skills of the robot, this study highlights that when designing robot as care providers, the robot's ability must go beyond the actual capability to perform the task. Robots caring for older adults must be able to adapt to the older adult's health conditions and specific sensitivities.

TABLE 8: Design recommendations for robot care providers.

Factors	Design recommendation
Professional skills	(i) Robots should be able to complete the task precisely, reliably, and in a way that the older adult expects (ii) As many of the older adults receiving care are suffering from various health conditions, it is important to that robots are able to be specialized for individuals for tasks such as transferring and bathing where certain areas might be more sensitive (iii) Design should ensure that appropriate pressure is applied for safety but still be gentle as to not incur harm on the older adult
Personal traits	(i) Need to be able to navigate the social environment in the home including family, friends, and pets of the older adult (ii) Able to demonstrate to older adult that tasks are being done for their well-being as primary motive (iii) Flexibility to demonstrate a set of values similar to each older adult (e.g., honesty, task values such as personal standards of cleanliness, or method in which task is performed) (iv) Material and temperature of robot for tasks that involve touch should be designed to be comfortable for older adult (e.g., not metal cold "hands" for bathing)
Communication	(i) Provide information while task is being performed to accurately describe the robots planned goals and steps for that task (ii) Create controls whether verbal or physical that allow the user to have clear input and output decisions so that they know the robot understood them and responds appropriately (iii) Ensure that user interface is usable by older adults (a) Audio: ensure that frequency is within common hearing range of older adults and volume is adjustable so that if hearing impaired the volume can be altered to ensure clear communication (b) Visual: make sure that guidelines are followed for older adults by providing high contrast, large print, and simple steps for navigation

For example, if the older adult fell and now has increased sensitivity due to bruising on their legs or hips, the robot must be able to adjust the way it performs the task of bathing or transferring to prevent incurring pain or discomfort to the older adult. If the robot care provider performs the task the same way as before and the older adult experiences discomfort, it may prevent the older adult from trusting the robot in the future.

When designing robot care providers, it is must be remembered that the robot will now be a part of the older adult's home. This requires not only designing for the older adult, but also taking in consideration the social environment such as family or pets. In addition, as robot care providers might be replacing a part of social interaction that the older adult was previously receiving, it is essential to the older adults that they still feel cared for. Designers should consider building in social components to these robot care providers so that older adults can still feel as though the care they are receiving is purely for their benefit. This study also demonstrated that designers should consider having some options for personalization of the robot in both social and capability aspects. For example, if prior to needing care an older adult would always clean the kitchen before cleaning the rest of the house, they might trust a robot care provider more if they could adjust its behavior to mirror what they previously did.

This study also highlights that clear communication between the older adult and the robot care provider is a key aspect of creating a trusting relationship. Designers should ensure that the robot care provider supports the older adult's knowledge of how the robot will perform the task and simple steps for navigation should be provided to access this information [33]. In addition, designer should pay specific attention to visual and auditory declines associated with aging, whether robots are designed to have a physical interface such as a screen or communicate via auditory displays. Recommendations for designing displays for older adults has been provided by prior literature such as ensuring that text is large and has high contrast of text [33]. For auditory communication, designers should consider having the frequency of communication in the typical audible range for older adults as well as the ability to adjust to volume.

This study provides initial insight into the variables that support trust between older adults and robot care providers, but it was not without limitations. Qualitative interviews allowed for obtaining in-depth information about the factors that support trust within this context, but as the older adults did not interact with a real robot or care technology these findings only provide insights about their attitudes, not their actual interaction behavior. We chose to not expose the older adults to specific robots as many of these technologies are either not in existence or still being developed (not commercially accessible) and we did not want to limit our study to a few specific robots. These initial factors should be further validated as these robots are developed. Another issue is that some individuals in this cohort of older adults have limited experience with technology. It is unclear how further experience with technology might impact these factors as well, but we do know that older adults with technology experience are more likely to adopt newer technologies [23]. Therefore, future studies should explore how various technology experiences influences individuals' perceptions of new technologies such as robot care providers. This study provided insights about what factors need further exploration in the future and what aspects older adults perceived as contributing to successful care between them and a robot care provider.

This study expanded the theoretical understanding of what factors support trust in the older adult and care provider context by solidifying that themes from other pieces of trust literature are pertinent to this context, but also by identifying factors that emerged specifically within this context such as gentleness. In addition, these findings give insight into how to design a care provider robot in a way that can support the older adult trusting it. By understanding what factors support trust in this context, we can improve not only the older adults' acceptance of the robot but improve the lives of older adults as well by creating successful interactions between the human and robot care providers.

Disclosure

NIDILRR is a center within the Administration for Community Living (ACL), Department of Health and Human Services (HHS). The contents of this publication do not necessarily represent the policy of NIDILRR, ACL, or HHS, and the reader should not assume endorsement by the Federal Government. Some details of the method—but none of the results—were presented at a workshop [34].

Conflicts of Interest

The authors declare that there are no conflicts of interest regarding the publication of this paper.

Acknowledgments

The contents of this publication were developed under a grant from the National Institute on Disability, Independent Living, and Rehabilitation Research (NIDILRR Grant no. 90RE5016-01-00) under the auspices of the Rehabilitation and Engineering Research Center on Technologies to Support Successful Aging with Disability (TechSAge; http://www.techsage.gatech.edu). The authors appreciate the insights of Charlie Kemp and Jamie Gorman on this project. They would also like to thank Emily McDonald and Lyndsie Koon for their help with the qualitative analysis.

References

[1] World Health Organization., "Research for the reorientation of national health systems," *World Health Organization. Technical Report Series*, 2015.

[2] R. C. Mayer, J. H. Davis, and F. D. Schoorman, "An integrative model of organizational trust," *Academy of Management Review*, vol. 20, no. 3, pp. 709–734, 1995.

[3] L. L. Couch and W. H. Jones, "Measuring levels of trust," *Journal of Research in Personality*, vol. 31, no. 3, pp. 319–336, 1997.

[4] L. Trojan and O. Yonge, "Developing trusting, caring relationships: home care nurses and elderly clients," *Journal of Advanced Nursing*, vol. 18, no. 12, pp. 1903–1910, 1993.

[5] L. Berg and E. Danielson, "Patients' and nurses' experiences of the caring relationship in hospital: An aware striving for trust," *Scandinavian Journal of Caring Sciences*, vol. 21, no. 4, pp. 500–506, 2007.

[6] R. Parasuraman and V. Riley, "Humans and automation: use, misuse, disuse, abuse," *Human Factors: The Journal of the Human Factors and Ergonomics Society*, vol. 39, no. 2, pp. 230–253, 1997.

[7] Y. Seong and A. M. Bisantz, "The impact of cognitive feedback on judgment performance and trust with decision aids," *International Journal of Industrial Ergonomics*, vol. 38, no. 7-8, pp. 608–625, 2008.

[8] P. Madhavan and D. A. Wiegmann, "Similarities and differences between human–human and human–automation trust: An integrative review," *Theoretical Issues in Ergonomics Science*, vol. 8, no. 4, pp. 277–301, 2007.

[9] J. D. Lee and N. Moray, "Trust, self-confidence, and operators' adaptation to automation," *International Journal of Human - Computer Studies*, vol. 40, no. 1, pp. 153–184, 1994.

[10] S. M. Merritt and D. R. Ilgen, "Not all trust is created equal: Dispositional and history-based trust in human-automation interactions," *Human Factors: The Journal of the Human Factors and Ergonomics Society*, vol. 50, no. 2, pp. 194–210, 2008.

[11] T. Sanders, K. E. Oleson, D. R. Billings, J. Y. C. Chen, and P. A. Hancock, "A model of human-robot trust: Theoretical model development," in *Proceedings of the 55th Human Factors and Ergonomics Society Annual Meeting, (HFES '11)*, vol. 55, pp. 1432–1436, USA, September 2011.

[12] P. A. Hancock, D. R. Billings, K. E. Schaefer, J. Y. C. Chen, E. J. De Visser, and R. Parasuraman, "A meta-analysis of factors affecting trust in human-robot interaction," *Human Factors: The Journal of the Human Factors and Ergonomics Society*, vol. 53, no. 5, pp. 517–527, 2011.

[13] N. Ezer, "Is a robot an appliance, teammate, or friend? Age-related differences in expectations of and attitudes towards personal home-based robots".

[14] D. R. Billings, K. E. Schaefer, J. Y. C. Chen, and P. A. Hancock, "Human-robot interaction: Developing trust in robots," in *Proceedings of the 7th Annual ACM/IEEE International Conference on Human-Robot Interaction, (HRI '12)*, pp. 109-110, USA, March 2012.

[15] W. A. Rogers and T. L. Mitzner, "Human-robot interaction for older adults," *Encyclopedia of Computer Science and Technology*, pp. 1–11, 2017.

[16] C.-A. Smarr, T. L. Mitzner, J. M. Beer et al., "Domestic robots for older adults: attitudes, preferences, and potential," *International Journal of Social Robotics*, vol. 6, no. 2, pp. 229–247, 2014.

[17] J. Beer, A. Prakash, C. Smarr et al., "Older users' acceptance of an assistive robot: Attitudinal changes following brief exposure," *Gerontechnology*, vol. 16, no. 1, pp. 21–36, 2017.

[18] K. E. Olson, *Understanding the construct of human trust in domestic service robots*, Georgia Institute of Technology, 2013.

[19] A.-S. Melenhorst, W. A. Rogers, and D. G. Bouwhuis, ""Older adults" motivated choice for technological innovation: evidence for benefit-driven selectivity," *Psychology and Aging*, vol. 21, no. 1, pp. 190–195, 2006.

[20] A. C. Costa, "Work team trust and effectiveness," *Personnel Review*, vol. 32, no. 5, pp. 605–672, 2003.

[21] D. Bernstein, K. Crowley, and I. Nourbakhsh, "Working with a robot: Exploring relationship potential in human-robot systems," *Interaction Studies*, vol. 8, no. 3, pp. 465–482, 2007.

[22] R. E. Stuck, *Understanding Dimensions of Trust between Older Adults and Human or Robot Care Providers*, Georgia Institute of Technology, 2017.

[23] S. J. Czaja, N. Charness, A. D. Fisk et al., "Factors predicting the use of technology: findings from the center for research and education on aging and technology enhancement (CREATE)," *Psychology and Aging*, vol. 21, no. 2, pp. 333–352, 2006.

[24] L. H. Barg-Walkow, T. L. Mitzner, and W. A. Rogers, "Technology Experience Profile".

[25] J. A. Sanford, P. C. Griffiths, P. Richardson, K. Hargraves, T. Butterfield, and H. Hoenig, "The effects of in-home rehabilitation on task self-efficacy in mobility-impaired adults: A randomized clinical trial," *Journal of the American Geriatrics Society*, vol. 54, no. 11, pp. 1641–1648, 2006.

[26] Z. S. Nasreddine, N. A. Phillips, V. Bédirian et al., "The montreal cognitive assessment, MoCA: a brief screening tool for mild cognitive impairment," *Journal of the American Geriatrics Society*, vol. 53, no. 4, pp. 695–699, 2005.

[27] D. Wechsler, *WAIS-III: Wechsler adult intelligence scale*, Psychological Corporation, 1997.

[28] C. T. K. C, R. W. Mitzner T, and C. T. K. C. R. W. Mitzner, "Identifying the potential for robotics to assist older adults in different living environments," *International Journal of Social Robotics*, vol. 6, no. 2, pp. 213–227, 2014.

[29] H. F. Hsieh and S. E. Shannon, "Three approaches to qualitative content analysis," *Qualitative Health Research*, vol. 15, no. 9, pp. 1277–1288, 2005.

[30] C. Pope, S. Ziebland, and N. Mays, "Qualitative research in health care Analysing qualitative data," *British Medical Journal*, vol. 320, no. 7227, pp. 114–116, 2000.

[31] J. Saldana, *The Coding Manual for Qualitative Researchers*, vol. 36, Sage Publications, England, UK, 2nd edition, 2014.

[32] T. L. Mitzner, R. Stuck, J. Q. Hartley, J. M. Beer, and W. A. Rogers, "Acceptance of televideo technology by adults aging with a mobility impairment for health and wellness interventions," *Journal of Rehabilitation and Assistive Technologies Engineering*, vol. 4, 2017.

[33] A. D. Fisk, W. A. Rogers, N. Charness, S. J. Czaja, and J. Sharit, *Designing for older adults: Principles and creative human factors approaches*, 2009.

[34] R. E. Stuck and W. A. Rogers, "Understanding older adult's perceptions of factors that support trust in human and robot care providers," in *Proceedings of the 10th International Conference on PErvasive Technologies Related to Assistive Environments (PETRA '17)*, pp. 372–377, Greece, June 2017.

8

Artificial Knee Joints Actuators with Energy Recovery Capabilities: A Comparison of Performance

Roberta Alò, Francesco Bottiglione, and Giacomo Mantriota

Dipartimento di Meccanica, Matematica e Management, Politecnico di Bari, Viale Japigia 182, 70126 Bari, Italy

Correspondence should be addressed to Roberta Alò; roberta.alo@poliba.it

Academic Editor: Shahram Payandeh

The human knee absorbs more energy than it expends in level ground walking. For this reason it would be useful if the actuation system of a wearable robot for lower limbs was able to recover energy thus improving portability. Presently, we recognize three promising technologies with energy recovery capabilities already available in the literature: the Series Elastic Actuator (SEA), the Clutchable Series Elastic Actuator (C-SEA), and the flywheel Infinitely Variable Transmission (F-IVT) actuator. In this paper, a simulation model based comparison of the performance of these actuators is presented. The focus is on two performance indexes: the energy consumed by the electric motor per gait and the peak torque/power requested to the electric motor. Both quantities are related to the portability of the device: the former affects the size of the batteries for a given desired range; the latter affects the size and the weight of the electric motor. The results show that, besides some well-explained limitations of the presented methodology, the C-SEA is the most energy efficient whereas the F-IVT allows cutting down the motor torque/peak power strongly. The analysis also leads to defining how it is possible to improve the F-IVT to achieve a reduction of the energy consumption.

1. Introduction

In the human normal gait cycle, the knee joint undergoes phases of positive and negative work. Therefore an optimally designed wearable device for lower limbs (orthoses, exoskeletons, and prostheses) should be able to store energy during the phases of negative work and to permit reusing it when required. Indeed, one of the greatest limits to the development of powered orthoses, exoskeletons, and prostheses deals with their portability that is often limited by the lack of actuators and energy sources enabling the development of devices with both reasonable weight and an acceptable range of locomotion [1]. The energy recovery during normal operation is a smart strategy to reduce the net energy consumption. It could extend the portability of wearable devices enlarging their operating range and reducing the power requested to the motor.

Most of the wearable devices for lower limbs rely on battery-powered electric motors [2]. Reversibility of the electric drives permits in principle the regeneration of electric energy to be stored in a battery pack. However, in the context of wearable robots for lower limbs, the charging/discharging cycles of the battery occur under variable tension/current conditions and their frequency is relatively high, resulting in a very large number of cycles during the lifetime of the device, which adversely affects the performance and the life of the battery [3, 4]. The relatively low efficiency of the mechanical to electrical energy conversion and of the battery charge/discharge process and the high frequency of the gait cycle are the reasons why the electric energy recovery has not usually been considered in this kind of application [5, 6] up to very recent times [7].

Great efforts are focused on developing mechanical energy storage devices, in which the lack of any energy conversion would result in a greater efficacy. SEAs (Series Elastic Actuators) include an elastic element between the motor and the load which stores energy and delivers peak power values greater than the maximum power of the motor, and it offers greater shock tolerance, lower reflected inertia, and more accurate and stable force control [8–11]. In SEAs, the linear torque-angle relationship of the early stance knee flexion-extension phase of human walking can be rendered only by a spring whose stiffness is equal to the slope of the torque-angle curve. VSAs (Variable Stiffness Actuators) which adapt their stiffness to match the torque-angle curve also during the swing-phase [12] are claimed to achieve greater energy saving and peak power decrease compared to SEAs. Furthermore VSAs are able to adjust the optimal value of their stiffness according to the load conditions, permitting obtaining greater benefits at more than one specific walking gait.

In SEAs, when only the spring powers the knee joint, the motor must provide only reactionary torque at negligible motor speed, thus working inefficiently [7]. This consideration motivated the improvement of the design of SEAs into the development of the C-SEA (Clutchable Series Elastic Actuator), where a clutch is activated or deactivated according to the phase of the gait cycle. The C-SEA is explained in detail in [7], where its performance is evaluated in powering knee prostheses. It was proved that the C-SEA reduces the net electric energy consumption also thanks to the electric energy recovery. However, a limitation of such device is that when the clutch is activated the motor is not able to provide additional torque; thus the stiffness of the spring limits the operation of the C-SEA mechanism to those activities whose torque demand is into the bandwidth of the device, which is defined by the stiffness value.

Another way to store mechanical energy involves kinetic energy storage in a high speed rotating mass (flywheel). The flywheel energy storage has been recently reconsidered in the automobile field under the denomination of mechanical KERS (Kinetic Energy Recovery System). It is one possible short-term solution to achieve an improvement of the fuel economy performance of automobiles. In mechanical KERS, the energy is moved from the load (vehicle) to the flywheel under negative work phases and from the flywheel to the load otherwise [13] with no energy transformation. Inspired by KERS, the F-IVT (flywheel-IVT) actuator has been presented by the authors [14] as a device leading to potential benefits when employed in powering knee bionic joints. The F-IVT includes a brushless DC motor, a flywheel, an IVT (Infinitely Variable Transmission), and HD (harmonic drive) gear. The flywheel acts as a kinetic energy storage device; energy is moved between the motor and the joint thanks to a proper change of the speed ratio of the IVT. The IVT, whose ratio changes between positive and negative values, also permits a nearly constant motor speed, while the desired joint speed is, in principle, always matched.

In [14] the performance of the F-IVT was predicted in one specific walking condition, for which it was optimally designed. The authors proved that the F-IVT architecture permits undersizing the electric motor, which in F-IVT delivers an almost constant power value roughly equal to the mean power requested and by far less than the peak power. Moreover, the electric motor can be sized for working at a nearly fixed point at the greatest efficiency values, with a resulting potential saving of electric energy consumption. The motor undersizing would be a great advantage in this kind of application because it facilitates the portability of the device, also enlarging its operating range.

However, the results described in [14] give a limited insight into the F-IVT potential because the performance is therein estimated in the nominal working condition, that is, the one taken as a reference to optimally design the F-IVT components. Indeed, the portable lower limb wearable robots are required to support the wearer in his daily activity; thus their actuators do not usually work under the nominal condition.

The aim of this analysis is to further investigate the F-IVT by predicting its performance when it does not work under the condition for which it was optimally designed, like different walking speeds and stairs climbing. The energy and power requirements of the F-IVT were estimated through a model based approach and compared with those of two very efficient actuators already used in biomechatronics devices, the SEA and the C-SEA, in order to evaluate if it could be a competitive actuator for artificial knee joints of wearable robots.

2. Materials and Methods

This study compares through a model based approach three different actuators for powering artificial knee joints: the F-IVT, a novel actuator presented very recently in [14], and two other high performance actuators, the SEA and the C-SEA, which were largely investigated in previous works [3, 7, 15, 16].

The actuators F-IVT, SEA, and C-SEA were compared in different working conditions: level ground walking at different speeds (0.5, 1.1, 1.6, and 2.1 m/s) and stairs climbing at different inclinations (24°, 30°, and 42°). These conditions often recur in daily life. Indeed, it would be desirable for any actuator of artificial human joints to support the subject in his daily activity, requiring both a large operating range and the portability of the device.

Two main performance indexes have been taken into consideration for comparison: the energy consumption (per gait) and the peak of electric power, both calculated at the motor electrodes. These are significant quantities for the development of lightweight and portable wearable robots because they affect the size and the weight of the actuator, as well as the operating range of the device. The energetic and power requirements of all the actuators were calculated through an inverse dynamic approach, that is to say, starting from the load characteristics (knee angle and torque versus time) given as input data to the model provided in Sections 2.1–2.3. Gait cycles data were taken from [17, 18] (Figure 1).

The following sections provide details on the working principle of all the actuators, as well as on their mathematical modeling and the sizing procedure. Both SEA and C-SEA were modeled and sized as already done in the literature.

2.1. The Flywheel-IVT (F-IVT) Actuator. The F-IVT is an innovative actuator with energy recovery capabilities to power artificial knee joints [14]. It is made up of the following devices: a brushless DC motor, a flywheel, an IVT, and a HD gear (Figure 2). The brushless DC motor and the HD gear are often employed in biomechatronics: the first one because of its great efficiency with respect to other electric drives [19], the second one because of its large torque capacity, high gear

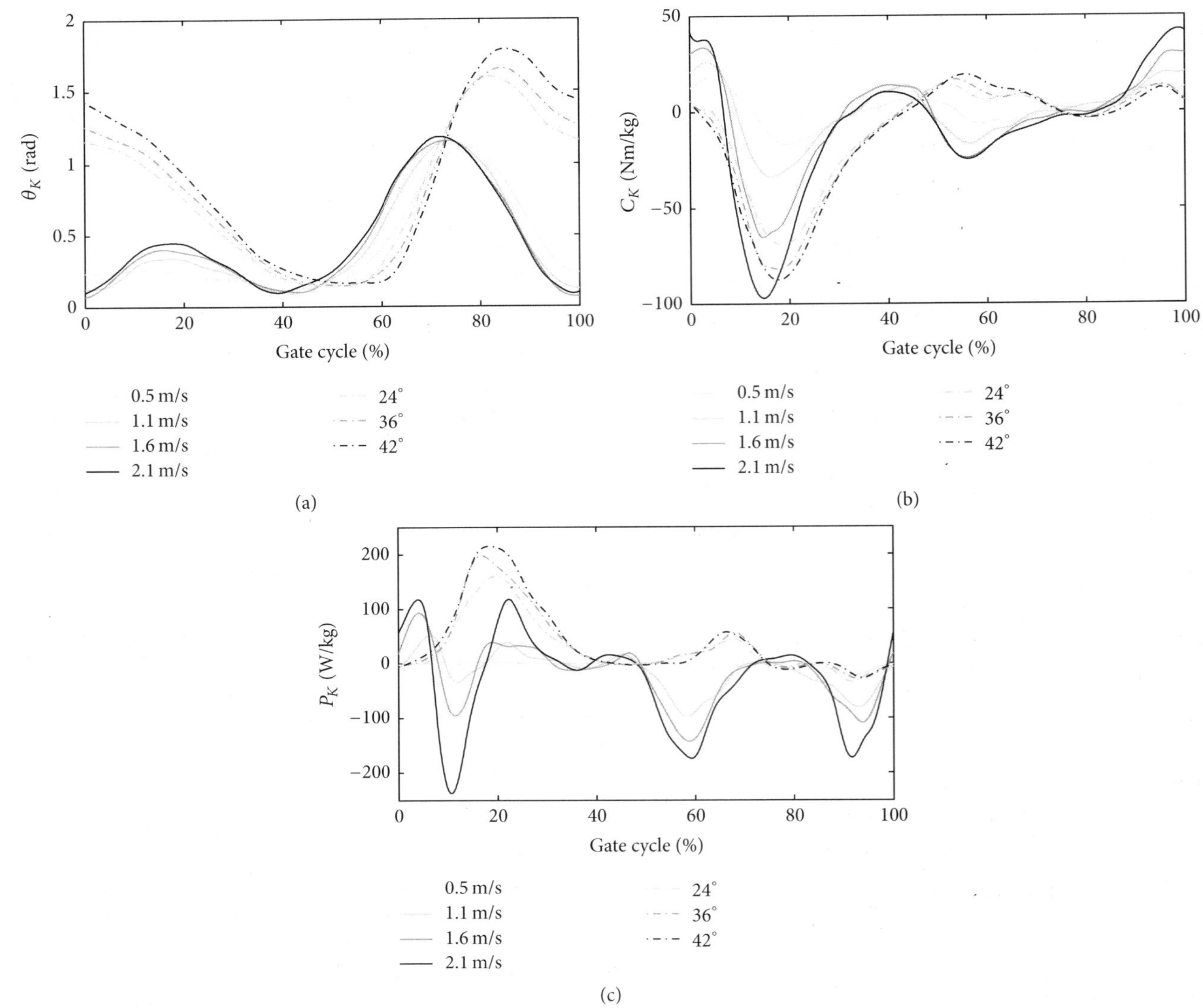

FIGURE 1: Gait cycle analysis data of the knee joint (angle, torque, and power) in level ground walking at different speeds and in stairs climbing at different inclinations. Data were adapted from [17, 18].

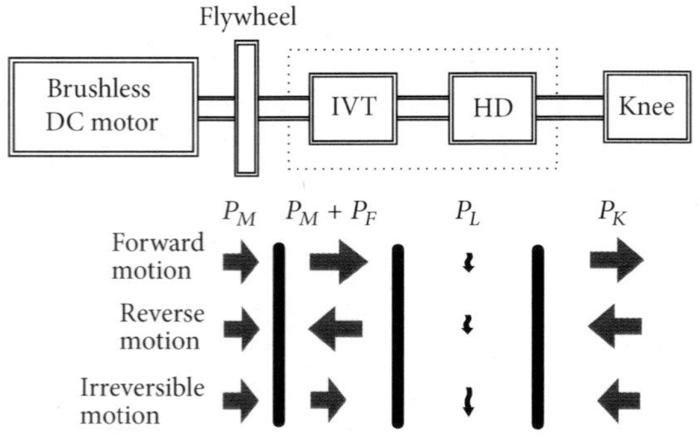

FIGURE 2: Fluxes of power in the F-IVT under forward, reverse, and irreversible motion conditions. The arrows indicate the power flows direction in each of them (P_M motor power, P_F rate of change of the kinetic energy of the flywheel, P_L power losses in the transmission devices, and P_K knee power).

ratios (τ_{HD}), and small size [19]. The flywheel and the IVT are introduced for the first time in an actuation system of artificial leg joints with the aim of recovering energy and reducing the irregularity of the motor working point.

As in a mechanical KERS (Kinetic Energy Recovery System) [13], in the F-IVT the flywheel operates as a kinetic energy storage device in which energy is stored when the power demand of the knee is negative (reverse motion) and it is released otherwise (forward motion), as schematically depicted in Figure 2. Energy is moved between the motor and the joint through a proper change of the speed ratio of the Continuously Variable Transmission. More specifically, an IVT is required because of the alternate motion of the knee joint. The speed ratio of the IVT (τ_{IVT}) is changed continuously between positive and negative values ($\tau_{IVT}^{min} \le \tau_{IVT} \le \tau_{IVT}^{max}$) permitting achieving the desired joint speed (ω_K) while the motor speed is nearly constant ($\omega_M = \omega_K/(\tau_{IVT}\tau_{HD})$) even when the output speed is zero thanks to the neutral gear condition of the IVT. An IVT with shunted CVT architecture was considered in the F-IVT actuator, because mathematical models of its efficiency are well established in the literature [20–22]. It is furthermore assumed that the speed ratio of the IVT is always perfectly regulated in order to be equal to the required value of τ_{IVT} at each instant of the gait cycle.

The operation of the F-IVT is such that the motor works at nearly fixed point with high efficiency and with a resulting possibility of undersizing the motor. Indeed, in the F-IVT the motor must provide an almost constant power, while the flywheel provides the variations of the requested power at the knee joint, permitting satisfying the following instantaneous power balance equation:

$$P_K = P_M + P_F + P_L, \tag{1}$$

where P_K is the knee power, P_M is the motor power, P_F ($P_F = -J_F\dot{\omega}_M\omega_M$) is the rate of change of the kinetic energy of the flywheel of inertia J_F, which rotates with the same motor angular speed (ω_M) and acceleration ($\dot{\omega}_M$), and P_L is the power lost in the transmission devices.

With the motor speed being nearly constant $\omega_M \approx \overline{\omega}_M$ ($\overline{\omega}_M$ is the average velocity of the motor), in order to achieve an almost constant power delivery, the motor is supposed to be torque controlled with constant torque C_M given by

$$C_M = \frac{1}{T\overline{\omega}_M}\int_0^T (P_K(t) - P_L(t))\,dt, \tag{2}$$

where T is the period of the gait cycle, $\overline{\omega}_M$ is the mean value of the motor angular speed, and $\int_0^T (P_K(t) - P_L(t))dt$ is the total amount of energy that the motor must provide to fulfill the knee power requirement and to overcome the power lost in the transmission devices.

Detailed efficiency models were adopted for the IVT and for the HD to realistically estimate the power losses. P_L (see (1)) can be written as a function of the instant values of the efficiency of the IVT (η_{IVT}) and the HD gear (η_{HD}) in forward ($P_L = -(1 - \eta_{IVT}\eta_{HD})(P_M + P_F)$) and reverse ($P_L = -(1 - \eta_{IVT}\eta_{HD})P_K$) motion. The adopted efficiency models also consider the nonreversible motion condition, which implies that P_M, P_F, and P_K are all delivered to the transmission and then dissipated as heat losses (Figure 2). This situation can involve the HD, the IVT, or both simultaneously, depending on the actual values of speed and torque. In particular, we adopted the model given in [20] (described in detail in Appendix A) for predicting η_{IVT} (Figure 3(a)), and we always referred to the guidelines of the data sheet of HD gears given by one manufacturer (AG Harmonic Drive) to estimate η_{HD} (Figure 3(b)) in this paper.

To determine the actual working conditions of the actuator, the equations given above are solved through an iterative routine (Figure 4) aimed at calculating the mechanical motor quantities ω_M, $\dot{\omega}_M$, C_M. The iterative calculation is necessary to find the actual value of the motor speed, which is not known a priori. Starting from a guess value of constant motor speed $\omega_M^{(0)}$, the corresponding values of C_M (see (2)) and then of $\dot{\omega}_M$ (from (1)) are calculated. $\dot{\omega}_M$ is then integrated through a numerical method (Newton-Raphson method) to find $\omega_M^{(1)}$. At integration step j, an error is defined as $\mathrm{ERR}_j = (1/T)\int_0^T (\omega_M^{(j+1)} - \omega_M^{(j)})^2 dt$. The iteration stops when $\mathrm{ERR}_j \le \mathrm{ERR}_{max}$, where ERR_{max} is the error tolerance, and it is requiring also that the constraint on the IVT ratio is satisfied ($\tau_{IVT}^{min} \le \tau_{IVT} \le \tau_{IVT}^{max}$).

After the mechanical motor requirements are defined, the electric power requested to the motor ($P_{EL} = v_M i_M$) is calculated as a function of the motor parameters (motor torque constant k_t, damping friction coefficient ν, Coulomb friction torque C_{fr}, motor resistance R, back EMF constant k_{emf}, and motor inertia J_M) and of the calculated requirements (motor torque C_M, angular motor speed ω_M, and angular motor acceleration $\dot{\omega}_M$):

$$\begin{aligned} i_M &= \frac{C_M}{k_t} + \frac{J_M\dot{\omega}_M}{k_t} + \frac{\nu\omega_M}{k_t} + \frac{\mathrm{sign}(\omega_M)C_{fr}}{k_t}, \\ v_M &= Ri_M + k_{emf}\omega_M, \end{aligned} \tag{3}$$

where i_M and v_M are, respectively, the electric motor current and voltage.

Finally, the electric energy to be supplied is calculated as the time integral of the motor electric power $P_{EL} = v_M i_M$ in the gait cycle; thus the energy consumption of all the electronic modules was neglected. According to [23], in the reverse operation it may happen that the electric motor is not able to generate electricity. In those cases, $P_{EL} = 0$. Consider

$$E_{EL} = \int_0^T P_{EL}\,dt. \tag{4}$$

The mathematical model of the F-IVT given above (see (1)–(4)) is implemented to estimate the performance of the F-IVT. In this work the F-IVT sizing was carried on in such a way that it is capable of powering the knee joint under different locomotion regimes, namely, the level ground walking at different speeds and the stairs climbing at different inclinations. In particular, the size of the HD gear and that of the motor were determined from the maximum power requested considering all possible walking regimes, whereas the most frequent one (level ground walking at 1.1 m/s) was used to optimally design the F-IVT in order to make it work very efficiently most of the time. More details about the criterion followed to size each component of the F-IVT, the nominal values of which are listed in Table 1, can be found in Appendix B.

2.2. Series Elastic Actuator (SEA). SEAs have been extensively employed in wearable devices for lower limbs. In SEAs a spring is put in series between the motor and the artificial powered joint (Figure 5(a)). The elastic element gives advantages to the actuation system such as increasing of shock tolerance and limited high frequency actuator impedance and energy storage [8–10]. Since the force/displacement characteristic of the spring is linear, the SEA is very efficient in powering the knee joint because it can be designed to take advantage of the linear torque-angle (or spring-like) relationship of the knee in the early stance flexion-extension

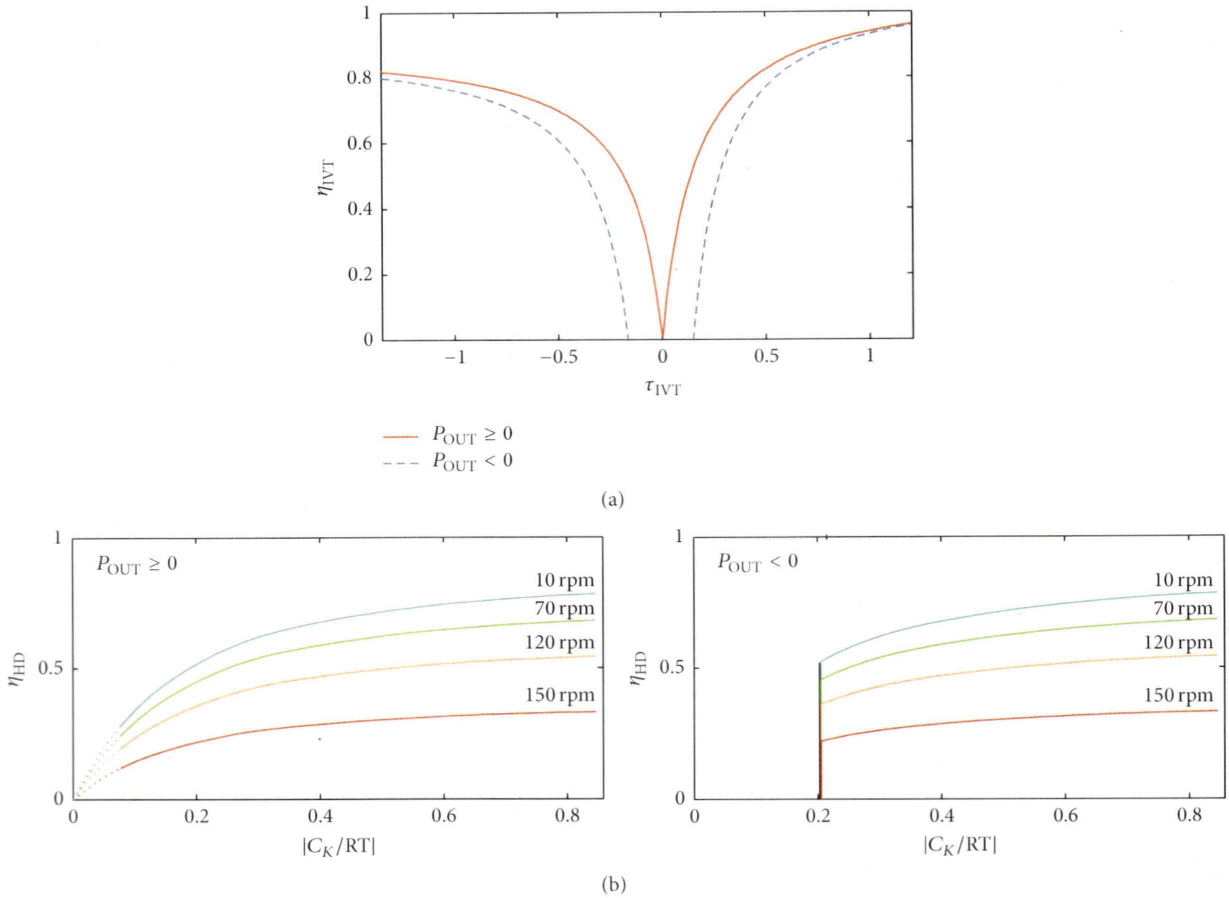

FIGURE 3: Efficiency maps of the IVT (a) and of the HD gear (b) obtained through detailed efficiency models which distinguish the performance of the transmission devices in forward ($P_{OUT} \geq 0$), revrse ($P_{OUT} < 0$), and irreversible motion conditions. In (a) η_{IVT} is given as a function of τ_{IVT} whereas η_{HD} is given as a function both of the output motor speed and of the torque at the output shaft normalized to the Rated Torque (RT) of the HD gear selected.

phase of level ground walking. In particular if the stiffness of spring (K) is chosen equal to the slope of the torque-angle curve, the spring alone would render the spring-like relationship, resulting in a great reduction of the motor mechanical energy requirement.

In this work the electric power and energy requirement of SEA to power knee joint in level ground walking at different speeds and in stairs climbing at different inclinations are calculated through (3)-(4) as a function of angle ($\vartheta_M = (C_K/K + \vartheta_K)(1/\tau_{GT})$, τ_{GT} being speed ratio of the gear train) and torque ($C_M = C_K\tau_{GT}/\eta_{GT}$, η_{GT} being efficiency of the gear train) requirements of the SEA motor.

As done for the F-IVT, the SEA was also sized to power the knee joint even under the most powerful condition examined and it was optimized according to the most frequent one. In particular the most powerful condition leads to the choice of both the motor and the HD, whereas the most frequent one leads to the choice of series stiffness which minimizes the energy consumption. In our calculations we used the same walking cycles of [17] so also the optimal value of the stiffness coefficient of the spring is the one determined in [17].

According to [24], the gear train of the SEA comprises a harmonic drive gear (HD). The value of gear ratio of the HD is larger for SEA than for the F-IVT to reduce the maximum torque value requested to the motor and thus to limit the motor size. This choice leads to an average efficiency of the HD that is smaller than that in the F-IVT. The selected 400 W brushless Maxon Motor (model: EC 60 p.n. 167131) is able to provide the maximum torque requested to the motor ($C_M^{max} = 1.494$ Nm) while working under the motor speed limit ($\omega_M^{max} = 7322$ rpm). We noticed that this motor is too heavy for a proficient use in a wearable device, but, on the other hand, it is the only one allowing the SEA to power the knee in all the conditions examined. The model parameters adopted are listed in Table 1.

2.3. Clutchable Series Elastic Actuator (C-SEA). The C-SEA (Figure 5(b)) was developed to overtake one limit of SEAs related to the inefficiency of the motor in the stance phase [7]. In fact, even if SEAs greatly reduce the motor mechanical energy requirement in the stance phase when the spring alone

TABLE 1: Model parameters of the actuators F-IVT, SEA, and C-SEA. The motor damping parameters and the Coulomb friction torque are not provided by data sheet, but they were estimated following the guidelines given by Maxon.

		F-IVT	SEA	C-SEA
Brushless motor	Model	EC-max 40 283871	EC 60 167131	EC 45 136209
	C_{fr} [mNm]	5.2	0.8	0.77
	J_M [g cm^2]	101	831	209
	k_{emf} [V/rad/s]	0.0628	0.147	0.0739
	k_t [mNm/A]	62.8	147	73.9
	R [Ω]	2.02	1.03	1.01
	ν [Nm/rad/s]	$8.66 \cdot 10^{-6}$	$1.4 \cdot 10^{-4}$	$2.7 \cdot 10^{-5}$
	i_C [A]	—		0.25
	v_C [V]	—		24
Flywheel	J_v [g cm^2]	$5 \cdot 10^3$	—	—
IVT	τ_{IVT}	[−1.35, 1.21]	—	—
Gear train	τ_{GT}	1/50	1/80	1/143
	η_{GT}	By model	By model	0.9
Spring	K [Nm/rad]	—	375.14	250

powers the knee joint, the electric motor works at very slow speeds and thus very inefficiently. The C-SEA improved the design of SEAs by including a clutch on the motor shaft that is conveniently activated or deactivated according to the cycle phase. In particular during the spring-like phase of the gait cycle the clutch is activated and the spring renders the complete torque-angle relationship while the motor is required to supply only reactionary torque; during the other phases of the cycle, the clutch is deactivated and the C-SEA works as a SEA.

Similar equations as those of SEAs govern the operation of the C-SEA. The complete set of equations given in [7] follows:

$$C_M = \begin{cases} \dfrac{C_K \tau_{GT}}{\eta_{GT}} & \text{clutch off} \\ 0 & \text{clutch on,} \end{cases}$$

$$\vartheta_M = \begin{cases} \left(\dfrac{C_K}{K} + \vartheta_K\right)\dfrac{1}{\tau_{GT}} & \text{clutch off} \\ 0 & \text{clutch on,} \end{cases}$$

$$i_M = \begin{cases} \dfrac{C_M}{k_t} + \dfrac{J_M \dot{\omega}_M}{k_t} + \dfrac{\nu \omega_M}{k_t} + \dfrac{\text{sign}(\omega_M) C_{fr}}{k_t} & \text{clutch off} \\ i_C & \text{clutch on,} \end{cases} \tag{5}$$

$$v_M = \begin{cases} R i_M + k_{emf}\omega_M & \text{clutch off} \\ v_C & \text{clutch on,} \end{cases}$$

where i_C is the clutch current and v_C is the clutch voltage when the clutch is activated.

In this work the authors estimated the electric energy and power consumption of the C-SEA in level ground walking at different speeds and in stairs climbing at different inclinations. The mathematical model and many of the parameters of [7] were adopted to calculate the performance of C-SEA. However, because of the different reference gait cycle data some characteristics have been changed and a bigger electric motor was selected. A 250 W brushless Maxon Motor (model: EC 45 p.n. 136209, parameters in Table 1) was chosen because it is able to provide a peak torque compatible with the maximum required ($C_M^{max} = 0.323$ Nm) in the locomotion regime explored and the maximum motor speed ($\omega_M^{max} = 11574$ rpm) is always below the motor speed limit. The optimal spring stiffness value (K) was also not the same as [7] because of the different gait cycles data considered in this work. In particular, the optimal spring stiffness in [7] is defined as the one which minimizes the kinematic similarity defined as

$$\Psi = \int (\vartheta_{CSEA} - \vartheta_K)^2 \, dt, \tag{6}$$

where ϑ_{CSEA} is the output displacement of the C-SEA and ϑ_K is the output displacement of the knee joint. Such criterion was adopted by Rouse et al. in [7] because they proved that the stiffness of the spring does not affect the electrical power profiles but determines very different knee kinematics. Indeed, when the stiffness affects the knee angle displacement, the clutch absorbs a nearly constant electrical power. Based on this minimization criterion, we calculated an optimal stiffness value of 250 Nm/rad, at which the kinematic agreement was of $8 \cdot 10^{-4}$ rad^2 s (Figure 6).

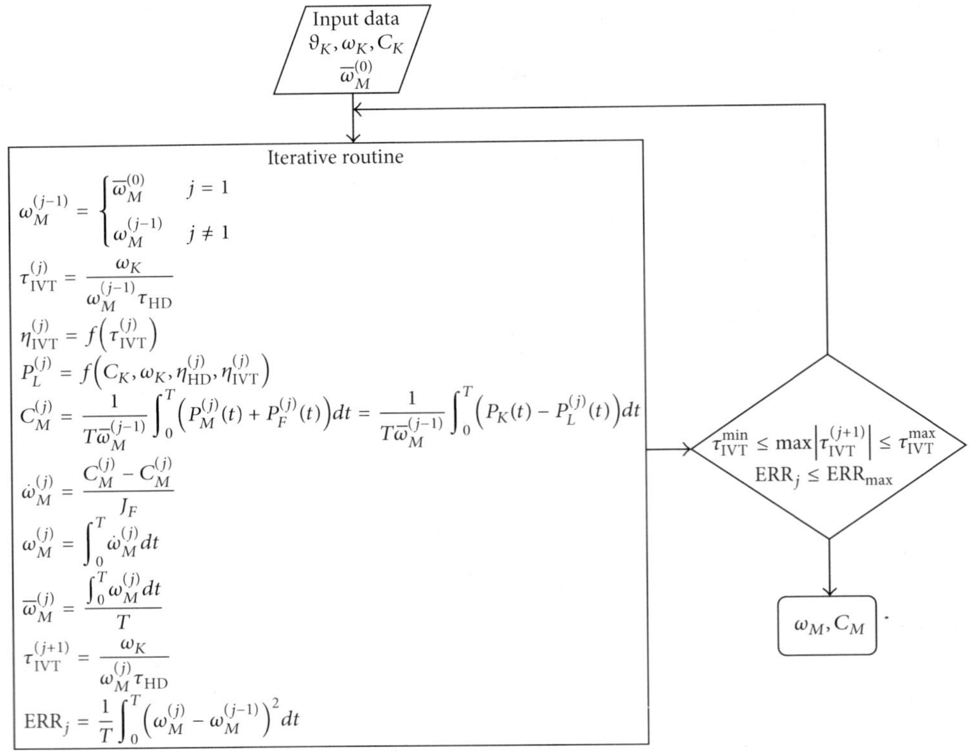

FIGURE 4: Schematic representation of the iterative routine to solve to calculate the mechanical motor requirements (ω_M, C_M).

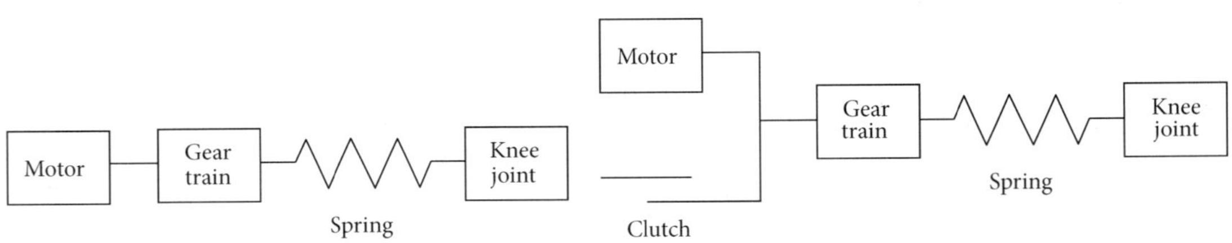

(a) SEA (b) C-SEA

FIGURE 5: Schematic pictures of the SEA (a) and C-SEA (b).

3. Results and Discussion

The electric power and energy consumption of F-IVT, SEA, and C-SEA actuators were estimated through the models described in Section 2, following an inverse dynamic approach. All the models parameters are listed in Table 1. Representative weight-normalized human locomotion data [17, 18] were given as input to the models to estimate the performance of all the actuators in powering the knee joint of a 75 kg person walking at different speeds and climbing stairs of different inclinations.

The electric energy demanded by the motor and the electric power profile result from the simulations done. These quantities are significant performance indexes for the development of lightweight and portable wearable robots because they affect the size and the weight of the actuator, as well as the operating range of the device.

The motor generation of electric energy has been usually neglected in the literature, because the efficiency of regeneration was considered to be too small. Only recently some researchers claimed that at the present state of the art the electric motors are enough efficient to permit a proficient use of the regenerative braking also in biomechatronic devices [7]. Indeed, although mathematical model of the brushless motor (Section 2.1) predicts very large values of the efficiency of the electric motor in the 2nd and 4th quadrant, experimental measurements are only in qualitative agreement with the model, since the measured values of efficiency are in the range 18–30% [25]. Therefore, two different estimations of the electric energy requirement were carried on in this work for SEA and C-SEA devices: in the first one the possibility of regenerating electric energy is considered; in the second one it is neglected. This distinction does not concern the F-IVT where the electric drive always works as a motor, providing

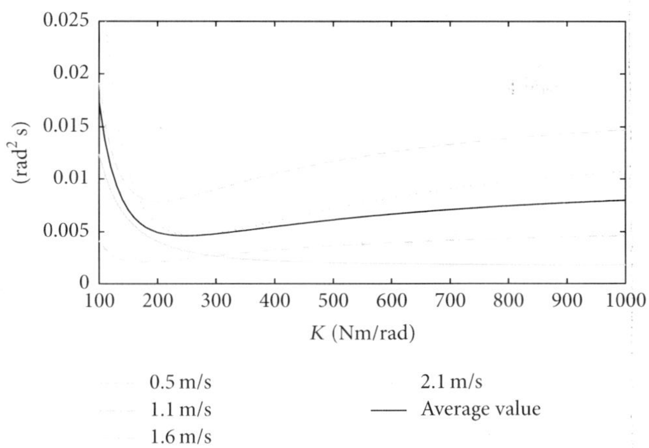

FIGURE 6: Kinematic similarity (Ψ) as a function of series stiffness averaged across walking speeds for subject weight of 75 kg. The minimum value of Ψ is kept approximately at the series stiffness value of 250 Nm/rad.

a positive amount of energy in the whole gait cycle at all the walking speeds and at all the stairs inclinations.

When the electric energy requirement of the F-IVT is compared with the electric energy requirements of SEA and C-SEA (Figure 7) two cases must be distinguished: SEA and C-SEA with regenerative braking and with no regenerative braking. In the former case (Figures 7(a) and 7(b)) the F-IVT is the most power-demanding device at all the conditions of walking and stairs ascending. The C-SEA is the only one actuator with a total negative energy requirement at all the level ground walking speeds, and this is due to the greater efficiency of the transmission devices when compared to efficiencies of the SEA. In fact, even if the motor of SEA generates electric energy when the motor power is negative, as a consequence of the low value of η_{HD}, which is often such to prevent the back drivability of the system, the total energy required by the motor is always positive for SEA.

In the no-regenerative braking case, the SEA has the greater electric energy consumption almost at all the walking speeds, whereas the C-SEA is still the most convenient one because of the low motor power absorbed when the clutch is activated (Figure 7(c)).

In stairs climbing the F-IVT always needs more electric energy than the SEA and the C-SEA if the regenerative braking of the motor is either considered or not (Figures 7(b) and 7(d)), whereas the C-SEA proves to be the most advantageous solution. It can be noticed that there is no evident difference of the values of energy required by SEA and C-SEA in stairs climbing between the regenerative and the nonregenerative cases. In fact, while the total energy required by the knee in level ground walking is negative, it is positive in stairs climbing and thus the possibility of recovering energy is reduced.

The peak of electric power is another important factor to estimate the performance of an actuator for lower limb joints, because it affects the size and the weight of the motor (Figure 8). The F-IVT permits greatly reducing the electric peak power at all the locomotion conditions, if the regenerative braking of the motor is either considered or not. In fact one of the achievements of the F-IVT is to stabilize the working point of the motor, which provides a nearly constant amount of power whose value is close to the mean value of the power request and by far less than the peak (Figure 9). On the contrary, the motor of both the SEA and C-SEA has to follow the torque request of the knee joint, supplying variable power. This is clearly shown in Figure 9 where the instant value of motor power is shown as a function of time for all the devices under analysis and in two walking regimes, for instance, walking at 2.1 m/s (Figure 9(a)) and stairs climbing at inclination of 30° (Figure 8(b)). Similar results have been obtained in all other walking regimes.

From these results the F-IVT seems to be the most effective solution for reducing the peak power of the motor, whereas the C-SEA is more energy efficient in all the conditions examined, if the regenerative braking is either permitted or not. On one hand, the F-IVT would permit a consistent electric motor downsizing; on the other hand, C-SEA would permit reducing the electric energy demanded by the motor for a given working range of the device. However, the gap between the C-SEA and the F-IVT in terms of energy consumption is considerably larger in the case of regenerative C-SEA, where our calculations may be affected by an overestimation of the electric energy recovered by regenerative braking with respect to the value achievable in practice [23].

Furthermore, the power losses in all electric modules may also alter the comparison of energy consumption of F-IVT and C-SEA (or SEA), and they have not been considered in this study. Indeed, all electronic modules and also the batteries would work differently in the F-IVT and in C-SEA (or SEA): in one case the electric power is almost constant with constant angular velocity of the motor; in the other cases both the power and the velocity of the motor vary periodically. Therefore, the total energy consumption of the actuators, that is, the energy requirement of the battery, may be quite different from the electric energy demanded by the motor.

4. Conclusions

In this work the authors compared the motor performance of three different actuators, F-IVT, SEA, and C-SEA, powering an artificial knee joint under different locomotion regimes very recurring in daily life: level ground walking at different speeds and stairs climbing at different inclinations. Detailed efficiency models of all the devices were developed to obtain realistic estimations of the motor consumptions following an inverse dynamic approach.

Two main performance indexes have been taken into consideration for comparison: the energy consumption (per gait) and the peak of electric power, both calculated at motor electrodes.

As a result of the simulations, it was found that the motor of the C-SEA has an electric energy consumption which is always lower compared to the F-IVT and the SEA at all the locomotion regimes considered. Thus the C-SEA would permit a larger operating range of the device. However, it is important to specify that in this work the energy requirement

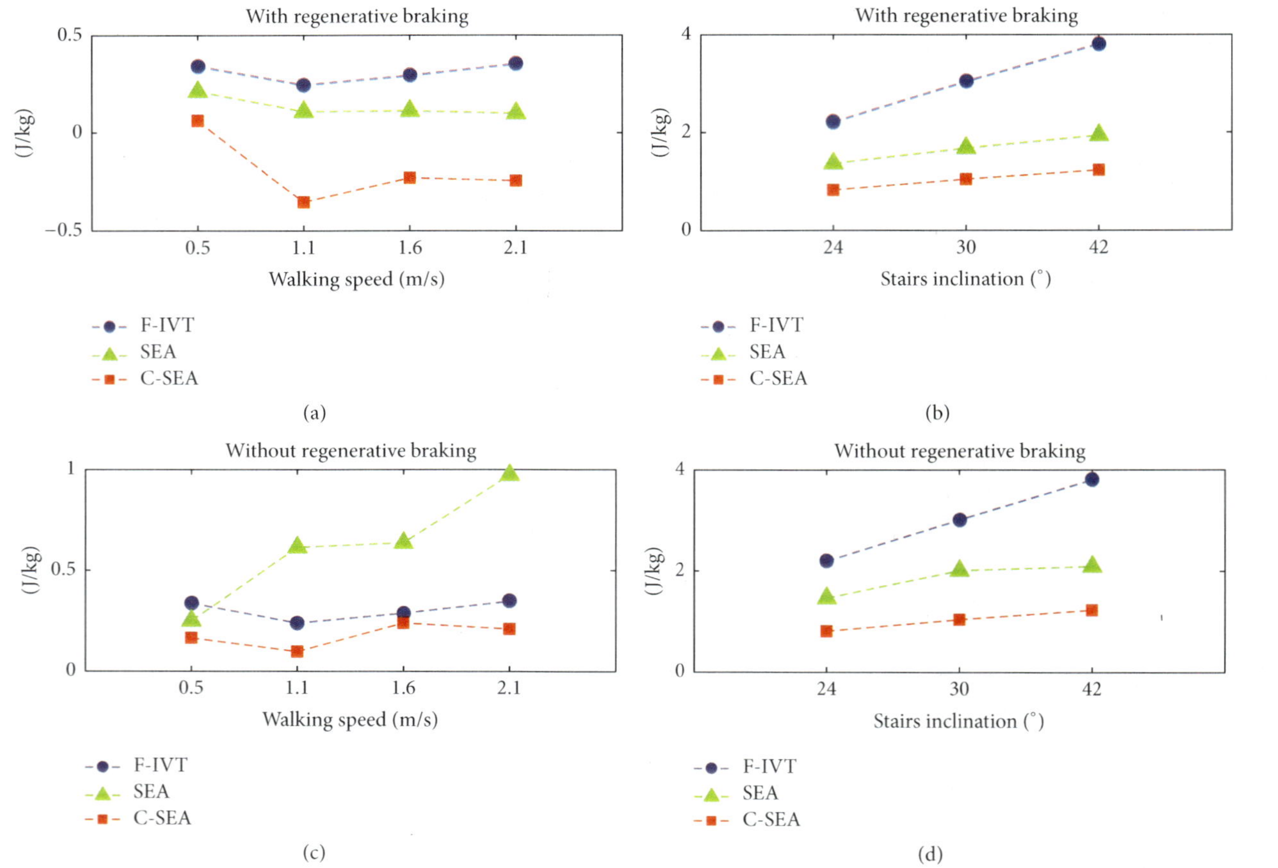

FIGURE 7: Motor electric energy requirement of the actuators compared (F-IVT, SEA, and C-SEA) in level ground walking at different speeds (a, c) and in stairs climbing at different inclinations (b, d), under the assumptions of regenerative braking admitted (a, b) and not admitted (c, d).

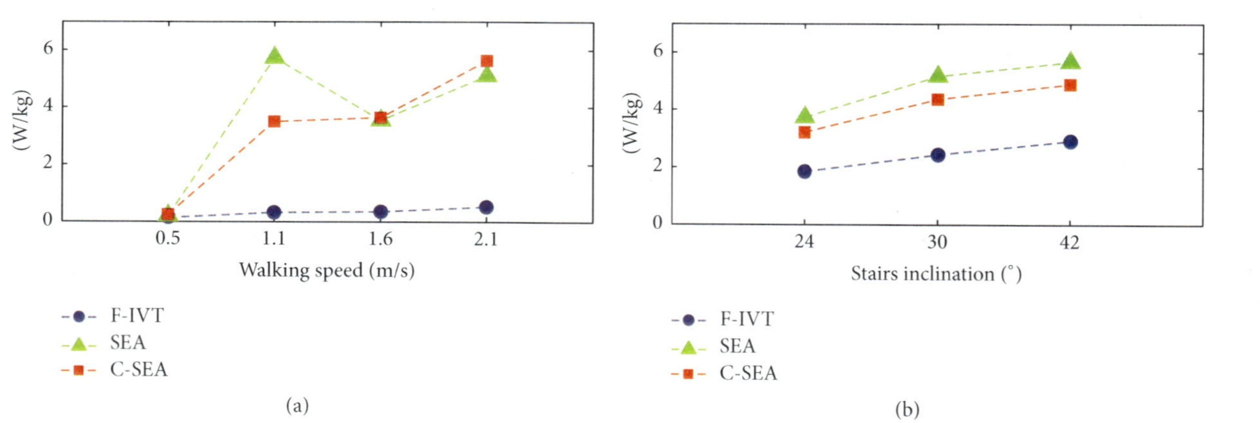

FIGURE 8: Electric peak power of the motors of the actuators compared (F-IVT, SEA, and C-SEA) in level ground walking at different speeds (a) and in stairs climbing at different inclinations (b).

of the motor has been calculated at the motor electrodes, not at batteries. There are at least two reasons why the effective energy requirement of the actuators, that is, the energy demanded by the battery, may be different with respect to the motor energy requirement, in particular in SEA and C-SEA. First, the working point of the motor is quickly and repeatedly variable, and this may affect the actual efficiency of the electronic devices between the batteries and the motor and the batteries themselves. Second, the electrical power generated by the motor could be smaller than the estimated, as the very low experimental values of the regeneration efficiency achieved in practice [23] would

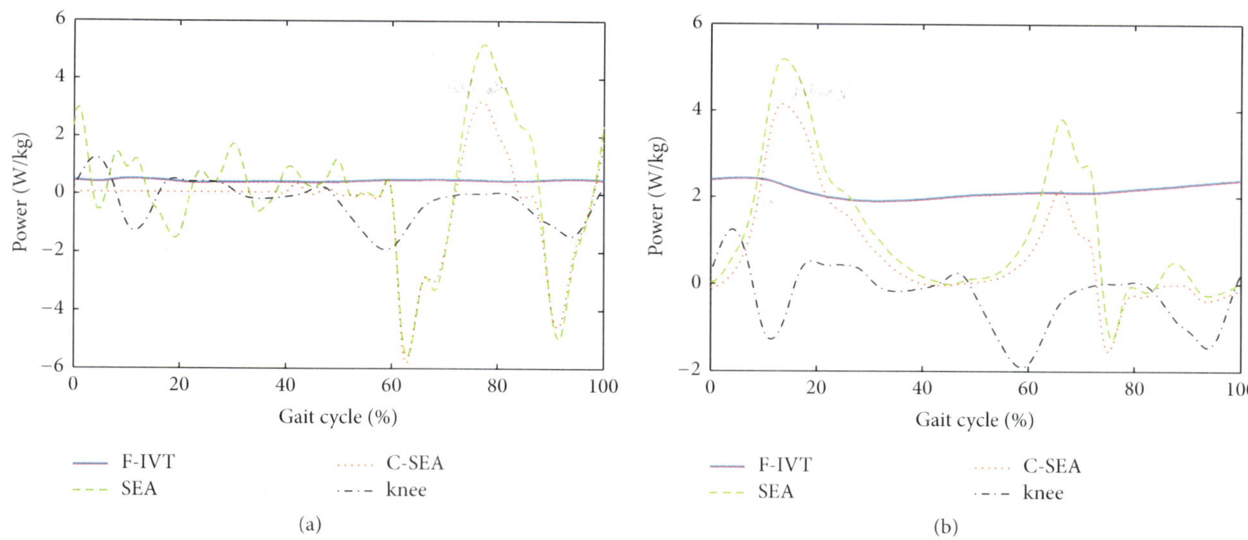

FIGURE 9: Instant power requirements of the knee joint and of the electric motor of the actuators compared (F-IVT, SEA, and C-SEA) in level ground walking at 2.1 m/s (a) and in stairs climbing at inclination of 30° (b).

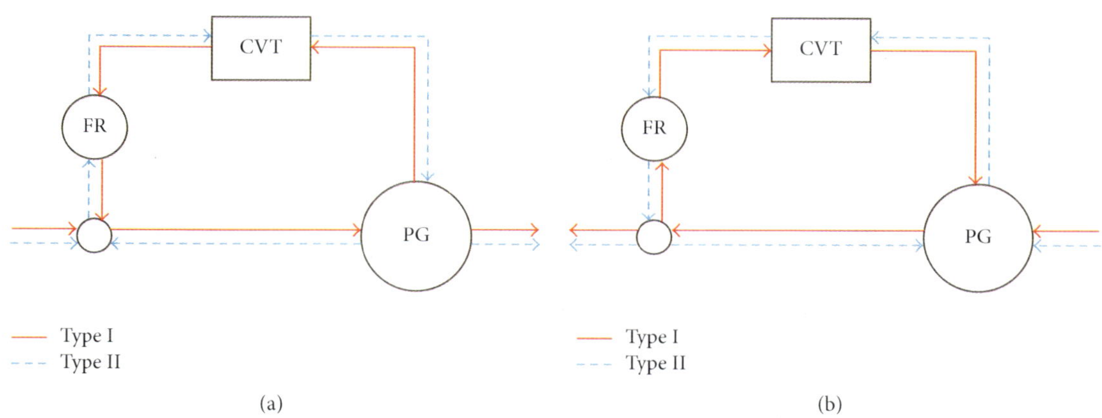

FIGURE 10: Schematic picture of the shunted CVT architecture of IVT, where power circulation of type I or that of type II can take place. Figures above depict the two types of power circulation in direct (a) and in reverse (b) operating modes.

demonstrate. Unfortunately, these issues are still debated, and no precise data are available in the literature.

Furthermore, in this study it was also proved that the F-IVT permit achieving a drastic reduction of the peak power at all the conditions examined. In fact, the flywheel and the IVT stabilize the working point of the motor, which always provides a nearly constant amount of power that is very lower than the peak power request. This achievement is very important in view of the possibility of downsizing the electric motor, which would also mean reducing the size and the weight of the actuator, with advantages in terms of portability. Of course, even if the motor mass is the most critical one, in order to make the F-IVT more lightweight than the other devices developed for the same purpose, the mass loss due mainly to the reduced motor size should overcome the additional masses of the micro-IVT and the flywheel.

Moreover, the motor of the F-IVT always works very efficiently, even when it does not work under the condition for which it is optimized, and this may be true for the power electronics and the batteries.

The comparison of the three actuators done in this paper aims at evaluating which are the potential benefits or the drawbacks of the F-IVT actuator, in order to evaluate if it could or could not be a competitive actuator for artificial knee joints of wearable robots. Furthermore, the results of the comparison are usefully supporting the development of a prototype of the F-IVT in which the authors are involved. For example, a careful analysis of the results obtained in this work reveals that at the present state of the art the weak point of the F-IVT is the overall efficiency of the driveline, the bottleneck of which is the harmonic drive. These results stimulate to find alternative possibilities in order to improve the efficiency of F-IVT to reduce the motor energy requirement, keeping its indisputable advantages safe.

Appendix

A. Shunted CVT Architecture

Among the different typologies of IVT proposed in literature [26–31], the most popular one is the shunted CVT architecture (Figure 10) which includes the following component devices: a microtoroidal CVT (Continuously Variable Transmission), PG (micro planetary gear train), and FR (fixed speed ratio drive). An IVT with shunted CVT architecture was considered in the F-IVT actuator, because mathematical models of its efficiency are well established in the literature [20].

An IVT is needed in the actuation system of lower limb joints because it is able to continuously change its speed ratio between positive and negative values, permitting the motor speed to be kept constant or nearly constant, while the joint always works at the desired speed value. The IVT speed ratio is given as a function of the transmission ratios of all the components of the IVT:

$$\tau_{\text{IVT}} = \frac{1 + \tau_{\text{CVT}}\tau_{\text{FR}}X}{1 + X}, \qquad \text{(A.1)}$$

where τ_{FR} is the speed ratio of the FR and X is the basic transmission ratio of the PG.

Very often the lower and upper bounds of τ_{CVT} are fixed, being an intrinsic geometric feature of the device. The IVT under consideration includes a microtoroidal CVT whose ratio range varies between 0.4 and 2.5. The speed ratio range of the IVT has also to be fixed to match the requirements of the load (Table 1). Once the lower and upper bounds of τ_{IVT} and τ_{CVT} are chosen, both τ_{FR} and X can be calculated through (A.1).

The IVT enlarges the transmission ratio range of the CVT (τ_{CVT}), but this increase involves a power circulation in the IVT device, which can be reverse or forward according to the relationship between τ_{IVT} and τ_{CVT}. The reverse power circulation, which is named power flow of type I, is characterized by a monotonic decrease of τ_{IVT} when τ_{CVT} increases; conversely the forward power circulation, named power flow of type II, is characterized by a monotonic increase of τ_{IVT} when τ_{CVT} increases. The IVT with type I power flow for positive value of the τ_{IVT} and with type II power flow for the negative ones was chosen as the most adequate for our purposes.

The efficiency of the IVT (η_{IVT}) was modeled as in [20], considering the different operating modes of the transmissions, that is, the direct and the reverse mode, and the conditions under which the IVT is not back drivable. η_{IVT} is affected by the type of power circulation both in direct and in reverse operating mode. The efficiency of the IVT is estimated through the following equations:

Direct mode is as follows:

$$\eta_{\text{IVT}} = \begin{cases} \dfrac{1 + \tau_{\text{CVT}}\tau_{\text{FR}}X}{1 + \tau_{\text{CVT}}\tau_{\text{FR}}X\eta_{\text{CVT}}\eta_{\text{FR}}} & \text{Type I } (\tau_{\text{IVT}} \geq 0) \\[2ex] \dfrac{\eta_{\text{CVT}}\eta_{\text{FR}}\left(1 + \tau_{\text{CVT}}\tau_{\text{FR}}X\right)}{\eta_{\text{CVT}}\eta_{\text{FR}} + \tau_{\text{CVT}}\tau_{\text{FR}}X} & \text{Type II } (\tau_{\text{IVT}} < 0). \end{cases} \qquad \text{(A.2)}$$

Reverse mode is as follows:

$$\eta_{\text{IVT}} = \begin{cases} \dfrac{\eta_{\text{CVT}}\eta_{\text{FR}} + \tau_{\text{CVT}}\tau_{\text{FR}}X}{\eta_{\text{CVT}}\eta_{\text{FR}}\left(1 + \tau_{\text{CVT}}\tau_{\text{FR}}X\right)} & \text{Type I } (\tau_{\text{IVT}} \geq 0) \\[2ex] \dfrac{1 + \eta_{\text{CVT}}\eta_{\text{FR}}\tau_{\text{FR}}X\tau_{\text{CVT}}}{1 + \tau_{\text{FR}}\tau_{\text{CVT}}X} & \text{Type II } (\tau_{\text{IVT}} < 0). \end{cases} \qquad \text{(A.3)}$$

In our calculations we assumed that $\eta_{\text{CVT}} = 0.93$ and $\eta_{\text{FR}} = 0.98$. η_{CVT} was assumed to be constant for the reasons given in [14].

The efficiency curves of the IVT shown in Figure 3(a) are obtained through (A.2)-(A.3). It can be noticed that η_{IVT} is larger in forward than in reverse operating mode in the whole range of τ_{IVT}. Moreover η_{IVT} gets its maximum value at the boundary values of the IVT ratio range and decreases when the absolute value of τ_{IVT} becomes small, close to the neutral gear condition ($\tau_{\text{IVT}} = 0$). This is why the ratio range was optimized according to the speed range of the knee joint to improve the global usage efficiency of the IVT. It can also be noticed that the back drive ability of the transmission is not permitted for very low values of $|\tau_{\text{IVT}}|$ ($\tau_{\text{IVT}} \in [-0.16, 0.14]$).

B. Sizing Procedure of the F-IVT

In this work the F-IVT was sized according to the task requirements of the actuator: walking at different speeds and stairs climbing at different inclinations. To design an efficient F-IVT, both the most powerful and the most frequent working conditions must be considered. In particular, the former defines the size of the HD gear and of the motor, whereas the latter (level ground walking at 1.1 m/s) is necessary to optimally design the F-IVT in such a way that it can work very efficiently as often as possible. The optimal design of the F-IVT leads to the choice of the HD gear, the ratio range of the IVT, and the related value of the motor working speed. Indeed, as it will be clarified later, the optimization of the IVT consists in the calculation of its optimal speed ratio range, which also has a direct influence on the actual working point of the motor.

The sizing procedure of the F-IVT started with the selection of the HD gear following the guidelines suggested by one manufacturer (Harmonic Drive AG), which mainly refers to the values of output torque and speed. Because the speed limit is by far respected (the knee speed is always lower than the maximum speed allowed), the HD gear was chosen constraining the greatest value of the output torque in all the gait cycles examined (i.e., stairs climbing at the maximum inclination), to be less than or equal to the Repeatable Peak

Torque (RPT) value of the HD gear ($\max |C_K| \leq$ RPT). The selection of the HD gear focuses on the smallest size and the greatest efficiency, mainly depending on the speed of the input shaft and on the torque of the output shaft, by means of a coefficient that is a function of the Rated Torque (RT) value of the device (Figure 3(b)). Among all the admissible options of HD gears (i.e., the ones that satisfy $\max |C_K| \leq$ RPT), the smallest one with the smallest RT value is chosen.

The second step is to choose the best motor for the required task. In the F-IVT the motor provides nearly the mean value of the power required by the load in a cycle, thanks to the effect of the flywheel. The flywheel inertia is determined in advance, as a compromise between its advantageous storage and regularizing function on one hand and its weight and size [14] on the other hand.

The motor nominal power must be approximately equal to the maximum value of the mean power demanded in all the working conditions explored. Unfortunately, the average power demanded per cycle is affected also by the efficiency of the IVT which has not yet been determined at this step of the design procedure. In order to overcome this problem, once the HD gear and the flywheel have been selected, the power requirement at the output shaft of the IVT is known and, thus, by assuming guess constant values of η_{IVT} both in direct ($\eta_{\text{IVT}} = 0.6$) and in reverse motion ($\eta_{\text{IVT}} = 0.5$), the mean value of the power requested to the motor in all the conditions was estimated. As it can be expected, the most powerful condition among those examined defines the size of the motor. In our case, we found the following motor for our application: 120 W brushless Maxon Motor, model EC-max 40 p.n. 283871. After having chosen a suitable motor for the application, the efficiency map of the motor is known, which is a necessary ingredient for the last step of the actuator design. The selected motor is able to provide greater torque and speed than the ones strictly necessary. Peak values of torque and speed are $C_M^{\max} = 0{,}165$ Nm and $\omega_M^{\max} = 10677$ rpm, respectively.

Design of the IVT proceeds in order to minimize the consumption of electric energy at the most frequent gait cycle (level ground walking at 1.1 m/s). The calculation of the optimal values of $\tau_{\text{IVT}}^{\min}$ and $\tau_{\text{IVT}}^{\max}$ needs an iterative procedure. Since the output speed of IVT is known for a given walking regime and for a given HD gear ratio, by assuming that the motor speed is constant and equal to an initial guess value $\omega_M^{(0)}$, which belongs to the motor speed working range, then τ_{IVT} can be calculated as a function of time as $\omega_K/(\omega_M^{(0)}\tau_{\text{HD}})$ and also $\tau_{\text{IVT}}^{\min}$ and $\tau_{\text{IVT}}^{\max}$ can be determined. It follows (see Appendix A) that also the efficiency of the IVT is known as a function of τ_{IVT}. Full simulation of the actuator with these parameters leads to the calculation of the actual trend of the motor speed $\omega_M^{(1)}$, which does not correspond to the initial guess value. An iteration error (at iteration 0) is defined as $\text{ERR}_0 = (1/T)\int_0^T (\omega_M^{(1)} - \omega_M^{(0)})^2 dt$. Then the procedure is repeated, starting from $\omega_M^{(1)}$ obtained at iteration 0 and so on. The error at iteration j is defined as $\text{ERR}_j = (1/T)\int_0^T (\omega_M^{(j+1)} - \omega_M^{(j)})^2 dt$ and the iterations are stopped when $\text{ERR}_j < \text{ERR}_{\max}$, where $\text{ERR}_{\max}$ is a proper error tolerance. After iterations are stopped, then the values of $\tau_{\text{IVT}}^{\min}$ and $\tau_{\text{IVT}}^{\max}$ are the ones calculated at the last iteration, and the consumption per cycle can also be calculated.

If the iterative routine is started from a different value of $\omega_M^{(0)}$, then different values of $\tau_{\text{IVT}}^{\min}$ and $\tau_{\text{IVT}}^{\max}$ and of the energy consumption are obtained. It follows that the optimal values of $\tau_{\text{IVT}}^{\min}$ and $\tau_{\text{IVT}}^{\max}$ are those which correspond to the minimum value of energy consumption achievable with all possible values of $\omega_M^{(0)}$ belonging to the working speed range of the electric motor.

Competing Interests

The authors declare that they have no competing interests.

References

[1] J. L. Pons, "Front matter," in *Wearable Robots: Biomechatronic Exoskeletons*, pp. 1–10, John Wiley & Sons, Chichester, UK, 2008.

[2] R. Bogue, "Robotic exoskeletons: a review of recent progress," *Industrial Robot: An International Journal*, vol. 42, no. 1, pp. 5–10, 2015.

[3] L. Mooney and H. Herr, "Continuously-variable series-elastic actuator," in *Proceedings of the IEEE 13th International Conference on Rehabilitation Robotics (ICORR '13)*, pp. 1–6, IEEE, Seattle, Wash, USA, June 2013.

[4] J. M. Donelan, Q. Li, V. Naing, J. A. Hoffer, D. J. Weber, and A. D. Kuo, "Biomechanical energy harvesting: generating electricity during walking with minimal user effort," *Science*, vol. 319, no. 5864, pp. 807–810, 2008.

[5] D. F. B. Haeufle, M. D. Taylor, S. Schmitt, and H. Geyer, "A clutched parallel elastic actuator concept: towards energy efficient powered legs in prosthetics and robotics," in *Proceedings of the 4th IEEE RAS & EMBS International Conference on Biomedical Robotics and Biomechatronics (BioRob '12)*, pp. 1614–1619, IEEE, Rome, Italy, June 2012.

[6] V. Luciano, E. Sardini, M. Serpelloni, and G. Baronio, "Analysis of an electromechanical generator implanted in a human total knee prosthesis," in *Proceedings of the IEEE Sensors Applications Symposium (SAS 2012)*, pp. 166–170, Brescia, Italy, February 2012.

[7] E. J. Rouse, L. M. Mooney, and H. M. Herr, "Clutchable series-elastic actuator: implications for prosthetic knee design," *The International Journal of Robotics Research*, vol. 33, no. 13, pp. 1611–1625, 2014.

[8] P. Cherelle, G. Mathijssen, Q. Wang, B. Vanderborght, and D. Lefeber, "Advances in propulsive bionic feet and their actuation principles," *Advances in Mechanical Engineering*, vol. 2014, Article ID 984046, 21 pages, 2014.

[9] D. Paluska and H. Herr, "The effect of series elasticity on actuator power and work output: implications for robotic and prosthetic joint design," *Robotics and Autonomous Systems*, vol. 54, no. 8, pp. 667–673, 2006.

[10] G. A. Pratt and M. M. Williamson, "Series elastic actuators," in *Proceedings of the IEEE/RSJ International Conference on Intelligent Robots and Systems, Human Robot Interaction and Cooperative Robots*, vol. 1, pp. 399–406, Pittsburgh, Pa, USA, August 1995.

[11] A. K. Raj, P. D. Neuhaus, A. M. Moucheboeuf, J. H. Noorden, and D. V. Lecoutre, "Mina: a sensorimotor robotic orthosis for mobility assistance," *Journal of Robotics*, vol. 2011, Article ID 284352, 8 pages, 2011.

[12] L. Flynn, J. Geeroms, R. Jimenez-Fabian, B. Vanderborght, N. Vitiello, and D. Lefeber, "Ankle-knee prosthesis with active ankle and energy transfer: development of the CYBERLEGs Alpha-Prosthesis," *Robotics and Autonomous Systems*, vol. 73, pp. 4–15, 2015.

[13] F. Bottiglione and G. Mantriota, "Effect of the ratio spread of CVU in automotive kinetic energy recovery systems," *Journal of Mechanical Design—Transactions of the ASME*, vol. 135, no. 6, Article ID 061001, 2013.

[14] R. Alò, F. Bottiglione, and G. Mantriota, "An innovative design of artificial knee joint actuator with energy recovery capabilities," *Journal of Mechanisms and Robotics*, vol. 8, no. 1, Article ID 011009, 8 pages, 2015.

[15] J. E. Pratt, B. T. Krupp, C. J. Morse, and S. H. Collins, "The RoboKnee: an exoskeleton for enhancing strength and endurance during walking," in *Proceedings of the IEEE International Conference on Robotics and Automation (ICRA '04)*, vol. 3, pp. 2430–2435, IEEE, May 2004.

[16] A. M. Dollar and H. Herr, "Design of a quasi-passive knee exoskeleton to assist running," in *Proceedings of the IEEE/RSJ International Conference on Intelligent Robots and Systems (IROS '08)*, pp. 747–754, Nice, France, September 2008.

[17] M. Grimmer, M. Eslamy, and A. Seyfarth, "Energetic and peak power advantages of series elastic actuators in an actuated prosthetic leg for walking and running," *Actuators*, vol. 3, no. 1, pp. 1–19, 2014.

[18] R. Riener, M. Rabuffetti, and C. Frigo, "Stair ascent and descent at different inclinations," *Gait &d Posture*, vol. 15, no. 1, pp. 32–44, 2002.

[19] A. Zoss and H. Kazerooni, "Design of an electrically actuated lower extremity exoskeleton," *Advanced Robotics*, vol. 20, no. 9, pp. 967–988, 2006.

[20] F. Bottiglione and G. Mantriota, "Reversibility of power-split transmissions," *Journal of Mechanical Design*, vol. 133, no. 8, Article ID 084503, 2011.

[21] F. Bottiglione, S. De Pinto, and G. Mantriota, "Infinitely variable transmissions in neutral gear: torque ratio and power re-circulation," *Mechanism and Machine Theory*, vol. 74, pp. 285–298, 2014.

[22] F. Bottiglione, S. De Pinto, G. Mantriota, and A. Sorniotti, "Energy consumption of a battery electric vehicle with infinitely variable transmission," *Energies*, vol. 7, no. 12, pp. 8317–8337, 2014.

[23] M. R. Tucker and K. B. Fite, "Mechanical damping with electrical regeneration for a powered transfemoral prosthesis," in *Proceedings of the IEEE/ASME International Conference on Advanced Intelligent Mechatronics (AIM '10)*, pp. 13–18, IEEE, Montreal, Canada, July 2010.

[24] F. Sergi, D. Accoto, G. Carpino, N. L. Tagliamonte, and E. Guglielmelli, "Design and characterization of a compact rotary series elastic actuator for knee assistance during overground walking," in *Proceedings of the 4th IEEE RAS & EMBS International Conference on Biomedical Robotics and Biomechatronics (BioRob '12)*, pp. 1931–1936, IEEE, Rome, Italy, June 2012.

[25] R. Rarick, H. Richter, A. Van Den Bogert, D. Simon, H. Warner, and T. Barto, "Optimal design of a transfemoral prosthesis with energy storage and regeneration," in *Proceedings of the American Control Conference (ACC '14)*, pp. 4108–4113, Portland, Ore, USA, June 2014.

[26] M. Bazyn, J. Carter, C. B. Lohr, C. Malone, L. T. McDaniel, and P. D. Poxton, "Continuously and/or infinitely variable transmissions and methods therefor," U.S. Patent 8,313,405, U.S. Patent and Trademark Office, Washington, DC, USA, 2002.

[27] L. G. Brown, G. A. Brown, and B. A. Brown, "Locked contact infinitely variable transmission," Patent US8419589 B1, 2013.

[28] C. J. Greenwood, A. D. De Freitas, and A. R. Oliver, "Drive mechanism for Infinitely Variable Transmission," Patent n. US7955210 B2, 2011.

[29] K. Kazerounian and Z. Furu-Szekely, "Parallel disk continuously variable transmission (PDCVT)," *Mechanism and Machine Theory*, vol. 41, no. 5, pp. 537–566, 2006.

[30] C. B. Lohr, J. W. Sherrill, B. P. Pohl, R. Dawson, and C. Pew, "Infinitely variable transmissions, continuously variable transmissions, methods, assemblies, subassemblies, and components therefor," Patent US8721485 B2, 2014.

[31] M. Douglas, "Infinitely Variable Transmission," Patent n. US7704184 B2, 2010.

9

A Social Robot, "Dali," for Aging in Place Technology

Dugan Um,[1] Jangwoon Park,[1] Jeongsik Shin,[2] and Woo Ho Lee[2]

[1] *School of Engineering and Computing Sciences, Texas A&M University-Corpus Christi, USA*
[2] *Phomatix LLC, Keller, TX, USA*

Correspondence should be addressed to Jangwoon Park; Jangwoon.Park@tamucc.edu

Academic Editor: Gordon R. Pennock

Healthcare has a trend of going hi-tech. With an aging population growing more than ever, researchers and health care providers are now relying on robots to ease the symptoms of dementia and help an aging population stay where they would like, at home. Several therapeutic robots such as Paro recently introduced in the markets are manifestation of such trends. In this paper, we propose a social robot missioned to autonomously capture images of people and feed multimedia contents to a social network or to a hospital for various social activities or for health monitoring purpose. The main technical barriers of such robots include autonomous navigation, human face detection, distance, and angle adjustment for clean and better shots. To that end, we study autonomous mapping/navigation as well as optimal image capturing technology via motion planning and visual servoing. To overcome the mapping and navigation at a crowded environment, we use the potential field path planning harnessed with two competitive potential update techniques. The robot is an agent navigating in a potential field where detected environmental significances provide sources of attractive forces, while previously occupied locations estimated by SLAM technique provide sources of repelling forces. We also study visual servo technique to optimize image capturing processes. This includes facial recognition, photographic distance/angle adjustment, and backlight avoidance. We tested several scenarios with the assembled robot for its usefulness.

1. Introduction

According to a new study from the Pew Research Center's Internet and American Life Project, four out of five Internet users participate in some kind of group in the "real" world, compared with just 56 percent of those who do not use the Internet regularly [1]. Those figures rise to 82 percent for users of social networks and to 85 percent for users of Twitter. This means that people being socially active online will more likely be social offline as well. However, the majority of modern professionals find incredibly constrained working environment with little or no chance of social networking in daily life. *The primary motivation of the paper lies in addressing the resolution as to how to help or promote social networking of those professionals so that they become more socially active not only in online but in offline domain as well.* Evidenced in the last decade is the evolution of social networking technologies and popular site such as Facebook and Twitter. Recently highlighted Instagram is another manifestation of such trend. However, taking photography of our own or family is, sometime, chore and we end up missing memorable moments during the daily life. Selfie becomes popular in the society and made a big impact on social networking as an enabling device. Again, the use of it sometimes never takes us to the level we all want to be in photographic quality or quantity. As a result, we propose a self-photographic device, namely, "Dali the photographer" as a breakthrough social networking device for our daily life (see Figure 1). As interests in social robots have increased rapidly, many researches have been conducted evaluations in socially assistive robots. For example, Fasola and Matarić [2] evaluated an effectiveness of spatial language based human-robot-interaction (HRI) with elders using a fully autonomous mobile robot. Fasola and Matarić [3, 4] designed, implemented, and evaluated a socially assistive robot to engage older adults in physical exercise. Tapus and Matarić [5] evaluated a music therapy feature of a social robot that aimed to maintain attention of people with cognitive impairments.

Many studies have addressed as to how social networking could affect people's daily life and the usage of social media, especially for women, older adults, and parents. Hampton et al. [3] found that women who use more social media report

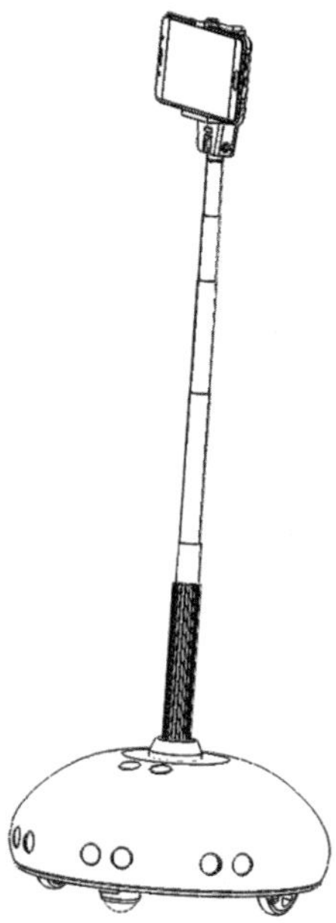

FIGURE 1: Dali social robot.

less stress than women who are nonusers. The article also states that people are more aware of events happening with their families and friends when they use social media. Pew Research Center [3] reported how seniors use social media. Nowadays, six out of ten seniors use social media, and higher-income and more highly educated seniors are more likely to use social media. However, around age 75 is when the use of social media drops and it is hard for them to use social media due to physical difficulties. Perrin [6] found that 65% of adults now use social media as a networking tool and it has been nearly tenfold increased in the past decade. Duggan et al. [3] reported how mothers use social media. They said that mothers are heavily engaged in social media, both giving and receiving support to their children. However, it has been also reported that posting multimedia contents in a social network such as Facebook is yet to be improved for mothers and elders. As a part of resolutions, Dali, the photographer, is proposed and developed to investigate the impact of a photographing social robot, not only to promote daily social networking, but also to offer a breakthrough as an aging in place technology. The primary function of Dali is to autonomously navigate around people and find human faces to take pictures. Several robotic photographers have been introduced in the market and demonstrated their usefulness. For instance, a robot photographer "Luke" was built based on Turtlebot utilizing Robot Operating System (ROS) open source autonomous navigation capabilities with the help of Kinect RGB-D camera [7]. Hsu [8] also implemented a photographer robot using Turtlebot platform and facial tracking feature based on facial landmark model. ROSBOT [9] is another ROS-based robot where a few social functions such as the following people, taking photos and uploading to photo sharing websites or online album. Microaerial vehicle (MAV) has been tested as a flyer photographer [10]. In this system, a Google's Yellowstone tablet was used as a main

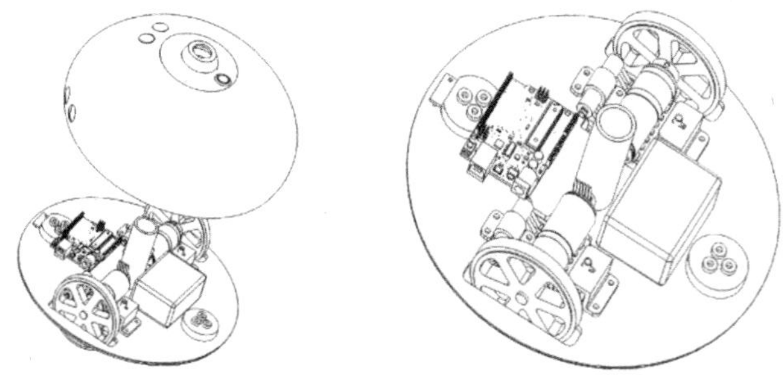

FIGURE 2: Dali internal layout and actuation mechanism.

control unit for autonomous navigation, pose estimation, and photographing. The success of those systems, however, is still in doubt due to several factors as listed below.

(1) Not cost effective

(2) Not human friendly

(3) Not able to navigate in crowded environment

The first factor is, in no doubt, the most important in creating desire for a product. Some of those aforementioned examples may have been successful for the cost effective solution, but not so for the second criteria. For instance, the "Luke" seems to be cost effective solution as a robotic photographer. However, due to its height limit, it is not as human friendly as other robots such as ROSBOT, which now causes the confliction with the first criteria. Navigation in crowded environments is also an important feature for social robots. In this paper, we propose a solution to satisfy all of those criteria at the same time so that more people can be involved in online and offline social activities for happier life style.

2. Robot Mechanism Design

A social robot Dali is designed and prototyped at Phomatix, LLC. Design considerations are size, low power consumption, maneuverability, and stability, to navigate indoor environments. Figure 1 illustrates Dali that is composed of a wheel based mobile platform, a selfie stick, and a smart phone. Different smart devices such as 3D or 360° cameras can be also mounted on the mobile platform. The size and weight of the mobile platform (the bottom part) are about 250 mm in diameter and 100 mm in height and 1 kg weight. Figure 2 illustrates major components inside of the mobile platform. Heavy components such as actuation mechanisms and battery are optimally distributed so that the center of mass of the robot is located at the geometric center of the robot.

The mobile platform has four wheels: two driving wheels and two ball caster wheels. The wheelbase of the mobile platform is designed to provide stability during the navigation. Two DC geared motor with encoders are used to drive the two driving wheels. A mounting boss is located at the center to hold a selfie stick (the middle part). Four sonar sensors are equipped at the upper case of the mobile platform to detect objects, and one distance sensor is equipped at

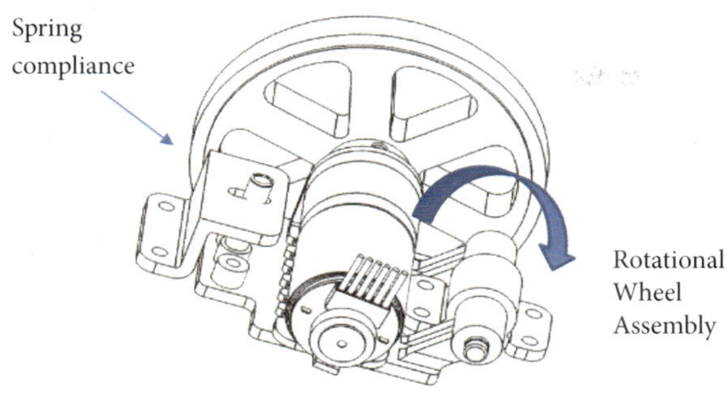

FIGURE 3: Compliant wheel mechanism for various surface conditions.

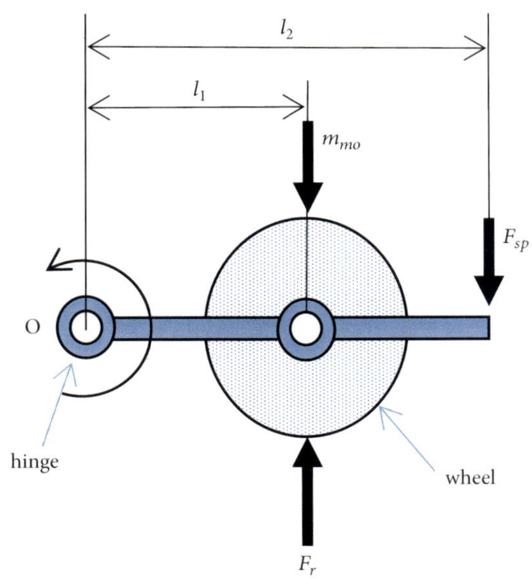

FIGURE 4: Free body diagram of a compliant wheel mechanism.

the lower case to avoid falling during the navigation. As shown in Figure 3, the compliant driving wheel mechanism is implemented to adapt various surface conditions including wood floors, carpets, and tiles.

A wheel directly coupled to a DC geared motor rotates with respect to a hinge joint and a linear compression spring at the opposite side provides a repulsive force. The main function of the spring compliance is to minimize the contact force on the front and rear ball caster to be able to move over small steps such as carpet or tiles. Figure 4 illustrates a free body diagram of a simplified compliant wheel mechanism. A static moment equation at O (hinge) is formulated as

$$F_r l_1 = m_{mo} g l_1 + F_{sp} l_2 \quad (1)$$

where F_r and F_{sp} are reaction forces exerting to the wheel from the ground and force generated by the deflection of the spring, m_{mo} is mass of the wheel and geared DC motor, g is gravity, and $l_1$and l_2 are the lengths between the center of mass of the wheel and the spring and hinge point O. Values of design parameters are m_{mo} = 0.15 kg, l_1= 33 mm, and l_2= 55 mm. As expressed in (1), the reaction force F_r can be determined by adjusting the spring force. A nominal deflection of a spring when all four wheels are on the ground is 4 mm, and each spring generates 1.16 N force from $F_{sp}= k_{sp}\delta$, where k_{sp} is a spring constant and δ is a nominal deflection. In this condition, the reaction force F_r becomes 3.4 N (≅ 0.3 kgf); that is, about 57% weight of the robot is applied to the driving wheel and 43% is applied to the ball caster wheel, where total robot weight is 1.2 kg. The weight balance between the driving wheel and the passive ball wheel can be determined by adjusting the spring constant, and the compliance of the wheel mechanism supports navigating uneven and bumped surfaces. Dali with this compliant wheel mechanism is validated while navigating indoor surfaces including wood floor, carpet, and tiles.

3. System Architecture

The basic philosophy of the design is to make the robot simple and affordable. To that end, a personal cellphone is used as the main brain of the robot, while the mobile platform is controlled by a low-cost microcontroller. The system hardware architecture of the proposed social robot is shown in Figure 5. An Arduino 101 by Intel functions as an embedded robot controller for low level motion control. The cellphone and robot controller work as a client and server model. The cellphone works as a client to request high-level motion requirements while the embedded computer works as a server and executes detailed motion control actions.

The cellphone software also processes camera images and provides the result of face detection to the embedded controller. This cooperative work allows patterned motions and visual servoing for robotic photographing.

Therefore, the robot itself autonomously moves and runs by a state machine (Figure 6). There are five different states: navigation, spin, manual, pattern, and ideal states. Each state is triggered either by sensor inputs or by manual commands. For instance, pattern state enables the robot to move in preprogrammed pattern such as linear, circular, arc, and rectangular patterns. Manual state allows a user to remote-control the robot using a cellphone app. Navigation state is the mode of potential field motion by sonar scanning and SLAM functions.

4. Localization

One of the most challenging functions of the Dali is to navigate in collision free fashion generating random motion in search of human face (or faces) to take photos. To that end, we use the potential field path planner whereby the Dali can find objects to avoid or approach. Our approach is simple in that the robot is an agent navigating in a potential field where detected environmental significance provides sources of attractive force, while previously occupied locations estimated by SLAM technique provide sources of repelling force (Figure 9). In order to install repelling sources for navigation and path planning, we use SLAM based localization technique. A seminal work in SLAM is the research of R.C. Smith and P. Cheeseman on the representation and estimation of spatial uncertainty in [11, 12]. Other pioneering works in

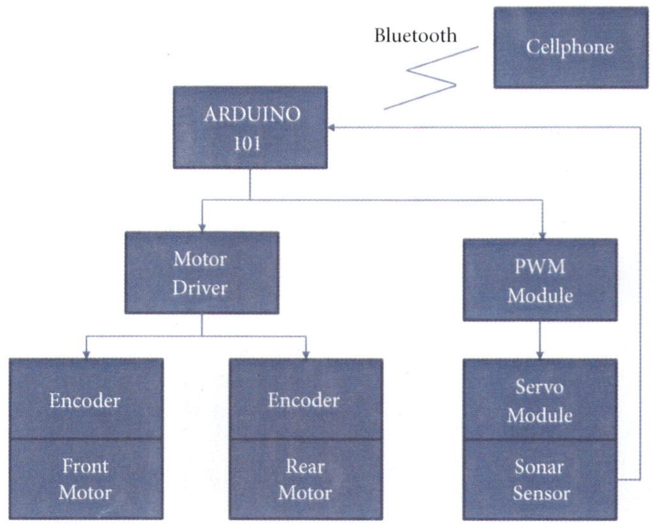

FIGURE 5: System architecture.

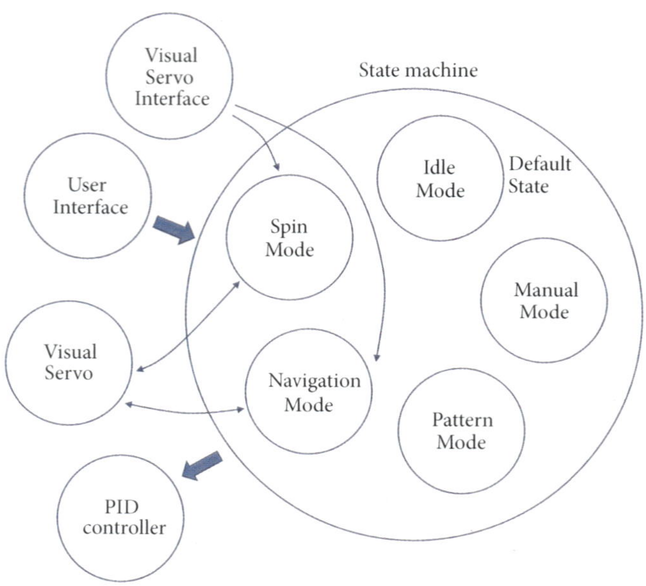

FIGURE 6: Software architecture.

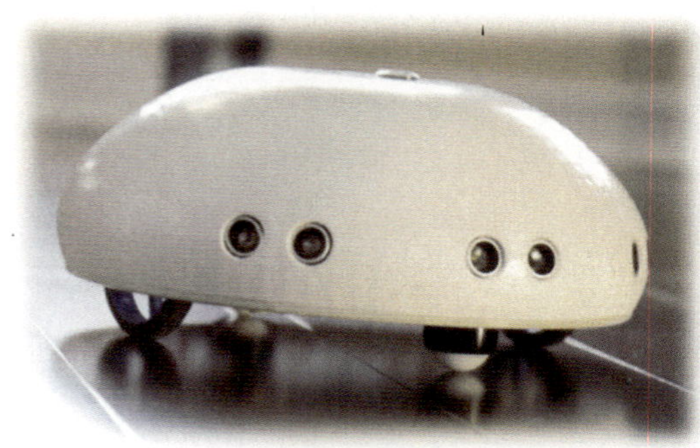

FIGURE 7: Nonholonomic robot.

this field were conducted by the research group of Hugh F. Durrant-Whyte in the early 1990s [13] by which tractable solutions to SLAM are demonstrated in the infinite data limit. We implement SLAM by using a linear system model. For the state model of a time invariant linear system such that

$$x_t = Ax_{t-1} + Bu_t + \varepsilon_t$$
$$z_t = Cx_t + \delta_t \tag{2}$$

where x is the state vector, z is the observation vector, and u is the control vector, ε and δ are Gaussian noise for the state estimation and observation process, respectively.

By using the nonholonomic platform as shown in Figure 7, for a given control input $u\ (v, w)$, where v is the translational velocity and w is the rotational velocity, the motion model or the odometry model becomes

$$\begin{pmatrix} x' \\ y' \\ \theta' \end{pmatrix} = \begin{pmatrix} x \\ y \\ \theta \end{pmatrix} + \begin{pmatrix} -\frac{v}{w}\sin\theta + \frac{v}{w}\sin(\theta + w\Delta t) \\ \frac{v}{w}\cos\theta - \frac{v}{w}\cos(\theta + w\Delta t) \\ w\Delta t \end{pmatrix} \tag{3}$$

Then SLAM becomes a state estimation problem of x. Recursive Bayes filter is the most common in state estimation. By definition, the belief of Bayes filter will be

$$bel(x_t) = p(x_t \mid z_{1:t}, u_{1:t}) \tag{4}$$

Using the conditional probability equation of $p(A \mid B) = p(B \mid A)p(A)/p(B)$,

$$\begin{aligned} bel(x_t) &= \eta \cdot p(x_t \mid z_{1:t}, u_{1:t}) \\ &= \eta \cdot p(z_t \mid x_t, z_{1:t}, u_{1:t}) \cdot \frac{p(x_t, u_{1:t})}{p(z_t, u_{1:t})} \\ &= \eta \cdot p(z_t \mid x_t, z_{1:t}, u_{1:t}) \cdot p(x_t \mid z_t, z_{1:t}, u_{1:t}) \end{aligned} \tag{5}$$

where η is a normalization factor for the probability in case the $bel(x_t)$ exceeds "1". To be more practical,

$$= \eta \cdot p(z_t \mid x_t, z_{1:t-1}, u_{1:t}) \cdot p(x_t \mid z_t, z_{1:t-1}, u_{1:t}) \quad (6)$$

It is known that a stochastic process has the Markov property if the conditional probability distribution of future states of the process (conditional on both past and present states) depends only upon the present state, not on the sequence of events that preceded it. Therefore, by applying Markov assumption of undependability on prediction of the current state observation to the past observation if the current state of the system is given, the belief will be simplified as

$$\eta \cdot p(z_t \mid x_t) \cdot p(x_t \mid z_t, z_{1:t-1}, u_{1:t}) \quad (7)$$

If we use the law of total probability, by which $P(A) = \int P(A \mid B) \cdot P(B) dB$, the above equation will be

$$= \eta \cdot p(z_t \mid x_t) \cdot \int_{x_{t-1}} p(x_t \mid x_{t-1}, z_{1:t-1}, u_{1:t}) \cdot p(x_{t-1} \mid z_t, z_{1:t-1}, u_{1:t}) \, dx_{t-1} \quad (8)$$

where $A = x_t$ and $B = x_{t-1}$. If we apply the Markov assumption again on both conditional probabilities on the right hand side of the equation, $z_{1:t-1}$ and $u_{1:t-1}$ can be ignored. Taking into consideration of u_t that cannot make any contribution to x_{t-1}, the above equation becomes

$$= \eta \cdot p(z_t \mid x_t) \cdot \int_{x_{t-1}} p(x_t \mid x_{t-1}, u_t) \cdot p(x_{t-1} \mid z_t, u_{t-1}) \, dx_{t-1} \quad (9)$$

or

$$= \eta \cdot p(z_t \mid x_t) \cdot \int_{x_{t-1}} p(x_t \mid x_{t-1}, u_t) \cdot bel(x_{t-1}) \, dx_{t-1} \quad (10)$$

Now the last equation is used as a prediction step such that,

$$\overline{bel}(x_t) = \eta \cdot p(z_t \mid x_t) \cdot \int_{x_{t-1}} p(x_t \mid x_{t-1}, u_t) \cdot bel(x_{t-1}) \, dx_{t-1} \quad (11)$$

In order to increase the likelihood of the predicted state to be more consistent to the observation by a sensor, by using (4), we propose a correction step such as

$$\begin{aligned} bel(x_t) &= \eta \cdot p(z_t \mid x_t, z_{1:t}, u_{1:t}) \cdot p(x_t \mid z_t, z_{1:t}, u_{1:t}) \\ &= \eta \cdot p(z_t \mid x_t) \cdot \overline{bel}(x_t) \end{aligned} \quad (12)$$

The probability of $p(x_t \mid x_{t-1}, u_t)$ in (11) is called a motion model, and the probability of $p(z_t \mid x_t)$ in (12) is called a sensor model. If we substitute the system state model to the motion model,

$$p(x_t \mid x_{t-1}, u_t) = \det(2\pi R_t)^{-1/2} \cdot \exp\left(-\frac{1}{2} E^T \cdot R_t^{-1} \cdot E\right) \quad (13)$$

where $E = x_t - (Ax_{t-1} + Bu_t)$ and R_t^{-1} is the Gaussian noise in state estimation. x_t here is the mean value of each state variable. In the same token, the linear observation model will become

$$p(z_t \mid x_t) = \det(2\pi Q_t)^{-1/2} \cdot \exp\left(-\frac{1}{2} F^T \cdot Q_t^{-1} \cdot F\right) \quad (14)$$

where $F = z_t - Cx_t$ and Q_t^{-1} is the Gaussian noise in measurement. z_t is again the mean value of the measurement. In state estimation, the challenge lies in minimizing the noise of R_t and Q_t for accurate state estimation. Kalman filter is known to be optimal in state estimation when the noise is in Gaussian distribution. First, we define the covariance of uncertainty, Σ_t, in the state estimation such as

$$\Sigma_t = A\Sigma_{t-1}A^T + R_t \quad (15)$$

In order to correct the system state due to the uncertainty in equation (13), we use the following Kalman gain:

$$K_t = \Sigma_t C^T \left(C\Sigma_t C^T + Q_t\right)^{-1} \quad (16)$$

Then, the system state, x_t, is corrected by the Kalman gain in (14) such that

$$x_t = Ax_{t-1} + Bu_t + K_t(z_t - Cx_{t-1}) \quad (17)$$

Equation (17) will replace the equation of E to $x_t - [Ax_{t-1} + Bu_t + K_t(z_t - Cx_{t-1})]$ in (13). The estimator extracts the most likely position from the possible location range matrix which corresponds to the highest probability from the probability matrix. The probability distribution is updated continuously during the recursive state estimation. Sonar sensor readings provide observer vectors to the Bayes filter. The state of the robot, then, is estimated for use in potential field motion planner as shown in Figure 8.

FIGURE 8: Localization by photometry.

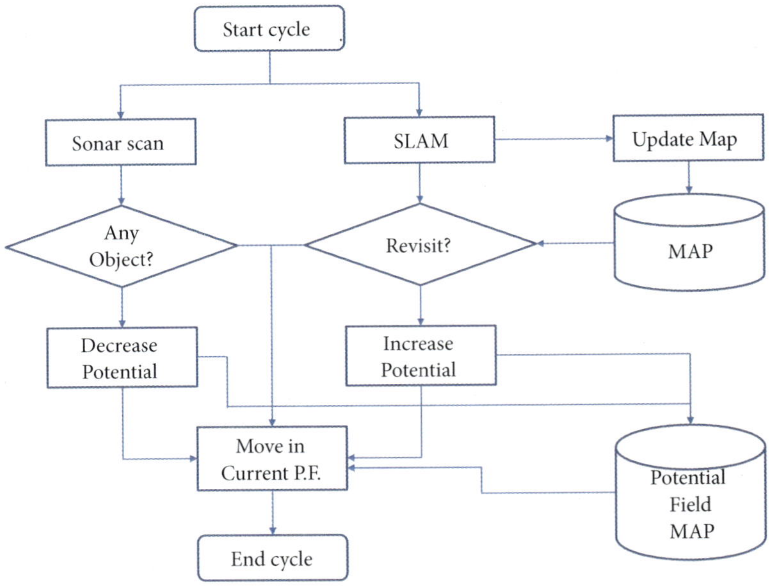

FIGURE 9: Potential field motion planning.

$U = U_{init}$ /*initialize potential field
do while $u_t > \varepsilon$
    $(x_t, \Sigma_t) = \text{SLAM}(x_{t-1}, u_t, z_t, \Sigma_t)$
    $v_t = \text{Visual\_scan}(F)$
    If $(v_t \neq 0)$
        $u_t = v_t$
    else
        $(U, n) = \text{Update\_potential}(n, u_t, x_t, z_t)$
        $u_t = -\blacktriangledown_{max} \text{U}(x_t)$
    end if
end do

ALGORITHM 1: Potential field motion planning.

$$\overline{x_t} = Ax_{t-1} + Bu_t$$
$$\overline{\Sigma_t} = A\Sigma_{t-1}A^T + R_t$$
$$K_t = \overline{\Sigma_t}C^T\left(C\overline{\Sigma_t}C^T + Q_t\right)^{-1}$$
$$x_t = \overline{x_t} + K_t(z_t - C\overline{x_t})$$
$$\Sigma_t = (I - K_tC)\overline{\Sigma_t}$$
Return x_t, Σ_t

ALGORITHM 2: SLAM$(x_{t-1}, u_t, z_t, \Sigma_t)$.

5. Motion Planning

Primary goal of the motion planning is to find environmental significance to autonomously navigate in a given environment. As mentioned earlier, we propose the potential field technique where detected environmental significance provides sources of attractive force, while previously occupied locations estimated by SLAM technique provide sources of repelling force. The original gradient is drawn from highest repelling on the left to the highest attraction on the right side. During the planning process, online attraction and repelling fields are updated depending on the conditions (Figure 9). Detail logics of each part in the planner shown in Figure 9 are in Algorithms 1–3.

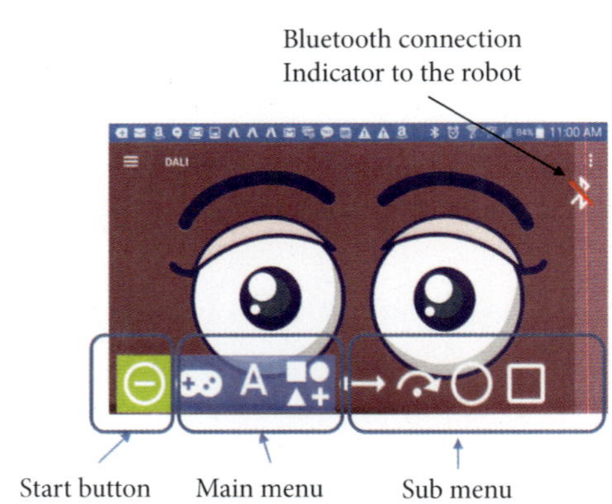

FIGURE 10: Screen shot of robotic photographer app.

6. Robotic Photography

Dali, the social robot, utilizes a cellphone as a main control unit for robotic photography. Cellphone App as shown in Figure 10 was developed for Android operating system. The app takes input from users regarding robot's required motions, detects human faces from video stream, and implements visual servoing in communication with the robot motion controller. The app provides several photographing modes (navigation, arc motion, circle motion, etc.) and animation features of eye for the events of searching and finding people.

The app uses Google Mobile Vision API [10] for face detection and tracking functions. The API is capable of detecting single or multiple faces. Therefore, the app can

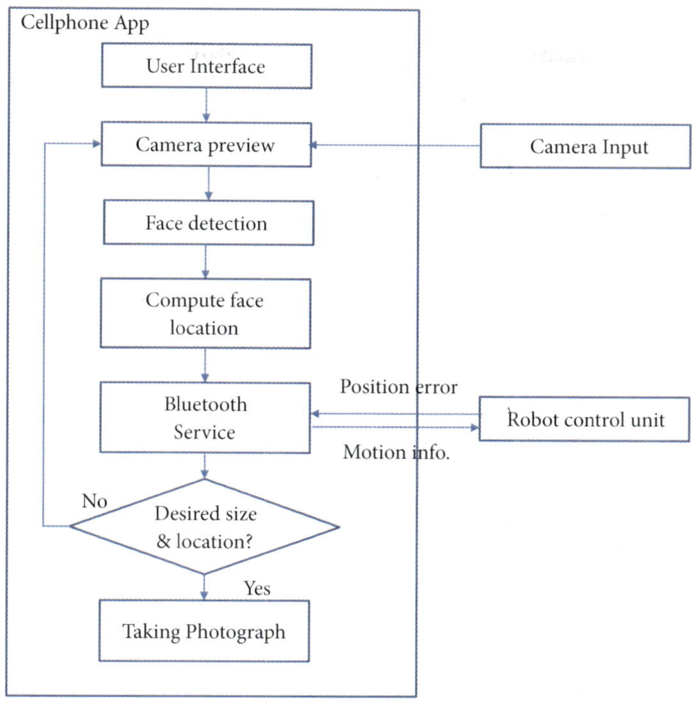

FIGURE 11: Flowchart of robotic photographing function.

/* update attractive object */
If $v_t \neq 0$
  $n = n + 1$
  $q_n = v_t$
end if
If $z_t \neq 0$
  $n = n + 1$
  $q_n = z_t$
end if
If $d(q, q_{1..n}) \leq d^*$
  $U_{att}(q) = \frac{1}{2}\zeta d^2(q, q_n)$
else
  $U^n_{att}(q) = d^*_{goal}\zeta d(q, q_n) - \frac{1}{2}\zeta d^{*2}$
end if
/* update repelling object */
$m = m + 1$
$Q^*_m = x_t$
If $D(q) \leq Q^*_m$
  $U^m_{rep}(q) = \frac{1}{2}\eta(\frac{1}{D(q)} - \frac{1}{Q^*_m})$
else
  $U^m_{rep}(q) = 0$
end if
$U(q) = \sum_{1..n} U^n_{att}(q) + \sum_{1..m} U^m_{rep}(q)$
Return $U(q)$

ALGORITHM 3: Update_potential(n, u_t, x_t, z_t).

apply different photographing strategies based on the number of people. The app allows the user to set photographing parameters on desired face size and location in photograph frame. This information is used as a basis when the robot generates desired visual servoing goals. For example, in case multiple faces are detected, the app can generate a desired input to include all of the detected faces in the camera, while requesting specific patterned trajectory. Figure 11 illustrates a flow chart of the robotic photographing algorithm in communication with robot controller.

User interface provides three control modes: manual, automatic, and patterned motion modes. Face detection is enabled in navigation and patterned motion modes. Faces are detected and tracked from any duration of appearance in video stream [10], and the face detections generate asynchronous events. In response to these events, the app processes the number of faces and sizes of each face compared with the user-predefined desired parameters. These parameters include individual face size ratio compared to camera frame, location of faces in camera coordinates, and their tolerable error ranges. Although suitable camera positions are determined by these geometrical parameters not by aesthetic point of view yet, the algorithm implements a closed loop visual servoing and photographing using asynchronous face detection events.

7. Experiments

In order to verify the performance of the proposed robot, we tested 10 times of actual photo taking tasks (Table 1). Three performance measurement parameters are identified: percent coverage of the given environment, number of photos taken, and convergence to the target point from the start point. For simplicity, percent coverage is calculated by adding the path length traveled multiplied by the sensing range divided by the actual area of the space. Number of photos

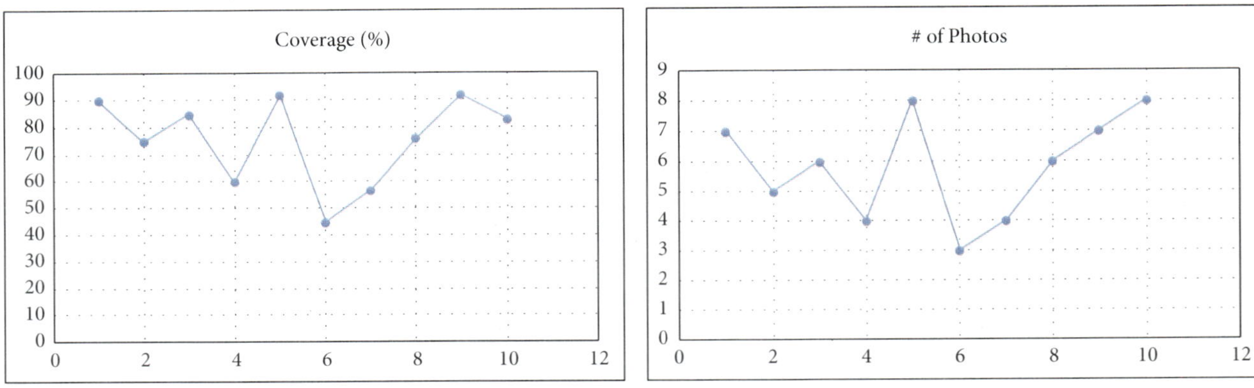

FIGURE 12: Percent coverage and photo taking results.

TABLE 1: Navigation, convergence, and photo taking experiments.

Test #	Coverage %	# of Photos	Convergence
1	90	7	yes
2	75	5	yes
3	85	6	yes
4	60	4	yes
5	92	8	yes
6	45	3	no
7	57	4	no
8	76	6	yes
9	92	7	yes
10	83	8	yes

is actual number taken by the robot during each navigation experiment. Convergence is the true or false Boolean value depending on the success or failure in reaching the goal point. Note in Figure 12 that there is a proportional trend between the percent coverage and the number of photos taken. Therefore, better coverage assures more photos in general. Two failures in convergence test was due to a local minima around a desk and chairs. While in local minima, the robot was not able to travel to the goal in a limited time (10min). Better strategies to resolve the confliction between increase and decrease of the potential will be necessary.

To investigate a social effect of the robotic photography by Dali, a user study was conducted in 3 steps. In the first step, a participant sat on a chair in a laboratory environment as shown in Figures 13 and 14. Second, Dali was set for navigation mode to move around in the space. Once the robot recognized the target participant's face, then the robot smoothly approaches to the participant until it is close enough to take a picture. After approaching, the robot automatically takes a picture of the seated participant's face. Third, the picture was uploaded to the participant's social network webpage (in this study, we used the participant's Facebook). This user study was conducted for 2 weeks, Monday through Friday, and responses were collected from the participant's online friends and family for the posted photos via the participant's Facebook. The collected responses with pros and cons of experiments are summarized in Table 2.

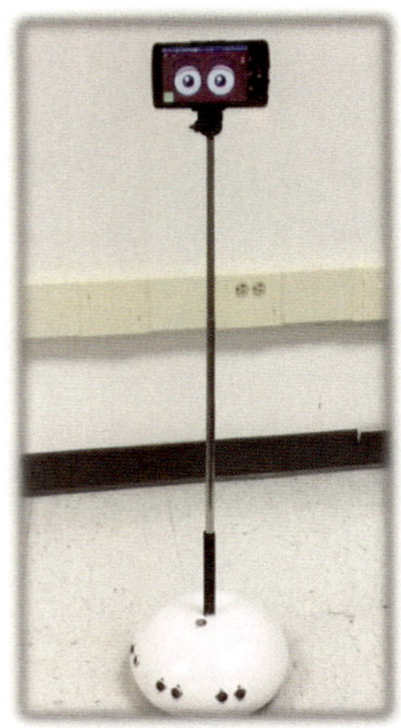

FIGURE 13: Dali the photographer.

8. Conclusion

In this paper, we proposed Dali, a social robot, whose mission is to autonomously capture images of users and feed pictures and live motion clips to social networks or to hospitals for health monitoring purposes. The robot autonomously navigates in indoor environments with collision avoidance capability. To implement full autonomy in navigation we proposed a potential field based path planning. A Kalman filter-based localization method generates a repulsive potential field, while sonar based object detection as well as visual servo based facial recognition generates an attractive potential field. We tested several scenarios with Dali including a user study conducted to evaluate the usefulness of the social robot. As a result of the user study, we found that the robotic photography has been very beneficial, especially for those who are apart from their family or friends. Some of the positive feedback from the participant's parent stated that they enjoyed being able to see their child's activities daily

TABLE 2: Collected pros and cons of robotic photography in terms of its social effect.

Pros
(i) Documentation of a daily life is much easier
(ii) People could see how their image develops. Also, people could see the difference in how they look with different routines. Such as they may look full of color one day because they ate a nutritious breakfast that morning, while another day they could look pale because they skipped breakfast.
(iii) By watching a person day by day, we can study the aging effect on users and their daily life's and how it affects aging.
(iv) Automatic photographing and posting on social network allows users concentrate on their work without worrying too much about reminding themselves to do social networking.
(v) You see yourself evolve and share with everyone
(vi) A memory in time is documented. Also, updating of people's daily life, especially about what you're doing each day help family in a close loop.
(vii) You can have your pictures of you instead of having someone take it for you
Cons
(i) Privacy issue due to some unwanted photographing and sharing.
(ii) If the picture is unexpected then the picture quality is not always the best.
(iii) You may not want pictures to be taken.
(iv) You have to be ready for the picture when the robot beeps for a good quality photographing, unless you want to share natural pose of you.

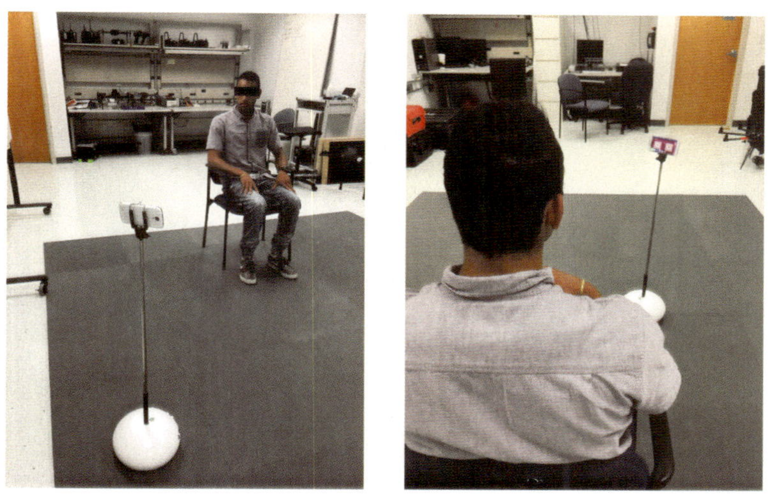

FIGURE 14: Experiment of the robotic photography using Dali with face recognition feature.

basis in far distance. It may result from the fact that real time up-to-date information in social media helps them feel more comfort, staying close, and intimate compared to the case of watching occasionally updated photos. The participant mentioned that the robotic photography made it easy for the user to keep up with social media, so that it is beneficial for a user who does not have time to manage their social media daily basis. It is addressed that autocharging capability seems to be a must-have feature to automatically activate the robot every day. In addition, conversation or Q/A function based on voice recognition and AI such as Amazon Echo will much enhance the experience of the social robot in daily life.

Conflicts of Interest

The authors declare that they have no conflicts of interest.

References

[1] L. Rainie, K. Purcell, and A. Smith, "The social side of the internet," Tech. Rep., PewInternet & American Life Project, 2011.

[2] J. Fasola and M. J. Mataric, "Evaluation of a spatial language interpretation framework for natural human-robot interaction with older adults," in *Proceedings of the 2015 24th IEEE International Symposium on Robot and Human Interactive Communication (RO-MAN)*, pp. 301–308, Kobe, Japan, August 2015.

[3] J. Fasola and M. J. Matarić, "Using socially assistive human-robot interaction to motivate physical exercise for older adults," *Proceedings of the IEEE*, vol. 100, no. 8, pp. 2512–2526, 2012.

[4] J. Fasola and M. J. Matarić, "Socially assistive robot exercise coach: motivating older adults to engage in physical exercise," in *Experimental Robotics*, vol. 88 of *Springer Tracts in Advanced Robotics*, pp. 463–479, Springer International Publishing, Heidelberg, Germany, 2013.

[5] A. Tapus and M. J. Matarić, "Socially assistive robotic music therapist for maintaining attention of older adults with cognitive impairments," in *Proceedings of the 2008 AAAI Fall Symposium*, pp. 124–127, Arlington, Va, USA, November 2008.

[6] A. Perrin, "Social networking usage: 2005-2015," Tech. Rep., Pew Research Center, 2015, http://www.pewinternet.org/2015/10/08/2015/Social-Networking-Usage-2005-2015/.

[7] M. Zabarauskas and S. Cameron, "Luke: An autonomous robot photographer," in *Proceedings of the 2014 IEEE International Conference on Robotics and Automation, ICRA 2014*, pp. 1809–1815, Hong Kong, China, June 2014.

[8] G.-S. J. Hsu and J.-W. Huang, "A photographer robot with multiview face detector," in *Proceedings of the IEEE International Conference on Industrial Technology, ICIT 2016*, pp. 2152–2156, Taipei, Taiwan, March 2016.

[9] G. Fu and X. Zhang, "ROSBOT: A low-cost autonomous social robot," in *Proceedings of the IEEE/ASME International Conference on Advanced Intelligent Mechatronics, AIM 2015*, pp. 1789–1794, Busan, Republic of Korea, July 2015.

[10] R. G. Valenti, Y. Jian, K. Ni, and J. Xiao, "An autonomous flyer photographer," in *Proceedings of the 2016 IEEE International Conference on Cyber Technology in Automation, Control, and Intelligent Systems (CYBER)*, pp. 273–278, Chengdu, China, June 2016.

[11] R. C. Smith and P. Cheeseman, "On the representation and estimation of spatial uncertainty," *International Journal of Robotics Research*, vol. 5, no. 4, pp. 56–68, 1986.

[12] R. C. Smith, M. Self, and P. Cheeseman, "Estimating uncertain spatial relationships in robotics," in *Proceedings of the 2nd Annual Conference on Uncertainty in Artificial Intelligence, UAI '86*, pp. 435–461, Raleigh, NC, USA, 1986.

[13] J. Leonard and H. Durrant-Whyte, "Simultaneous map building and localization for an autonomous mobile robot," in *Proceedings of the IEEE/RSJ International Workshop on Intelligent Robots and Systems (IROS '91)*, pp. 1442–1447, Osaka, Japan, November 1991.

A Control Architecture of Robot-Assisted Intervention for Children with Autism Spectrum Disorders

Yongli Feng, Qingxuan Jia, and Wei Wei

School of Automation, Beijing University of Posts and Telecommunications, No. 10 Xitucheng Road, Haidian District, Beijing 100876, China

Correspondence should be addressed to Yongli Feng; fengyongli@bupt.edu.cn

Academic Editor: Ali Meghdari

Robot-assisted intervention has been successfully applied to the education and training of children with autism spectrum disorders. However, it is necessary to increase the autonomy of the robot to reduce the burden on the human therapists. This paper focuses on proposing a robotic architecture to improve the autonomy of the robot in the course of the interaction between the robot and the child with autism. Following the model of perception-cognition-action, the architecture also incorporates some of the concepts of traditional autism intervention approach and the human cognitive model. The details of the robotic architecture are described in this paper, and in the end, a typical scenario is used to verify the proposed method.

1. Introduction

Autism, or Autism Spectrum Disorder (ASD), describes a broad range of developmental disorders whose main symptoms include impairment in social communication and restricted or repetitive patterns of behavior, interests, or activities [1]. Previous studies reveal that behavioral intervention is one of the major approaches to stimulate children with ASD to reduce their symptoms and improve their abilities of social interaction [2]. In the field of ASD intervention, some of the major challenges are to maintain motivation of the children, while long-term and repetitive therapies require therapists to spend a lot of time and energy. To address these challenges, the novel techniques and devices have been studied to ensure effective intervention of ASD while reducing the workload of therapeutic professionals [3, 4].

In these technologies, Socially Assistive Robotics (SAR) is considered as support tool for autism therapy through social interaction [5]. In the current studies, these robots have shown high efficiency in the intervention in autism, and the social skills of children with ASD have been positively improved [6, 7]. The approach of robot intervention for ASD is expected to obtain a better assist to traditional intervention procedure. In many systems, the robot is controlled remotely by the operator to make appropriate responses. This control method limits the long-term, large-scale use of robots in ASD interventions. In the field of ASD intervention, robots should be positioned as support tools for the human therapists rather than as alternatives. During the intervention, robots require collaborating with people who lack knowledge of ASD intervention or robotic. In order to achieve this goal, it is a major trend that to develop autonomous robots to interact with children with ASD in the course of intervention sessions. In the field of ASD intervention, the robot could play the role of human therapists, and the function of the robot is to stimulate and encourage the social skills of children with ASD (as show in Figure 1).

In the unstructured environment, it is a complex process to realize autonomous interaction between the robot and the child. In order to improve the autonomy of the robot in the intervention in autism therapy, the appropriate control architecture design is an indispensable part. In particular, a well-designed robotic architecture is an essential part to adjust the robot's action by interpreting interaction signals reliably [8]. The main purpose of this paper is to present a control architecture that enables the robot to intervene in autism therapy with higher autonomy. The remainder of this paper is organized as follows: Section 2 provides an

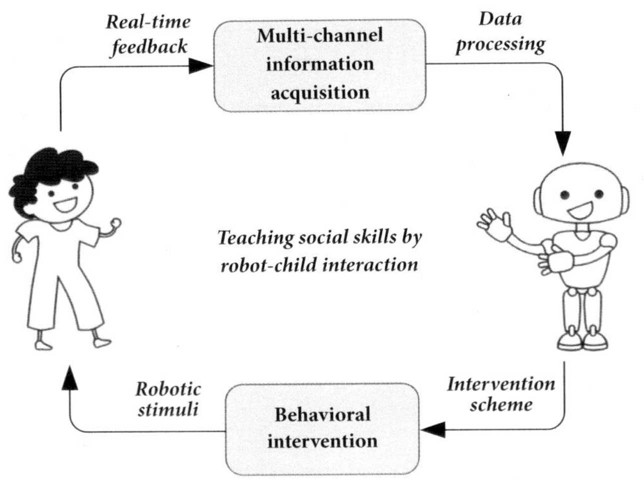

FIGURE 1: A basic overview of the robot application for autism intervention.

overview of related work; Section 3 presents the detail of the control architecture of the robot in each part and explains the intervention process of the robot for autism therapy; Section 4 describes the experiment to validate proposed architecture; in Section 5, we conclude our study and present future work.

2. Related and Previous Work

At present, the pathogenesis of autism is not clear and the early intervention is an effective means to relieve symptoms of children with ASD to promote their skills. In this section, we will introduce the traditional methods of intervention and the application of robots for autism therapy.

2.1. Traditional Methods of Autism Intervention. With years of development, researchers have presented various methods and models to help children with ASD, for example, Denver Model, Comprehensive Programs, TECCH Program, Applied Behavioral Analysis, and DIR/Floortime. Among these methods, Applied Behavioral Analysis (ABA) and DIR/Floortime are widely used and accepted. In the following sections, we will introduce the two methods, respectively.

(i) Applied Behavioral Analysis for Autism. Applied Behavior Analysis (ABA) which is based on the principles of learning and motivation is a general term for behavior modification [9]. ABA therapy is usually defined as a systematic approach to improve people's social behavior based on the mixture of psychological and educational techniques. Over the past few decades, the intervention methods based on the ABA principle have been applied to the behavior and learning of children with special needs. Particularly, in the intervention of children with ASD, ABA therapy is one of the most effective means of intervention for children with ASD [10].

In the course of intervention of autism, Discrete Trial Teaching (DTT) is a concrete implementation method based on principle of ABA therapy [11]. DTT was developed by psychologists Ivar Lovaas in the 1970s and its core elements include instruction, individual response, auxiliary, reinforcement, and pause [9]. In practical execution, DTT includes the following main contents. Firstly, the target task is decomposed into a series of smaller or mutually independent steps in a certain manner and sequence. Subsequently, each small step is trained and educated by the appropriate reinforcement method. Through this process, children with autism can master all the steps and fulfill the task independently. The ultimate goal of DTT teaching for children with ASD is to allow them to apply the knowledge and skills in other situations. DTT is a highly structured teaching model and this type of teaching is generally one-on-one intervention of children with ASD. The basic process of DTT is shown in Figure 2.

(ii) The DIR/Floortime. The Developmental, Individual-difference, and Relationship-based (DIR) mode is an outstanding developmental approach for autism intervention. The mode of DIR was developed by Stanley Greenspan and its core is Floortime [12].

Different from the ABA method, DIR/Floortime emphasizes children's emotional experience, imagination training, and interpersonal interaction. According to children with ASD's characteristics and stages of development, the ladder of capability development is set up which consists of six milestones. These milestones include (1) attention and interest, (2) engagement and intimate relationship, (3) two-way communication, (4) continuous problem solving, (5) creative thinking, and (6) abstract and logical thinking [13]. Based on specific developmental stages of children with ASD, adopting the different strategies will achieve better effect of intervention. For children with autism, the time and place of DIR/Floortime intervention are flexible, and intervener can use this method in drawing rooms, bedrooms, schools, or training institutions at any time [14]. The intervention process of DIR/Floortime for autism can be summarized as shown in Figure 3.

2.2. Application of Robots in ASD Intervention. Researchers have done a lot of studies on the use of robots in the intervention of children with autism. The results of these studies demonstrate that it is an effective way to use the robot in autism intervention.

In the course of intervention of children with autism, robots play a role of complementary therapy. In previous studies, researchers always use robots to assist human therapists for mediated interaction in tasks of action imitation, joint attention, turn talking, and so on [15]. These tasks can improve the social skills of children with ASD and help them integrate into society. Nevertheless, in the current approaches, almost all robots which are usually controlled by human operators remotely lack autonomy. In this mode, the operators need specialized knowledge of autism intervention and robots operation which is unsustainable for use of robots in long-term and large scale. Therefore, it is necessary to make robots have a certain degree of autonomy. In order to achieve this goal, the research of suitable control architecture of robots for intervention of children with autism requires developing preferentially.

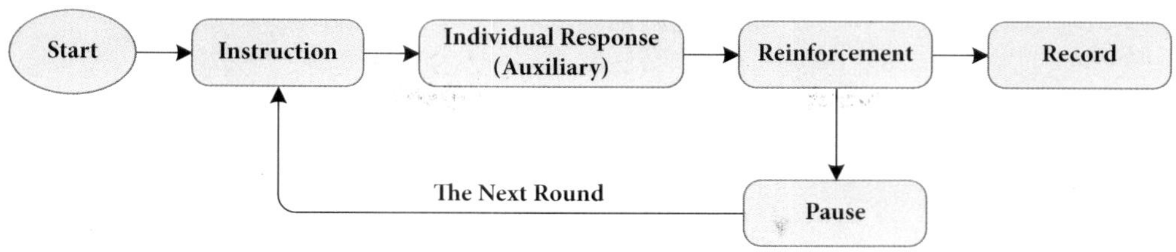

FIGURE 2: The basic process of DTT for children with ASD.

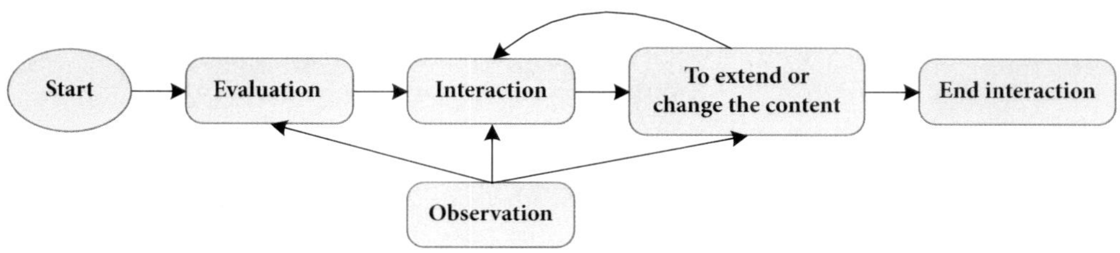

FIGURE 3: The basic process of DIR/Floortime for children with ASD.

At present, the researches of robots' control architectures for autism intervention mainly focus on the control of robots, and rarely take into account the traditional methods of intervention. Feil-Seifer and Mataric [16] proposed a kind of behavior-based control architecture which is named B^3IA. The B^3IA is composed of sensor and interpretation module, activity history module, behavior network, and effector module. In the task of playing video games, Wainer et al. [17] designed the control architecture of detection-planning-action, which makes the robot complete the task with a certain degree of autonomy. Sang-Seok et al. [18] developed control architecture based on four modules including human perception, interaction manager, user input, and robot effector. In their study, the DTT protocol was used. In addition, Gonzalez et al. [19] proposed a three layer planning architecture to carry out a rehabilitation therapy for patients with physical impairments. Pour A G et al. [20] proposed human-robot facial expression reciprocal interaction platform. The reciprocal interaction is composed of two main phases: nonstructured and structured interaction modes. In this platform, the imitation task of facial expression between the robot and the children with autism is realized. In their study, the traditional rehabilitation procedure was considered, which gives us some inspiration.

The intervention of autism therapy has its particularity. It is a promising research direction to integrate the ideas of traditional approaches of intervention into the control architecture design of robot. In this paper, we carried out the exploratory study.

3. The Design of Control Architecture for the Robot in Autism Intervention

3.1. The Feasible Procedure of the Robot for Autism Intervention. In autism intervention, it is necessary to improve the autonomy of the robot which can reduce the workload of the human (e.g., therapists, educators, and parents). Furthermore, people who lack professional knowledge of autism intervention or robotics can use robots to achieve their aims. In this way, the scope of use of the robot can be expanded, and the robot can play better roles for autism intervention. In order to achieve this goal, it is necessary to design the control architecture for robot based on traditional methods of autism intervention.

As mentioned in the preceding section, ABA and DIR/Floortime are two traditional methods for autism's early intervention commonly. DTT is the specific method of ABA program in autism therapy. This method has the advantages of structured process and easy to apply and has been used by robots in some studies for autism intervention [21]. Although DTT is very effective in the early intervention for children with ASD, quite a few scholars criticize that this method is hidebound and mechanical and is not conducive to the development of children with ASD [13]. Compared to DTT, DIR/Floortime attaches great importance to the developments, individual differences, and relationships of autism children. It advocates the intervention into the child's daily life, according to the child's development stages. Nevertheless, for the robot in autism intervention, this method requires the robot has high-level ability of natural interaction, but the existing technology cannot support that.

Therefore, as shown in Figure 4, we proposed a feasible procedure of the robot for autism intervention to improve the autonomy of the robot.

In the course of autism intervention, the robot awakes children's interest by dancing, singing, dialogue, and so on. In this way, children can focus their attention on the robot which is the basis for efficient interaction. Simultaneously, the robot collects the environmental information and evaluates the behavior of children with ASD. The content of assessment includes engagement, the stage of development of child which refers to the method of DIR/Floortime, requirement analysis of the child, and the evaluation of task feedback. And afterward, the robot assign tasks based on the results of evaluation and the tasks include DTT teaching and other ones (e.g., storytelling, singing, and dancing). Following this,

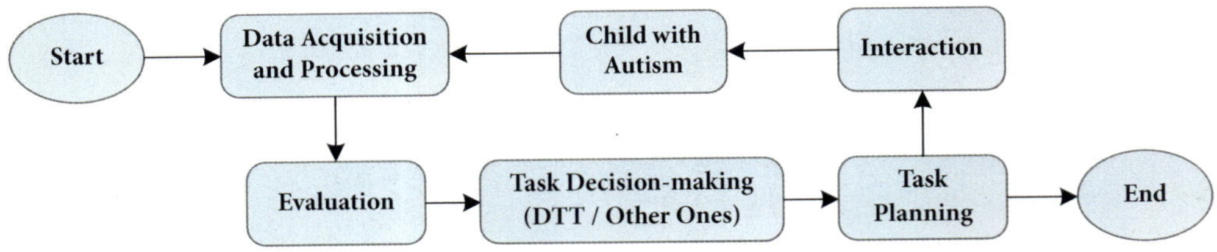

FIGURE 4: The flowchart of the robot intervention for children with autism.

the robot decomposes the task which has been decided and planning to perform. Finally, the robot interacts with the child or else ends the interaction.

In the process of interaction between robots and children with ASD, robots play the role of human therapists. Therefore, the cognitive mechanism of human should be considered in the design of robotic architecture. Among cognitive architectures, ACT-R (adaptive character of thought rational) has been studied for a long time and has good evaluation. ACT-R system is a hybrid cognitive architecture, which consists of two parts: symbolic system and subsymbolic system. ACT-R is a general theory of cognition and provides a framework for information processing [22]. The system of ACT-R is composed of several different modules. The production module connects other modules into a whole through buffers. In this framework, the symbolic system is driven by production system and the buffers of different modules are operated through the production rules. Besides, the subsymbolic system runs in the external structure of the ACT-R, and it controls the operation of symbolic system through a series of mathematical methods. One of the important features of ACT-R which are different from other similar theories is to apply a large number of experimental information directly to the research work. Therefore, some ideas of ACT-R were used in our study.

In the next section, we will propose the control architecture of the robot for autism intervention based on ACT-R and the flowchart which we have presented in this section.

3.2. The Control Architecture of the Robot for Autism Intervention. In order to improve the autonomy of the robot in the intervention of children with autism, we designed the control architecture of the robot for autism intervention (CARAI), which is shown in Figure 5. Following the perception-cognition-action model, the CARAI comprises several modules and submodules with special functions which are represented by rectangles. In the architecture, the arrows express the direction of information transmission and the dependency between modules. How each module works is described as follows.

3.2.1. Perception Module. The perception module is an interface that the robot obtains environment information and maps the information to the internal representation. In our study, the perception module is divided into two submodules, data acquisition module and data processing module. Among them, in the data acquisition module, the robot collects the data through sensors and transfers them to the data processing module. Besides, the data processing task is carried out in three steps. Firstly, according to the needs of the interaction task, the data processing module obtains the instruction from upper process of CARAI through production rules; then, the data processing module transfers the focus of the attention to the corresponding position according to the instruction, obtains the detailed information of the object, and carries on the data processing; finally, the data processing module transmits the results to the high-level module. Some general algorithms are applied to this module. For example, the robot locates the child with ASD through HOG-Linear SVM and detects the face of the child based on Haar feature. The library based on Deep Neural Network (DNN) is used for natural language processing.

3.2.2. Intention Module. Goal making is a very important process in human-robot interaction. Because the ability of the robot is always limited, the robot should revolve around its purpose when accepting input and processing and making output. In the CARAI, we use intention module to store the goals and aim to generate therapy plans for children with autism. In addition, the goal buffer is used as an interface to implement the interaction between the intention module and the robot core processing process. In the goal buffer, the goal can be created, temporarily stored, and modified. When the robot interacts with the child, a goal can be decomposed into subgoals, which are managed in a stack way.

As shown in Figure 6, the ultimate goal for the robot to achieve is G_0, which is the initial state of the goal stack. However, in order to achieve the goal G_0, some subgoals $G_1 \sim G_m$ need to be completed. Therefore, these subgoals are pushed into goal stack. In this way, the goal G_m is at the top of the goal stack and is popped to execute firstly. During the robot's execution of the current goal, the new goal may be created. At this point, the new goal will be pushed onto the top of the goal stack. This process is looped until the ultimate goal is completed.

3.2.3. Memory Module. The function and running rule of memory module are similar to the declarative module in ACT-R. In this module, the knowledge of intervention therapy for autism is stored. The knowledge, which is converted into production rules, comes from the experienced therapists and is filtrated according to the characteristics of the robot intervention. During the intervention, the central processing system of the CARAI retrieves the rules from memory module in real time and updates the knowledge (e.g., modify, remove, or add rule). In addition, the information of users (e.g., names, genders, and symptoms), robot's behavior set, and task information are also stored in this module.

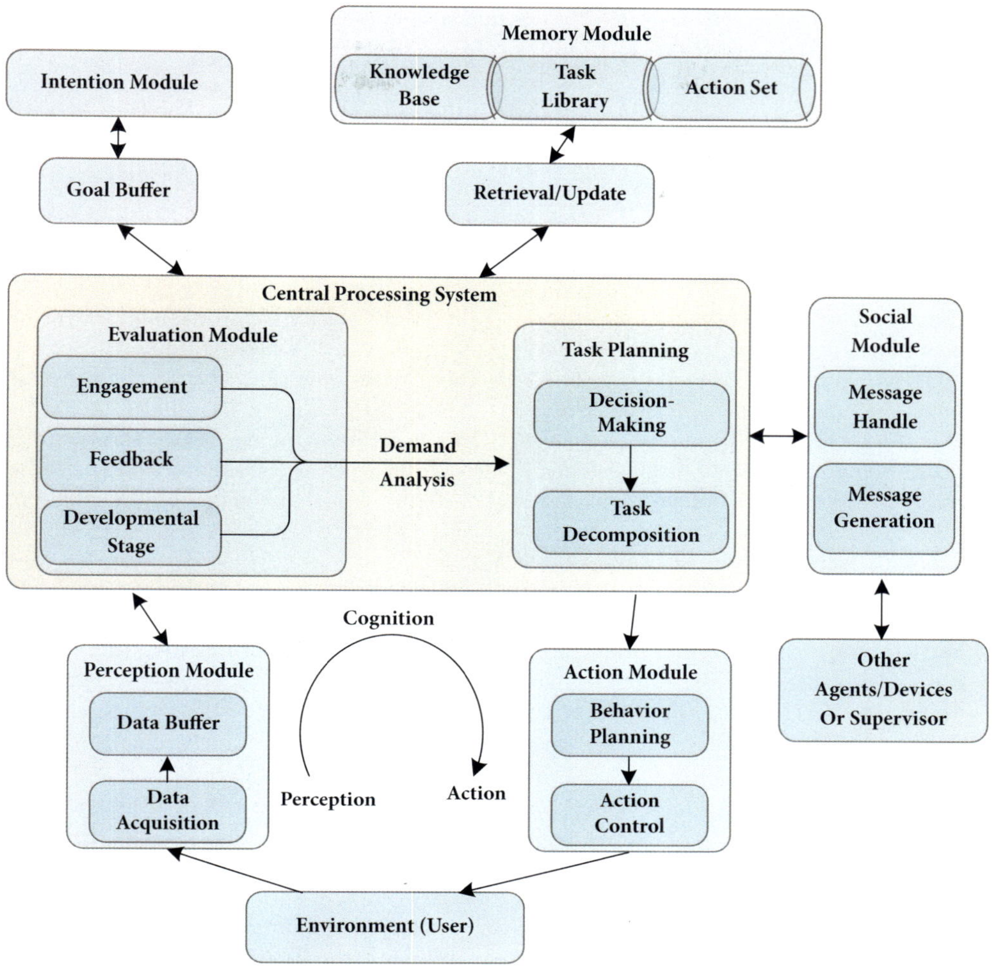

FIGURE 5: The control architecture of the robot for autism intervention (CARAI).

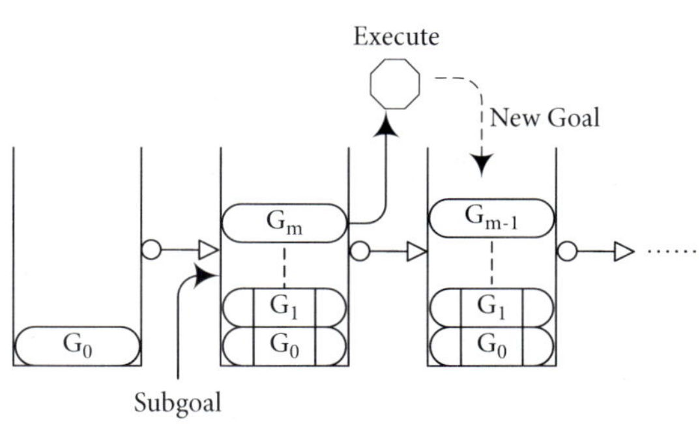

FIGURE 6: The schematic diagram of goal management.

Production rules can generally be expressed as "$P \leftarrow Q$". In this expression, Q represents a set of preconditions; P denotes several conclusions or actions. Its meaning is "if the premise Q is satisfied, the conclusion P (or action P should be performed) can be introduced". The knowledge represented by a production rule is an ordered set of productions. The syntax can be represented by (Backus Normal Form) as follows: ⟨symbol⟩::= ⟨name of rule⟩: IF "Q" THEN "P". In this study, two of the robot production rules are shown as follows:

⟨Rule 1⟩::=⟨*visual_location*⟩: IF *visual_goal* | *visual_state* | *location_state* THEN *move-attention*;

⟨Rule 2⟩::=⟨*action_leftarm*⟩: IF "*action_goal* | *action_state* | *leftarm_state*" THEN "*joint_angle_1* | *joint_angle_2* | *joint_angle_3* | *joint_angle_4*";

3.2.4. Evaluation Module. The evaluation module appraisals the state or situational context using data received from perception module. The evaluation module appraisals the state of the child based on several features and variables. In the CARAI, we used a parallel mechanism to evaluate the user's state. The evaluation module contains three submodules: engagement recognition, feedback appraisal, and assessment of developmental stage. Among them, the engagement recognition module mainly assesses the user's degree of participation in the interaction task currently; the feedback appraisal module is used to evaluate the completion of the interaction object for the current task; in the development stage module, the development of children with ASD is divided into several levels, and the level of the

TABLE 1: The description of the task.

Event	Content		
The difficulty of the task	Level 1	Level 2	Level 3
Task overview	Imitates a single arm of the robot	Imitates both arms of the robot at the same time	Ask child to imitate and answer questions
Score	1	3	5

user is determined by his performance in the human-robot interaction. The evaluation results of these modules will serve as the basis for goal setting and task planning of the system.

In the evaluation module, the engagement evaluation is the important step for using the robot to interact with the children with ASD. The evaluation results of the engagement can reflect the degree of acceptance of intervention tasks by children with ASD. In order to classify the degree of participation of the child, we developed an engagement evaluation model based on dynamic Bayesian networks and domain experts' knowledge. The inputs (evidence variables) of the model are child's features which are face orientation, interpersonal distance, and acoustic state. The outputs (query variables) of the model are the state of the child when he/she interacts with the robot. The expression of engagement evaluation is shown as

$$\begin{aligned} \boldsymbol{P}\left(S^t \mid \boldsymbol{e}^t\right) &= k_1 \boldsymbol{P}\left(S^{t-1} \mid \boldsymbol{e}^{t-1}\right) \boldsymbol{P}\left(S^t \mid S^{t-1}\right) \\ &\quad \circ \sum_{i=1}^{3} \left(P\left(H_i^t \mid S^t\right) \boldsymbol{P}\left(H_i^t \mid \boldsymbol{e}^t\right)\right) \\ \boldsymbol{P}\left(H_i^t \mid \boldsymbol{e}^t\right) &= k_2 \boldsymbol{P}\left(H_i^{t-1} \mid \boldsymbol{e}^{t-1}\right) \boldsymbol{P}\left(H_i^t \mid H_i^{t-1}\right) \\ &\quad \circ \left(\boldsymbol{P}\left(e_{i1}^t \mid H_i^t\right) \circ \boldsymbol{P}\left(e_{i2}^t \mid H_i^t\right)\right) \\ & \qquad i = 1, 2, 3 \end{aligned} \tag{1}$$

where k_1 and k_2 are normalization factors. Under the influence of the evidence variables $\boldsymbol{e}^t$, $\boldsymbol{P}(S^t \mid \boldsymbol{e}^t)$ denotes the probability value set of query variables, and $\boldsymbol{P}(H_i^t \mid \boldsymbol{e}^t)$ represents the calculated values of hidden variables in time t. Transition relations between two adjacent time slices are represented by 3-order confused matrices, which are $\boldsymbol{P}(S^t \mid S^{t-1})$ and $\boldsymbol{P}(H_i^t \mid H_i^{t-1})$. The symbol "$\circ$" indicates the dot product between matrices. Detailed content of engagement evaluation can be found in [23].

The feedback evaluation can measure whether the child's response to the robot is correct. For imitation task, we use "PyOpenPose" to evaluate the posture of the child. PyOpenPose is an implementation of OpenPose (an algorithm for human posture recognition) based on Python. In the application, the key points of the human body are first detected. Then, based on the coordinates of these points, a threshold is set to evaluate the child's feedback.

The evaluation result of developmental stage is the important basis for robot decision-making. In our application, we divided the development of children with ASD into three levels which include focused attention, two-way communication, and continuous interaction. During the interaction, the robot asks the child to complete the tasks with different difficulty which is assigned different score. For example, the robot raises a hand to ask the child to imitate, or the robot raises a hand, asking the child to answer which hand it is. The description of the task is shown in Table 1. When the children with ASD complete the tasks, the robot evaluates their developmental stages by the addition and subtraction of the score.

3.2.5. Task Planning Module. The process of task planning is divided into two stages: decision-making and task decomposition. In decision-making, the system fully integrates information which is from evaluation module, memory module, intention module, and social module. The learning mechanism and selection mechanism are also formulated in this step. In the task decomposition, the target task which is set in the previous step is decomposed into simple primitive tasks based on the robot's ability. In this way, these primitive tasks can be performed through the robot's actuators directly.

During the interaction between robots and children with ASD, the role of the supervisor is indispensable [24]. Therefore, we considered the role of the supervisor in the decision-making of the robot and achieved the purpose through the method of interactive reinforcement learning which can be defined by

$$a_t \longleftarrow \underset{a \in \mathbf{A}(s)}{\arg\max} \left((1-\beta) \times Q(s,a) + \beta \times E(s,a)\right) \tag{2}$$

where a_t is the result of decision which expresses the robot's action. $\boldsymbol{A}(s)$ represents the action set in state s. $Q(s,a)$ is the value of Q-learning and $E(s,a)$ is the result of supervisor's reinforcement. The symbol β represents the confidence level for supervisor's reinforcement. $Q(s, a)$ can be obtained by (3) which is the expression of Q-learning and the $E(s,a)$ can be received through prior agreement.

$$\begin{aligned} Q(s,a) &\longleftarrow Q(s,a) \\ &\quad + \alpha \left(r + \gamma \max_{a'} Q\left(s', a'\right) - Q(s,a)\right) \end{aligned} \tag{3}$$

where $0 \leq \alpha \leq 1$ is the learning rate, which defines the extent of new information covering old information. $0 \leq \gamma \leq 1$ is the discount factor and reflects the importance of future rewards in the learning process. r represents the reward value of executing action a at the state s. s' and a' denote the state of the next step and the action to be performed. The detailed derivation process of the algorithm will be discussed in another paper.

In the CARAI, the task decomposition is achieved through hierarchical task network. The simple primitive tasks are comprised of the basic tasks that the robots can accomplish, like the joints of robot rotating to the specific angles, robot voice output, setting the LED colour of the robot eyes, and so on.

3.2.6. Action Module. When the task planning is finished, the CARAI translates the primitive tasks into the robot's behavior through the action module. First, the action sequence of the robot is formulated by the behavior planning module, and then the movements of the actuators are controlled by the action control module.

3.2.7. Social Module. In the therapeutic task, we expect that the robot could infer and understand the intentions of the child and take appropriate behavior to meet the child's individual needs. Usually, the capabilities of a robot are limited. In intervention schemes, sometimes, the robot needs to use other agents (e.g., glowing ball which can be controlled by programming) or sensors to achieve the intervention goals. Besides, it is unrealistic that the robot is fully automated which means the robot can adapt to any event during the intervention in unstructured environments. Therefore, in our study, while the robot interacts autonomously with the child, it needs to accept and prioritize the special instruction which given by the supervisor (the therapist, the teacher, or parents). In other words, when the robot's behavior does not correspond to the interactive content, the supervisor should be able to command the robot to adjust its behavior via special instruction. In order to achieve the above targets, we design the social module in the CARAI. The social module includes two submodules: one module is the message handling and the other is message generation. Through the social module, the robot can interact with other agents, devices or supervisors for information and data.

3.3. Intervention Process of the Robot. In this section, we will describe the robot intervention process based on the robotic architecture which is proposed in the previous section.

In the robot-assisted behavioral intervention for children with ASD, the overall process can be divided into two steps. As shown in Figure 7, the first step is to develop intervention plan and the second step is to execute the plan.

The intervention procedure starts with the individual information of the interactive object. Based on the evaluation result, the robot develops an initial scheme. This process is accomplished by the intention module of robotic architecture. The robot updates the intervention plan according to the environmental information along with the interaction.

The execution of the intervention plan includes three parts: real-time evaluation, decision support, and robot execute. Among them, the real-time evaluation is realized by evaluation module of the robotic architecture. The decision support function of the robot is achieved by task planning module of the robotic architecture. The function of robot execution belongs to the basic planning of robots which can be realized by action module of robotic architecture. At this stage, the robot converts the decision result into the sequence of instructions which can be executed directly, including the angle of the robot joints, voice, and LEDs control.

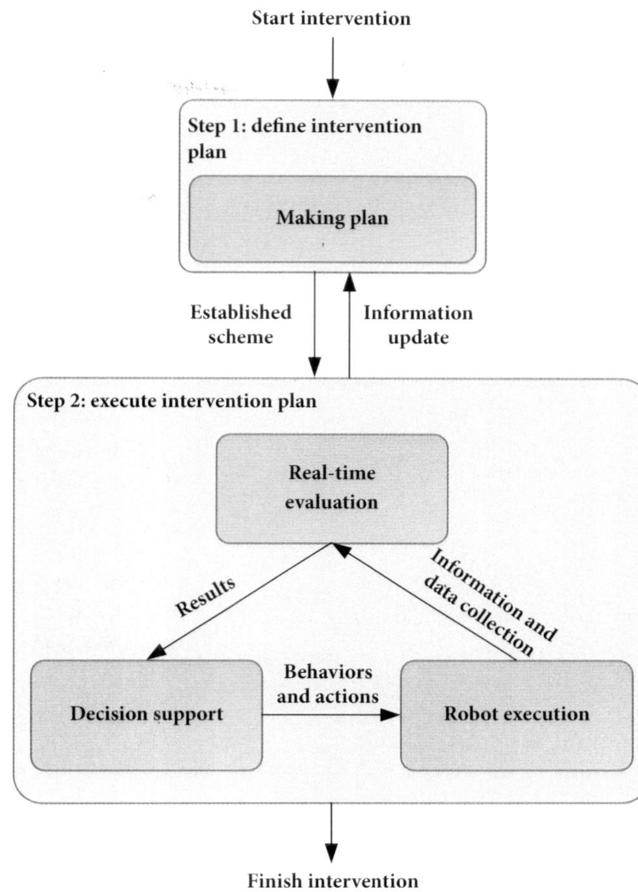

FIGURE 7: Steps of robot-assist behavioral intervention.

4. Experimental Scenarios

4.1. The Robot Platform. We prepared the robot platform NAO to support our study, as shown in Figure 8. The NAO robot is a humanoid robot, it has an appealing appearance and is easy to be accepted by children. In previous studies, the NAO robot has been successfully applied to training tasks for children with ASD and achieved ideal results [25–28]. With the height of 574 mm and the weight of 5.4 kg, the NAO robot integrates various sensors (i.e., video cameras, microphones, inertial unit, tactile sensors, and joint position sensors) and actuators (i.e., loudspeakers, joint motors, and LEDs). Therefore, the NAO robot can collect environmental information and interact with people conveniently. The supported programming languages of the NAO robot include Python, C++, Java, and JavaScript. Based on this, we can be flexible for program development. In the experiment, we developed the application based on Python SDK of the NAO robot.

4.2. Interaction Session Design. In the study, we designed a scenario where the robot guided the child with ASD to imitate

Table 2: Overview and explanation of the basic event.

Phase	Basic event	Explanation
Initialization	Detect child	The robot detects the child and adjusts its position and poses for the child
	Identify child	The robot identifies the child and loads the data, including name, gender, stage of development, preference, etc.
	Greet child	The robot greets the child with body movement and speech
Arousing child's interest	Play audio	The robot plays music or other sounds, such as birds singing
	Dancing	The robot performs the dance
	Dialogue	The robot tries to talk to the child
	Body movement	The robot swings its limbs and head
	Facial expression	The robot attracts the child through changes in the eyes' LEDs
Training	Introduce training	The robot gives an explanation of what will to be done
	Correct mistake	If the child makes a wrong action, the robot will point out the mistake and ask the child to do it again. If the child fails three times, this action will be abandoned
	Pause session	The robot makes a short break for the child
	Claim attention	If the child is distracted, the robot reminds him to focus attention
	Guidance	The robot guides the child to imitate its action by language. For example, the robot says, "raise your right hand."
	Adjust the degree of tasks	When the child imitates correctly or mistakenly three times, the robot will adjust the difficulty level of the next action
	Encourage the child	When the child completes the task correctly, the robot praises him; when the child completes the task incorrectly, the robot encourages him
Finishing session	Say goodbye	The robot says goodbye to the child
	Update file	The robot updates the child's information and stores them in the database

Figure 8: The NAO robot.

its action in the session. The training session was divided into four phases: initialization, arousing child's interest, training, and finishing session. In Figure 9, the rectangular box is used to represent each phase, where the basic events are indicated in a brace. The explanations of basic events are shown in Table 2.

4.3. Results. A 4-year-old boy and a 6-year-old girl were invited to participate in our study, which were supported by their parents. Figures 10 and 12 show sequences of the interaction between the robot and the child A and the child B, respectively. The session was divided into five rounds, and each round lasted 3 minutes. When the robot output the action, it waits for the child to perform the task for 5 seconds. The degree of difficulty of action imitation was divided into 3 training levels, and the reaction of the robot was counted. In each round, the distribution rate of the training levels and the robot's reaction on children's performance are shown in Figures 11 and 13. Figures 11(a) and 13(a) show the proportion of the feedback that the robot outputs when the child performs a task of imitation, including encouragement, praise, and claim attention. Figures 11(b) and 13(b) show the proportion of three different difficult movements output by the robot in each round. These percentages are obtained by dividing the number of corresponding items by the total number.

Figure 11 shows the distribution of training levels and robot's feedback when the robot interacted with the child A. For the training levels, the difficulty of the robot's action increases gradually from level 1 to level 3. According to Figure 11, different legends are used to distinguish the distribution of the robot's feedback and training levels in each round. In round 1, the robot asked the child to imitate the actions with a higher proportion (64.3%) of level 1, and the feedback of the robot consisted of encouragement (38.3%), praise (43.2%), and claim attention (18.5%). In round 2,

TABLE 3: The times of the supervisor intervenes during interaction.

Event	Content				
Rounds	Round 1	Round 2	Round 3	Round 4	Round 5
Total A	25	26	24	25	25
Times A	2	3	4	3	4
Total B	26	24	24	24	25
Times B	3	4	3	5	3

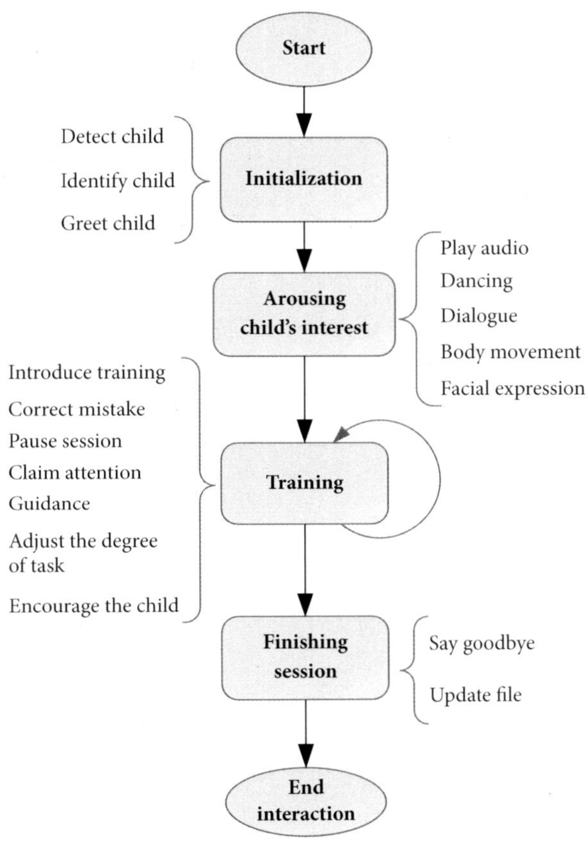

FIGURE 9: Phases of the interaction session.

the difficulty in training level was improved. According to Figure 11(b), the proportion of training of level 1 decreased while the proportion of training of level 2 and level 3 increased. In this round, more praises (57.5%) were given by the robot. With the child skilled in the task, the robot guided the child more to imitate actions of level 3 in round 3. However, in this round, the child did not perform well, so the robot gave more reaction of encouragement, and the robot reduced the difficulty of training level in round 4. In round 5, the child was tired of the imitation task, and the robot had to spend more time asking the child to keep his attention.

Similar to Figure 11, Figure 13 shows the distribution of training levels and robot's feedback when the robot interacted with the child B. Compared with the child A, the child B has the stronger ability of imitation. Therefore, the robot performed higher-level actions during the interaction. As shown in Figure 13(b), the proportion of training of level 2 and level 3 steadily increased, while the proportion of training of level 1 was decreased from round 1 to round 5. However, during the interaction, the child B showed the state of inattention. Therefore, compared with the interaction with the child A, the robot's feedback showed a higher proportion of claim attention (as shown in Figure 13(a)).

In the experiment, the robot can adjust its action according to the change of interaction environment and quickly meet the individual needs of the interactive object under the guidance of the supervisor. Sometimes, the robot's perception and cognition of the environment were not accurate, and the robot's behavior output was not reasonable. At this time, it is necessary for the supervisor to intervene in the interactive process to ensure the smooth progress of the task. This process is implemented through social module of the control architecture. The Table 3 shows how many times the supervisor gives the suggestion to the robot during interaction.

As shown in Table 3, total A represents the total number of imitative actions that the robot outputs in each round when it interacts with child A. Times A indicates that when the robot interacts with the child A, the number of the supervisor gives advice. Total B and times B indicate the same meaning when the robot interacts with the child B. During the interaction, when the supervisor thinks the robot's decision is inappropriate, he will intervene in the robot's decision through language. The robot adjusts its output through its decision-making algorithm (based on interactive reinforcement learning). In the experiment, the overall rate of intervention of the supervisor on the decision-making of the robot is about 14%. Through analysis, it is found that the error rate of the robot feedback evaluation to children is the main factor that influences the robot to make inappropriate decisions. Therefore, in the future research, the supervisor's intervention in the robot-child interaction process can be reduced by improving the robustness of the related algorithm.

5. Conclusion

Robot-assisted intervention is an effective way to facilitate social skills for children with ASD. In this study, we proposed a control architecture to improve the autonomy of the robot, when the robot is used in the intervention of children with ASD. Following the perception-cognition-action model, the architecture is designed based on some ideas of traditional intervention (DTT and DIR/Floortime) and ACT-R. In the paper, the operating mechanism and some algorithms of the proposed architecture are described in detail. Finally, through

(a) (b) (c)

FIGURE 10: The robot is interacting with the child A.

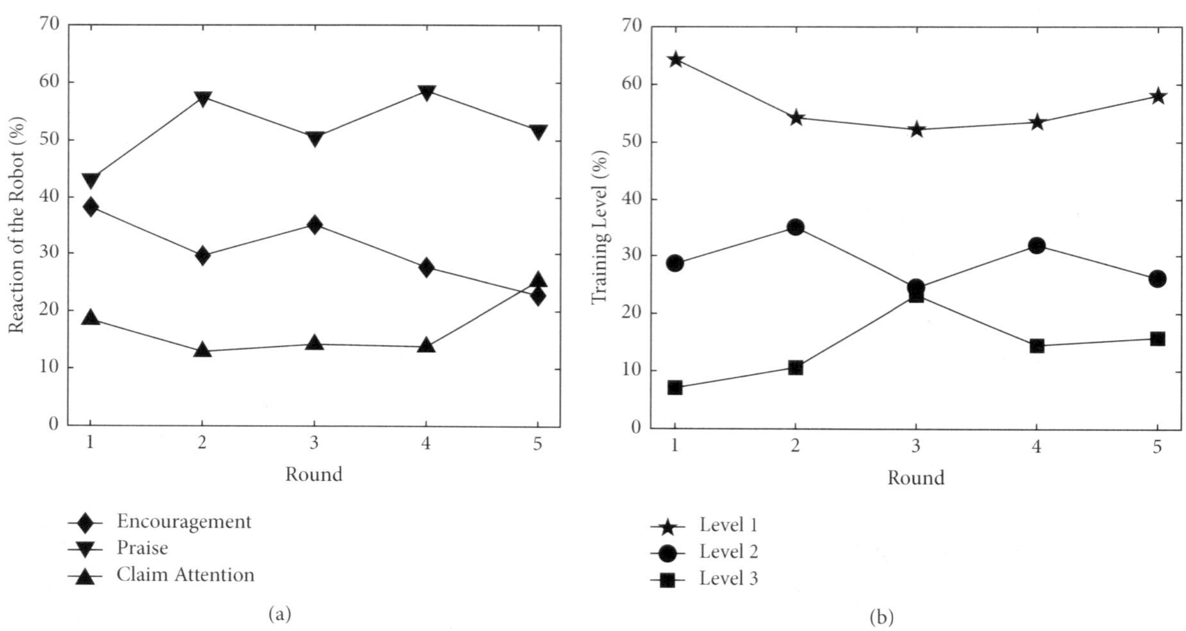

FIGURE 11: The distribution rate of training levels and the robot's feedback for the child A in the session.

experimental verification, the proposed control architecture can improve the autonomy of the robot in the intervention of children with ASD and reduce the burden on supervisors.

It should be noted that some limitations of this study still exist. For example, two children participated in the validation in this study. However, children with ASD have large individual differences and there are many situations in the intervention sessions. Therefore, more participants are needed to verify the advantages and disadvantages of the architecture. In addition, only one robot platform (NAO robot) was used in our study, and the application of some algorithms was limited. Moreover, the tasks of robot intervention for children with ASD need to be further enriched.

In a near future, we are planning to work in three aspects: (1) according to the characteristics of interactive objects, the collection, analysis, and interpretation of sensory information are carried out and improve the robustness of the robot by perfecting existing algorithms; (2) we will design interactive tasks which suitable for robot expression based on experience of traditional intervention; (3) we will expand the scope of clinical application, which can help us improve the study and make our study more meaningful.

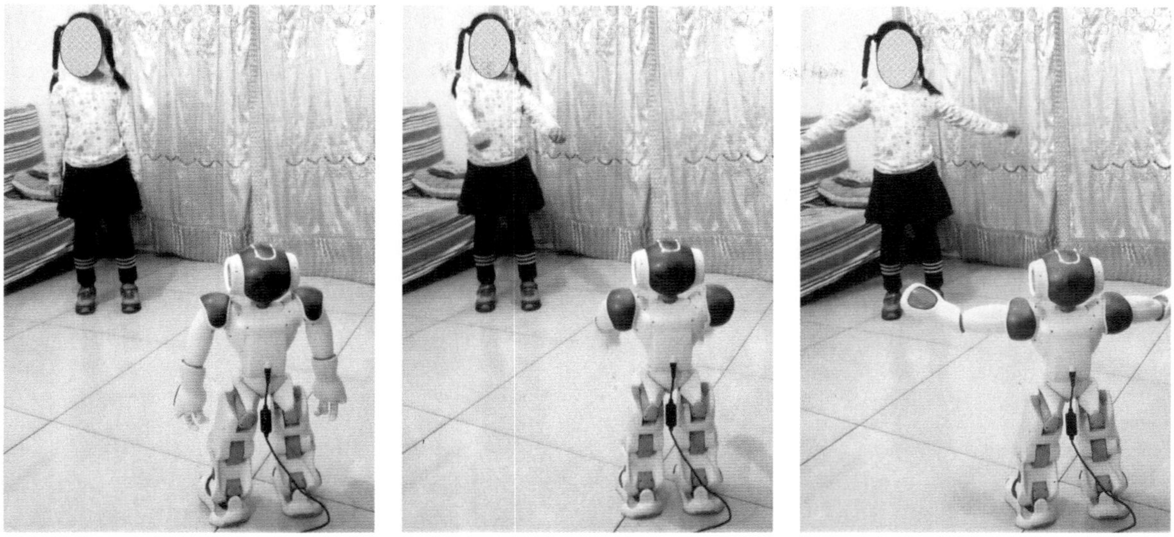
(a) (b) (c)

FIGURE 12: The robot is interacting with the child B.

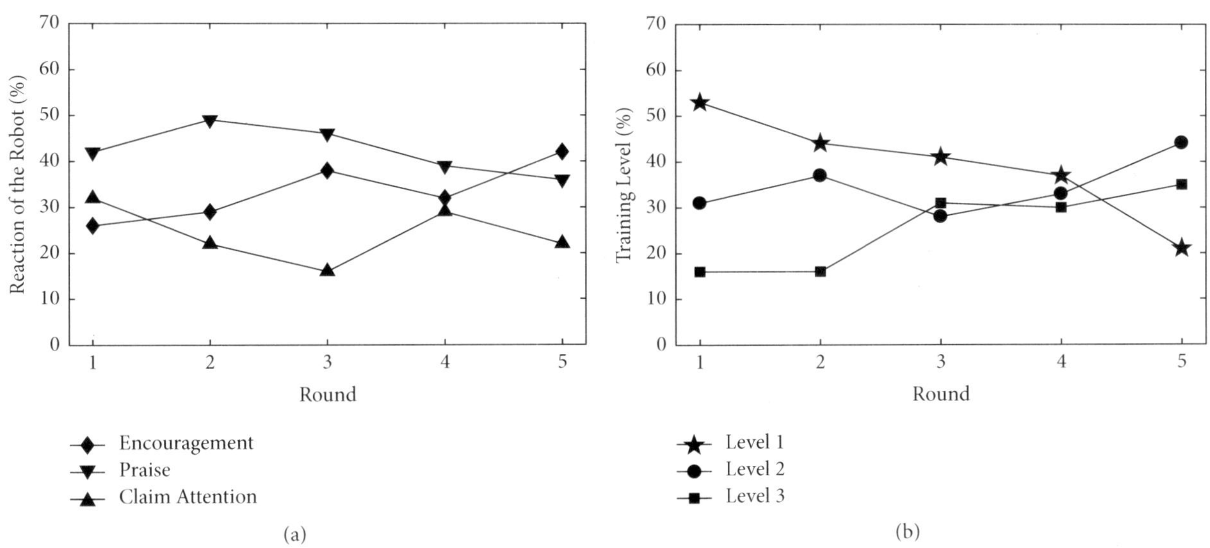

FIGURE 13: The distribution rate of training levels and the robot's feedback for the child B in the session.

Conflicts of Interest

The authors declare that there are no conflicts of interest regarding the publication of this paper.

Acknowledgments

The research presented in this paper is supported in part by the National Natural Science Foundation of China (61573066 and 61327806).

References

[1] J.-J. Cabibihan, H. Javed, M. Ang Jr., and S. M. Aljunied, "Why Robots? A Survey on the Roles and Benefits of Social Robots in the Therapy of Children with Autism," *International Journal of Social Robotics*, vol. 5, no. 4, pp. 593–618, 2013.

[2] B. Ingersoll, "Brief report: Effect of a focused imitation intervention on social functioning in children with autism," *Journal of Autism and Developmental Disorders*, vol. 42, no. 8, pp. 1768–1773, 2012.

[3] S. Boucenna, A. Narzisi, E. Tilmont et al., "Interactive Technologies for Autistic Children: A Review," *Cognitive Computation*, vol. 6, no. 4, pp. 722–740, 2014.

[4] L. Escobedo, C. Ibarra, J. Hernandez, M. Alvelais, and M. Tentori, "Smart objects to support the discrimination training of children with autism," *Personal and Ubiquitous Computing*, vol. 18, no. 6, pp. 1485–1497, 2014.

[5] A. Peca, M. Coeckelbergh, R. Simut et al., "Robot Enhanced Therapy for Children with Autism Disorders: Measuring Ethical

Acceptability," *IEEE Technology and Society Magazine*, vol. 35, no. 2, pp. 54–66, 2016.

[6] C. A. G. J. Huijnen, M. A. S. Lexis, and L. P. de Witte, "Matching Robot KASPAR to Autism Spectrum Disorder (ASD) Therapy and Educational Goals," *International Journal of Social Robotics*, vol. 8, no. 4, pp. 445–455, 2016.

[7] P. Pennisi, A. Tonacci, G. Tartarisco et al., "Autism and social robotics: A systematic review," *Autism Research*, vol. 9, no. 2, pp. 165–183, 2016.

[8] B. Scassellati, H. Admoni, and M. Matarić, "Robots for use in autism research," *Annual Review of Biomedical Engineering*, vol. 14, no. 1, pp. 275–294, 2012.

[9] R. P. Hastings, "Behavioral adjustment of siblings of children with autism engaged in applied behavior analysis early intervention programs: The moderating role of social support," *Journal of Autism and Developmental Disorders*, vol. 33, no. 2, pp. 141–150, 2003.

[10] J. Virués-Ortega, "Applied behavior analytic intervention for autism in early childhood: meta-analysis, meta-regression and dose-response meta-analysis of multiple outcomes," *Clinical Psychology Review*, vol. 30, no. 4, pp. 387–399, 2010.

[11] M. W. Steege, F. Charles Mace, L. Perry, and H. Longenecker, "Applied behavior analysis: Beyond discrete trial teaching," *Psychology in the Schools*, vol. 44, no. 1, pp. 91–99, 2007.

[12] S. I. Greenspan and S. Wieder, "A Functional Developmental Approach to Autism Spectrum Disorders," *Research and Practice for Persons with Severe Disabilities*, vol. 24, no. 3, pp. 147–161, 1999.

[13] C. M. Corsello, "Early intervention in autism," *Infants & Young Children*, vol. 18, no. 2, pp. 74–85, 2005.

[14] S.-T. Liao, Y.-S. Hwang, Y.-J. Chen, P. Lee, S.-J. Chen, and L.-Y. Lin, "Home-based DIR/Floortime™ intervention program for preschool children with autism spectrum disorders: Preliminary findings," *Physical & Occupational Therapy in Geriatrics*, vol. 34, no. 4, pp. 356–367, 2014.

[15] S. Thill, C. A. Pop, T. Belpaeme, T. Ziemke, and B. Vanderborght, "Robot-Assisted Therapy for Autism Spectrum Disorders with (Partially) Autonomous Control: Challenges and Outlook," *Paladyn, Journal of Behavioral Robotics*, vol. 3, no. 4, pp. 209–217, 2012.

[16] D. Feil-Seifer and M. J. Matarić, "B3IA: A control architecture for autonomous robot-assisted behavior intervention for children with autism spectrum disorders," in *Proceedings of the 17th IEEE International Symposium on Robot and Human Interactive Communication, RO-MAN*, pp. 328–333, deu, August 2008.

[17] J. Wainer, B. Robins, F. Amirabdollahian, and K. Dautenhahn, "Using the humanoid robot KASPAR to autonomously play triadic games and facilitate collaborative play among children with autism," *IEEE Transactions on Autonomous Mental Development*, vol. 6, no. 3, pp. 183–199, 2014.

[18] S.-S. Yun, H. Kim, J. Choi, and S.-K. Park, "A robot-assisted behavioral intervention system for children with autism spectrum disorders," *Robotics and Autonomous Systems*, vol. 76, pp. 58–67, 2016.

[19] J. C. González, J. C. Pulido, and F. Fernández, "A three-layer planning architecture for the autonomous control of rehabilitation therapies based on social robots," *Cognitive Systems Research*, vol. 43, pp. 232–249, 2017.

[20] A. Ghorbandaei Pour, A. Taheri, M. Alemi, and A. Meghdari, "Human–Robot Facial Expression Reciprocal Interaction Platform: Case Studies on Children with Autism," *International Journal of Social Robotics*, vol. 10, no. 2, pp. 179–198, 2018.

[21] M. Salvador, A. S. Marsh, A. Gutierrez, and M. H. Mahoor, "Development of an ABA Autism Intervention Delivered by a Humanoid Robot," in *Proceedings of the International Conference on Social Robotics*, vol. 9979, pp. 551–560, Springer International Publishing.

[22] G. Trafton, L. Hiatt, A. Harrison, F. Tanborello, S. Khemlani, and A. Schultz, "ACT-R/E: an embodied cognitive architecture for human-robot interaction," *Journal of Human-Robot Interaction*, vol. 2, no. 1, pp. 30–55, 2013.

[23] Y. Feng, Q. Jia, M. Chu, and W. Wei, "Engagement Evaluation for Autism Intervention by Robots Based on Dynamic Bayesian Network and Expert Elicitation," *IEEE Access*, vol. 5, pp. 19494–19504, 2017.

[24] P. G. Esteban, P. Baxter, T. Belpaeme et al., "How to Build a Supervised Autonomous System for Robot-Enhanced Therapy for Children with Autism Spectrum Disorder," *Paladyn, Journal of Behavioral Robotics*, vol. 8, no. 1, pp. 18–38, 2017.

[25] S. Shamsuddin, H. Yussof, L. I. Ismail, S. Mohamed, F. A. Hanapiah, and N. I. Zahari, "Humanoid robot NAO interacting with autistic children of moderately impaired intelligence to augment communication skills," *Procedia Engineering*, vol. 41, no. 2012, pp. 1533–1538, 2012.

[26] L. I. Ismail, S. Shamsudin, H. Yussof, F. A. Hanapiah, and N. I. Zahari, "Robot-based Intervention Program for autistic children with Humanoid Robot NAO: Initial response in stereotyped behavior," *Procedia Engineering*, vol. 41, no. 41, pp. 1441–1447, 2012.

[27] A. Tapus, A. Peca, A. Aly et al., "Children with autism social engagement in interaction with Nao, an imitative robot: A series of single case experiments," *Interaction Studies*, vol. 13, no. 3, pp. 315–347, 2012.

[28] M. A. Miskam, M. A. C. Hamid, H. Yussof, S. Shamsuddin, N. A. Malik, and S. N. Basir, "Study on social interaction between children with autism and humanoid robot NAO," *Applied Mechanics and Materials*, vol. 393, pp. 573–578, 2013.

11

Biomechanic and Energetic Effects of a Quasi-Passive Artificial Gastrocnemius on Transtibial Amputee Gait

Michael F. Eilenberg, Ken Endo, and Hugh Herr

Biomechatronics Group, MIT Media Lab, Massachusetts Institute of Technology, Cambridge, MA 02139, USA

Correspondence should be addressed to Hugh Herr; hherr@media.mit.edu

Academic Editor: Kazuo Kiguchi

State-of-the-art transtibial prostheses provide only ankle joint actuation and thus do not provide the biarticular function of the amputated gastrocnemius muscle. We develop a prosthesis that actuates both knee and ankle joints and then evaluate the incremental effects of this prosthesis as compared to ankle actuation alone. The prosthesis employs a quasi-passive clutched-spring knee orthosis, approximating the largely isometric behavior of the biological gastrocnemius, and utilizes a commercial powered ankle-foot prosthesis for ankle joint functionality. Two participants with unilateral transtibial amputation walk with this prosthesis on an instrumented treadmill, while motion, force, and metabolic data are collected. Data are analyzed to determine differences between the biarticular condition with the activation of the knee orthosis and the monoarticular condition with the orthosis behaving as a free-joint. As hypothesized, the biarticular system is shown to reduce both affected-side knee and hip moment impulse and positive mechanical work in both participants during the late stance knee flexion phase of walking, compared to the monoarticular condition. The metabolic cost of walking is also reduced for both participants. These very preliminary results suggest that biarticular functionality may provide benefits beyond even those of the most advanced monoarticular prostheses.

1. Introduction

The loss of a leg below the knee can have a formidable impact on one's quality of life [1–3]. The widespread, passive ankle-foot prostheses on the market today provide only a rudimentary approximation to the function of a human ankle joint [4, 5]. Instead of providing net mechanical work to the wearer during walking, these passive devices act at best in a spring-like manner; they can only provide as much mechanical energy return as is provided to them by the wearer, and they do not provide the articulation normally seen in the biological ankle-foot complex during walking. As evidence of this technological limitation, transtibial amputees display a variety of pathological features of their walking gaits. Specifically, transtibial amputees naturally select a 30–40% slower walking speed than those without amputation, and when walking at the same pace as a nonamputee, these amputees require 20–30% more metabolic power than their nonamputee counterparts [4, 6–8]. Further, these amputees exhibit increased levels of hip positive power during late stance phase. This increased hip power may be a compensatory response to lack of calf muscle function and could contribute to the aforementioned increase in metabolism while walking [4].

In the last several years, robotic advances in prosthetic technology have led to the introduction of powered ankle-foot prostheses (EmPower, BionX Medical Technologies, Inc., Bedford, MA), which, unlike the passive conventional devices, provide levels of mechanical work comparable to those provided by the human ankle-foot complex. As a result of this functional improvement, many of the aforementioned gait pathologies have been drastically reduced; amputees using the powered prostheses have preferred walking speed, metabolic cost at a given speed, and contralateral limb impacts that are not significantly different from those of nonamputees [9, 10]. These improvements are thought to stem from the propulsive net mechanical work provided by these devices to the wearer [9], as this propulsion helps to redirect the center-of-mass, thereby reducing collisions of the contralateral limb [11–14]. This ankle propulsion may also assist swing-initiation of the affected-side limb [15–18].

These new prosthetic devices are, however, limited to emulating the function of the ankle-foot complex alone and consequently cannot restore the full function of the powerful gastrocnemius muscle. The gastrocnemius provides not only a plantar flexion moment at the ankle, but also a flexion moment at the knee. Without the knee-flexing function, compensatory mechanisms are required. Indeed, transtibial amputees exhibit higher hamstring muscle activity during level-ground walking than nonamputees [19], possibly as an attempt to stabilize or flex the knee in place of the nonfunctional gastrocnemius muscle. This higher muscle activity is still apparent when amputees walk with the powered ankle-foot prostheses, indicating that a monoarticular intervention at only the ankle-foot complex may not be sufficient to restore biological function. It is possible that this pathological muscle activity has detrimental effects on amputee gait. However, little work has been done to develop devices for restoring this missing gastrocnemius functionality.

Since those with transtibial amputation maintain their biological knee joints, a direct assistance of these joints requires the use of an exoskeletal device such as an orthosis that provides assistance in parallel to the existing joint. A variety of robotic orthoses have been developed to assist human knee joints for locomotion. Many of these devices utilize active components such as electric motors [20–22] or pneumatic actuators [23–26], but these elements tend to require a large power source or otherwise require off-board tethered actuation, limiting the scope of their applicability. Alternatively, some researchers have developed quasi-passive knee exoskeletons that do not provide net mechanical power to the wearer but instead require at most the minimal input power to operate the electronics and mechanical components [27–30]. In one such study, a spring-loaded, custom toothed clutch was designed for assisting extension moment during human running [29]. This device locked to support the knees of nonamputee participants during the stance phase and unlocked to be a free joint during the swing phase. Using this quasi-passive device, large knee moments of 190 Nm could be supported with an exoskeletal joint weighing only 710 grams. This, and other exoskeletons have been built to assist with knee extension in nonamputees, but no such devices to date have been built for knee flexion assistance, particularly for the gastrocnemius, in those with transtibial amputation.

In order to develop an assistive device for the gastrocnemius, it is important to first understand how this muscle functions. In vivo ultrasonography shows that the gastrocnemius muscle fascicle length is largely isometric during the early and mid-stance phases of level-ground walking [31], indicating that the passive tissues such as tendons are responsible for much of the power delivery from the gastrocnemius. Hence, the gastrocnemius may be approximated by a clutched spring, with the spring representing the compliant tendinous structures, and the clutch representing the isometrically acting muscle fibers. Indeed, a model of human gait by Endo and Herr with only spring-clutch structures at the knee joint was able to achieve human-like metabolic economy, while capturing the dominant kinetic features of human walking [32–34]. It is therefore possible that a physical clutch in series with a spring can provide much of the missing knee function of the gastrocnemius in transtibial amputees.

Researchers have begun to study the effects of a robotic device at the affected-side knee to provide the knee flexion moment of the missing gastrocnemius. In the first such study [35], a quasi-passive knee orthosis, dubbed an artificial gastrocnemius (AG) was built, based on the spring-clutch representation of the gastrocnemius in the aforementioned Endo-Herr model. The AG was physically realized as clutchable rotary spring at the orthosis knee joint. Two copies of this AG, worn along with a powered-ankle-foot prosthesis (PAFP), was tested on both legs of one bilateral transtibial amputee, with promising results: the amputee's metabolic energy expenditure was reduced with the AG-PAFP condition, compared to the amputee walking at the same speed with only conventional leaf-spring-type ankle-foot prostheses. However, since the PAFP was not tested independently of the AG, it is not clear what incremental effects the AG had over the PAFP alone. A need, therefore, exists to evaluate the incremental biomechanical and metabolic effects of the AG.

In this study, we design and evaluate a new version of the AG unit, based on the exoskeletal knee clutch design [29]. We evaluate the effects of this new AG-PAFP on the walking gait of transtibial amputees. We focus on the late stance knee flexion phase of walking, during which the knee flexes to prepare for the swing phase, so to detect improvements on the pathologies thought to be most likely affected by the lack of calf muscle function [4]. We expect that the energy stored in the spring of the AG early in the gait cycle will be returned during late stance knee flexion to replace positive muscle work. Specifically, we hypothesize that including a clutched-spring AG at the affected-side knee joint of transtibial amputees in conjunction with a PAFP will reduce both biological flexion moment impulse and positive mechanical work of the affected-side knee and hip joints during late stance knee flexion, as compared to the same conditions with only the PAFP. We further hypothesize that these changes will provide a corresponding reduction in metabolic cost of walking. We expect that this metabolic reduction will stem from reduced joints moments and net work in the affected knee and hip joints. We evaluate these hypotheses by analyzing the kinematics, kinetics, and metabolism of two transtibial amputees walking with the AG-PAFP combination on an instrumented treadmill. We compare these data between the two conditions where the AG is active versus acting as a free-joint.

2. Methods

Two devices were used for the intervention in this study. A powered ankle-foot prosthesis provided ankle function in the sagittal plane, and a quasi-passive artificial gastrocnemius (AG), that was a clutched-spring joint, was mounted on a knee orthosis on the affected-side knee. Together, these two devices represented a biarticular transtibial prosthesis that could actuate both joints independently.

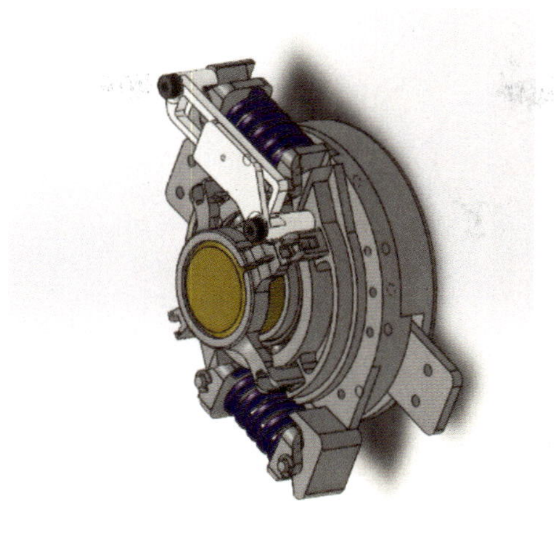

(a)

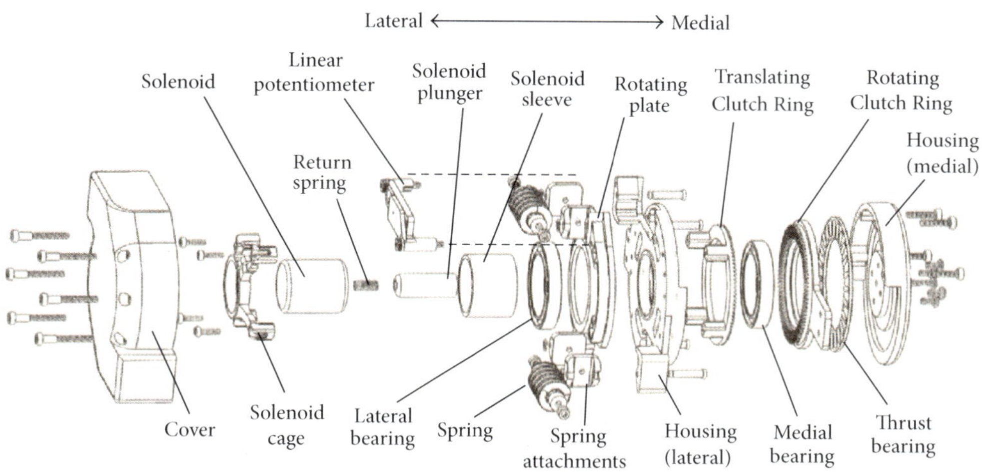

(b)

FIGURE 1: Joint mechanism of the quasi-passive artificial gastrocnemius. The exploded view of the joint design (b) is shown along with a rendering of the assembled joint without the cover (a). The *Housing* was attached to the proximal part of the orthosis, and the *Rotating Clutch Ring* attached to the distal part of the orthosis.

2.1. Powered Ankle-Foot Prosthesis. An EmPower powered ankle-foot prosthesis (BionX, Bedford, MA) was used as the prosthesis for all clinical trials. This prosthesis had the capability of providing positive net power at levels comparable to the human ankle-foot complex [9]. At the core of the prosthesis was a series elastic actuator, comprising a brushless DC motor, ball screw, and carbon fiber leaf spring. For dorsiflexion angles, a rigid hard-stop engaged, both to reduce the torque required by the motor in dorsiflexion and to act as a safety feature in the event of a power failure. The total mass of the prosthesis and battery was 1.8 kg.

The prosthesis controller employed a positive force feedback strategy that served to approximate the biological muscle dynamics and neural reflexes. A wireless communication link enabled real-time tuning of the control parameters.

2.2. Clutch-Spring Joint. The AG, shown in Figure 2, was a knee orthosis, comprising a pair of polycentric hinges that connected a thigh cuff to an 1/8″-thick aluminum bracket made of aluminum sheet stock (6061-T6). The bracket connected to the base of a participant's socket, between the prosthetic pylon attachment and the socket. The proximal side of the brace was strapped to the participant's thigh as a typical knee orthosis. A custom, dog-tooth rotary clutch spring with series compliance was attached lateral to the knee, in parallel with the hinges.

This clutch-spring joint was based on a previous design [29] but had the addition of series compliance and reversed the direction of the clutch teeth so that the teeth engaged to provide a knee flexion moment, rather than an extension moment to the wearer. In addition, our device lacked a planetary gear train that was present in the previous work. The machined clutch components were made of 6061-T6 aluminum alloy and included two rings of dog teeth, brought together by the action of a solenoid (LT8x9, Guardian Electric, Woodstock IL). As in Figure 1, the clutch controlled the relative rotation of the *Housing* to the *Rotating Clutch Ring*. A solenoid was built into the middle of the joint and translated

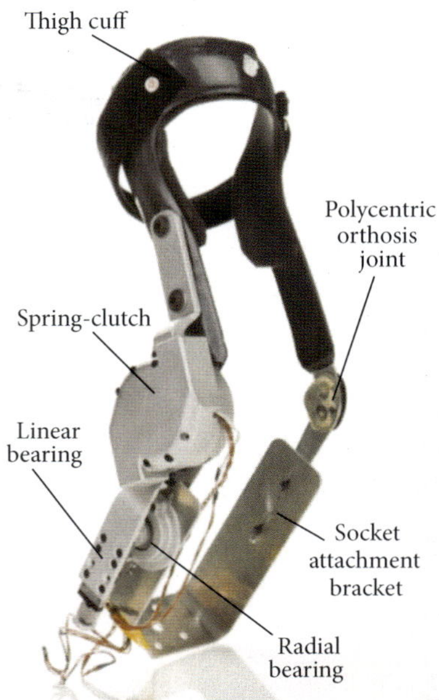

Figure 2: Quasi-passive artificial gastrocnemius (photo credit: Chris Conti Photography).

in the medial-lateral direction under action of its actuation power and a return spring. When the solenoid was energized, the *Translating Clutch Ring was engaged with the Rotating Clutch Ring*. The *Translating Clutch Ring* constrained the rotation of the *Rotating Plate* that, in turn, was connected to the *Housing* through two tangential linear compression springs. Thus, when the solenoid was energized, the *Housing* was coupled to the *Rotating Clutch Ring* through these springs. When the solenoid was inactive, the return spring separated the *Translating Clutch Ring* from the *Rotating Clutch Ring* and, hence, they were then free to rotate with respect to each other. The return spring stiffness was chosen experimentally so that when the clutch developed more than 0.5 Nm of torque, the tooth friction prevented the two rings from separating. This tooth-binding effect acted as a safety mechanism and was included as a part of the control algorithm, as described in Section 2.8. This aforementioned friction maintained the clutch in a clutched state until the torque level dropped back to the safe value of 0.5 Nm. As a result, the bulk of the stored spring energy could only be released gradually into the wearer's leg, rather than suddenly into the mechanism.

The *Housing* of the clutch was bolted to the thigh cuff of the knee orthosis. The *Rotating Clutch Ring* connected to a distal output link with a linear ball bearing and radial ball bearing, acting in series. These bearings accommodated for the kinematic difference between the polycentric hinges of the orthosis and the single-axis rotation of the clutch joint. Thus, only forces tending to flex or extend the joint could be transmitted through this set of bearings.

2.3. Sensing. The AG had onboard sensing for use in the control algorithm. Knee angle of the AG was measured using a 10 kOhm rotary potentiometer. Knee moment provided by the spring-clutch element to the wearer was estimated by measuring the deflection of one of the two identical tangential springs with an 8 kOhm linear potentiometer. The voltage output from the potentiometer was scaled with an experimentally determined gain to provide an estimate of torque applied by the AG to the wearer. This gain was found by applying various tangential forces to the distal end of the brace using a force gauge (gauge moment arm = 28 cm), while the clutch was engaged. With this force and the known moment arm of the gauge, the applied knee moment was computed and compared to voltage readings from the linear potentiometer to achieve a scaling factor for conversion from voltage to Nm/rad. The resulting measured spring-clutch knee torque reading is subsequently referred to as the AG knee moment. A similar calibration procedure was performed using the AG angle-sensing rotary potentiometer and an infrared camera system (model T40s, Vicon Motion Systems Ltc, Oxford, UK). In this calibration procedure, reflective markers were placed on the center of the joint and on the distal and proximal ends of the orthosis. Marker-based joint angle was computed by obtaining the relative angles of lines connecting the distal and proximal markers to the joint center marker. These angles were then compared to voltage readings from the angle-sensing potentiometer. Prosthesis-side ground contact was detected using a resistive pressure sole footswitch (model: FSW, B&L Engineering, Santa Ana, CA), inserted into the shoe between the prosthetic foot and the insole.

2.4. Failure Analysis. Unlike the titanium toothed rings in the exoskeletal clutch [29], the clutch in the present study was made from aluminum. Thus, it was necessary to reevaluate the load-bearing limits of the teeth. Finite element analysis (FEA) was performed using Solidworks (DS Solidworks Corp, Waltham, MA) to ensure this clutch could support the required knee moments. Tangential loads were simulated symmetrically on the tips of all teeth to simulate worst-case loading scenario.

2.5. Modeling and Spring Stiffness Selection. The selection of the spring constants for the tangential springs allowed for the control of rotary stiffness of the clutch-spring element. It was desirable for the stiffness of the clutch-spring joint to be such that the spring-clutch behavior of the artificial gastrocnemius would most closely reproduce the gastrocnemius muscle behavior of a healthy gastrocnemius. To this end, walking data from healthy nonamputees was fed as input to a modified version of the Endo-Herr sagittal-plane human leg model [32–34] during the development of the artificial gastrocnemius in [35].

2.6. Control. The quasi-passive artificial gastrocnemius produced no positive net mechanical work to the wearer but still required a control system to determine the appropriate times to engage and disengage the clutch. This controller took knee joint angle and stance information as input and engaged the clutch appropriately through each gait cycle. The clutch was controlled to engage at maximum stance knee flexion angle, enabling the spring to store energy during the subsequent

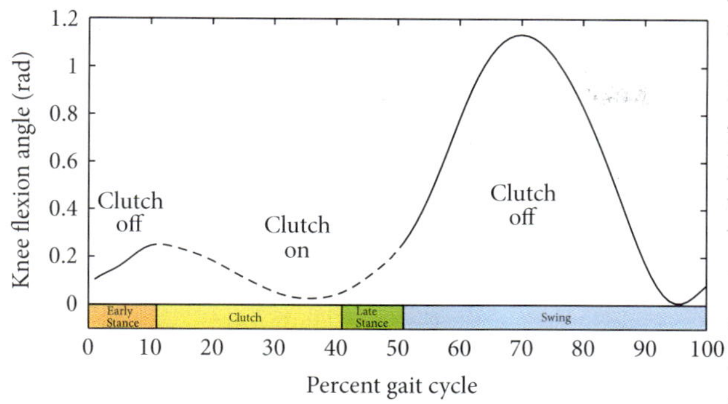

FIGURE 3: Example action of the clutch engagement. A typical knee flexion angle profile is shown for level-ground walking. The dashed red section indicates the region in the gait cycle when the clutch in the AG would be engaged. Colored bars show the sequence of intended finite-state-machine state transitions.

knee extension and flexion into swing phase, as shown in Figure 3.

2.7. Control Electronics. The computer platform for controlling the AG was a commercial single-board computer (model: Raspberry Pi Version B, Raspberry Pi Foundation, Cambridgeshire, UK). The computer was equipped with an 800 MHz ARM11 processor, 512 MB SDRAM, with Linux Debian. The system was powered by a 6-cell lithium polymer battery (nominal voltage of 22.2 V).

2.8. Control Algorithm. High-level control was implemented using a finite state machine, implemented in Python. The gait cycle was divided into four states, shown in Figure 4: (1) *Swing*; (2) *Early Stance*; (3) *Clutched*; and (4) *Late Stance*.

The *Swing* state was triggered from any other state when the affected-side foot left the ground, as detected by a drop in footswitch signal, FSW, to less than a fraction, foot_{SW} of the maximum possible signal. During *Swing*, the clutch was disabled, allowing the knee to swing freely.

The *Early Stance* state was triggered from the *Swing* state at foot contact of the affected-side with the ground, defined as FSW increasing to the stance threshold, foot_{ST}, provided that the time elapsed in the current state, t_{state}, was at least the minimum time required for the *Swing* state, t_{swing}. During the *Early Stance* state, the clutch was disabled, and the controller monitored the knee angle of the brace for maximum stance knee flexion, at which point the *Clutched* state would be engaged. A least squares algorithm, similar to the one used previously [29], continually predicted the time remaining before maximum knee flexion angle. This prediction provided time to initiate the engagement of the clutch, so that the clutch would be fully engaged as close as possible to the time of maximum knee flexion. In addition to this knee flexion detection algorithm, two safety features were in place to ensure that the motions detected were resulting from a walking gait. The *Clutched* state could only be enabled if the following two conditions were also met: (1) the knee was flexed a minimum angle of θ_{ES} from the angle at heel strike, θ_{HS}; (2) the maximum knee flexion angular velocity, $\dot{\theta}_{\max}$, measured during the *Early Stance* state was at least $\dot{\theta}_{\text{ES}}$. The values θ_{ES} and $\dot{\theta}_{\text{ES}}$ were experimentally determined during early testing as the lowest values that successfully prevented false-triggering of the *Clutched* state outside of a steady, level-ground walking gait (Table 1).

The *Clutched* state served to activate the clutch near the maximum knee flexion during stance phase of walking. This state was activated when the following criteria were met: (1) the maximum stance flexion angle was predicted to occur within the time t_{delay} of the current time, (2) the knee angle was flexed more than a threshold, θ_{ES}, and (3) the maximum knee flexion angular velocity, $\dot{\theta}$, during the *Early Stance* state exceeded a threshold $\dot{\theta}_{\text{ES}}$. Conditions (2) and (3) were used to differentiate a walking gait with slow, nongait motions, the latter of which did not warrant activation of the clutch.

The *Late Stance* state served to turn off the clutch after it engaged. Once the clutch spring began to develop force, the clutch teeth would bind, preventing the clutch from disengaging until the spring force dropped sufficiently. Therefore, the clutch solenoid was deactivated by entering the *Late Stance* state when the time elapsed in the *Clutched* state, t_{state}, exceeded the clutch timeout threshold, t_{cl}.

In order to maximize spring energy storage and return, it was desirable to engage the clutch as near as possible to the moment of peak knee flexion in the *Early Stance* state. In fact, the clutch solenoid needed to be engaged slightly before the desired clutching time, as to account for the time required to close the gap between the two sets of clutch teeth. To achieve the necessary prediction of peak stance flexion, a look-ahead algorithm was used.

First, the knee angle during the *Late Stance* state was approximated as parabolic, given this assumption, the location of the vertex of the parabola may be found by performing a linear fit to the knee angular velocity data, via running sums, and solving for the zero-crossing. The algorithm used here was similar to one described by [29]. However, the earlier algorithm assumed a fixed time step that was not applicable with the system in this study, as the computer platform was not a truly real-time system. Hence, the algorithm proposed here did not presume a fixed time step.

The parameter $\boldsymbol{\beta} = [\beta_0 \ \ \beta_1]^T$ that minimizes the error in the least squares sense of linear model $\hat{y}(t) = \beta_0 + \beta_1 x(t)$ to time-series data $(x(t), y(t))$ is

$$\boldsymbol{\beta} = \left(\mathbf{X}^T\mathbf{X}\right)^{-1}\mathbf{X}^T\mathbf{y}, \tag{1}$$

where the matrix $\mathbf{X}$ has elements

$$X_{i,j} = \frac{\partial \hat{\mathbf{y}}}{\partial \beta_j} = \begin{cases} 1, & j = 0 \\ x_i, & j = 1 \end{cases} \tag{2}$$

and x_i is the ith element in the discretely sampled $x(t)$ vector. Expanding (1) yields

$$\boldsymbol{\beta} = \frac{\begin{bmatrix} \sum x_i^2 \sum y_i + \sum x_i \sum x_i y_i \\ \sum x_i \sum y_i + W \sum x_i y_i \end{bmatrix}}{W \sum x_i^2 - \left(\sum x_i\right)^2}, \tag{3}$$

Table 1: Parameter values for the AG controller.

Parameter	Description	Value
$foot_{ST}$	Footswitch threshold for entering the *Early Stance* state; fraction of the maximum possible value	0.2
$foot_{SW}$	Footswitch threshold for entering the *Swing* state	0.15
t_{swing}	Minimum time in the *Swing* state before it is possible to exit *Swing*	200 ms
θ_{ES}	Minimum knee flexion angle (rad) during the *Early Stance* state before the clutch can be engaged	0.087 rad
$\dot{\theta}_{ES}$	Minimum knee flexion angular velocity that must be observed during the *Early Stance* state before the clutch can be engaged	1.2 rad/s
t_{delay}	Delay time required between activation of the solenoid and the full engagement of the clutch	50 ms
t_{cl}	Maximum time to energize the solenoid	400 ms

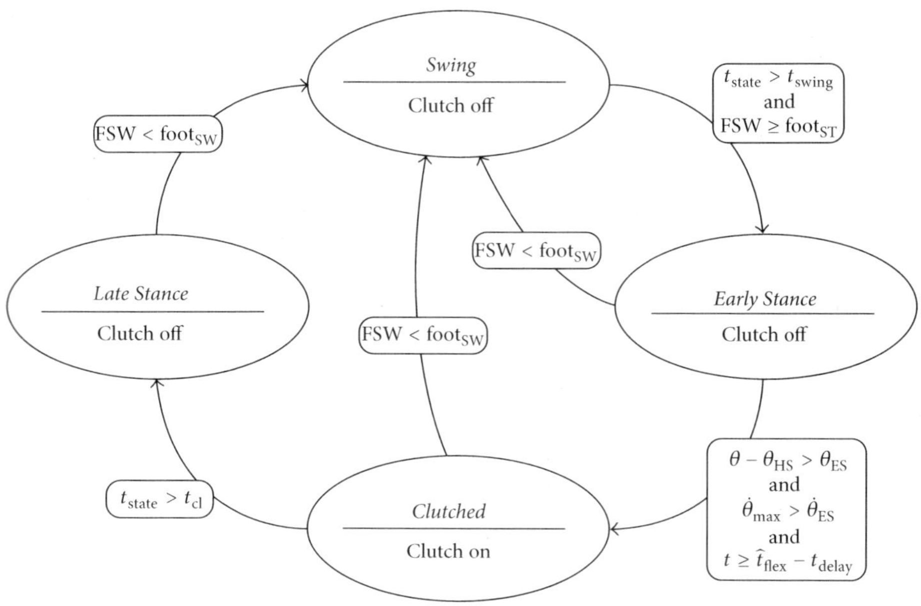

Figure 4: Finite state machine for the quasi-passive artificial gastrocnemius.

where all summations in (3) span i from 1 to W. For the task of estimating the angular velocity over time, the values of x_i were timestamps, and the values of y_i were knee angular velocity values. For the window of size W, the most recent W values of x and y were maintained in a queue. At each new timestep, the oldest values of x and y were popped from the queue, while the current x and y values were added to the queue, maintaining the 0th element as the first in the queue. This method allowed the sums to be updated each timestep without needing to store all values of the computed sums. Finally, the estimated time of the angular velocity zero-crossing was $\hat{t}_{flex} = -\beta_0/\beta_1$. It was found experimentally that the minimum error in clutching time between the initiation of the development of clutch torque and the peak knee flexion angle occurred when the clutch was engaged 50 ms prior to the predicted maximum knee angle. As a result, the time delay parameter, t_{delay}, was set to 50 ms.

2.9. Experimental Protocol. Two participants with below-knee amputation were involved in this study (Table 2). Both participants had right-side unilateral amputation and were of generally good health. The clinical evaluation was conducted at MIT (Cambridge, MA) and was approved by MIT's Committee on the Use of Humans as Experimental Subjects (COUHES). Each participant provided written, informed consent that was obtained before data collection was initiated.

Table 2: Amputee participant body parameters.

	Height (cm)	Weight (kg)
Subject 1	180	93
Subject 2	193	94

An infrared camera system (model T40s, Vicon Motion Systems Ltc, Oxford, UK) was used to track the three-dimensional motion, recorded at 100 Hz, of reflective markers, placed at 47 anatomical locations on the participants' bodies, based on the Helen Hayes marker model. Ground reactions forces and center of pressure locations were measured using a dual-belt instrumented treadmill (Bertec Corporation, Columbus, OH) with a sampling rate of 1 kHz. The net metabolic cost of walking during each condition was estimated using standard open-circuit gas exchange techniques (model: K4b2, COSMED, Rome, Italy).

At the beginning of each session, the participant was asked to don the powered prosthesis in place of their conventional prosthesis. The knee orthosis of the AG was then affixed to the prosthesis and donned by the participant.

For each participant, the prosthesis controller's power setting was adjusted using the commercial tuning app as the participant walked over a treadmill at 1.25 m/s so as to achieve net work from the prosthesis per step that was within one standard deviation of the mean for nonamputees walking at the same walking speed [36]. It was verified that the level of prosthesis net work remained within the desired range for both walking conditions for a given participant (0.045 to 0.16 J/kg) and that this level of net work stayed reasonably consistent across walking conditions for each participant.

A short time period of approximately 15 minutes was given to ensure that both the prosthesis and the artificial gastrocnemius orthosis were functioning properly. This acclimatization was evaluated by verifying that the clutch was engaging and disengaging at appropriate points in the gait cycle (as in Figure 3) and that the PAFP net mechanical work was reasonable as compared to biological values.

Both the powered prosthesis and artificial gastrocnemius were worn for all trials. The participants were asked to perform one standing trial to measure standing metabolism. Walking trials were performed on the treadmill at a speed of 1.25 m/s.

Two walking conditions were tested: (1) a baseline condition (*Baseline*) in which the AG acted as a free-joint at the knee with the spring-clutch disabled, and (2) an active condition (*Active*), in which the AG was controlled as a clutched spring at the knee with the described control algorithm. For both conditions, the powered ankle-foot prosthesis was active. The *Baseline* condition represented a monoarticular transtibial prosthesis, as the knee joint was a free-joint when the clutch was inactive. Yet the mass distribution of the device was identical to the *Active* condition. Thus, a direct comparison could be made to determine the incremental effects of the clutched-spring knee joint.

2.10. Data Processing. Fourth-order Butterworth filters were used to filter the marker position and ground reaction force data with 6 Hz and 25 Hz cutoff frequencies, respectively. The marker and force data were postprocessed through the SIMM (Musculographics Inc., Evanston, IL) inverse dynamics module to produce total joint moments and angles in three dimensions. However, only sagittal-plane dynamics were considered. Affected-side biological knee moment contribution was computed by subtracting the AG knee moment from the total knee moment estimated from the SIMM-based inverse dynamics.

Gait events were determined using vertical ground reaction force data from the embedded force plates. Approximate event timing was found by determining the times when the force increased beyond a 40 N threshold. Exact heel strike and toe-off times were found by progressing backward and forward in time, respectively, until the force value dropped to zero. Data were then cut to gait cycles based on heel strike times and resampled to 101 points. Gait cycles were discarded for the beginning and end of each trial, during the speed transients of the treadmill. Gait cycles in which the stride times were below 0.7 seconds or above 1.3 seconds or in which a foot crossed the midline of the treadmill were also discarded.

Joint powers were computed as the product of joint moments and joint velocities from SIMM-derived joint kinematics, where positive power was defined as that produced by the joint on the environment. Positive joint work during late stance knee flexion was computed as the positive contribution of the time-integral of joint power from maximum stance knee extension to toe-off. This region of the gait cycle was chosen for analysis because, as the knee flexes from the maximum extension angle, it provides an opportunity for the AG to provide positive power to the wearer. Joint flexion moment impulse was computed using the same integral for joint flexion moment. Net prosthesis work was computed by integrating the SIMM-derived joint torque over joint angle for the entire gait cycle.

Metabolic cost for each walking speed was computed by taking average oxygen and carbon dioxide data over a two-minute window at the end of each six-minute trial. The metabolic power was computed using the equation

$$P = K_{O_2}\dot{V}_{O_2} + K_{CO_2}\dot{V}_{CO_2}, \tag{4}$$

where P is the metabolic power in Watts, $\dot{V}_{O_2}$ is the volume flow rate of oxygen inhaled, $\dot{V}_{CO_2}$ is the volume flow rate of carbon dioxide exhaled, and K_{O_2} and K_{CO_2} are constants with values from literature [37], given as $K_{O_2} = 16{,}580$ W/L and $K_{CO_2} = 4{,}510$ W/L. The above equation is only valid for conditions when the metabolism is primarily aerobic. As a verification of this condition, the respiratory exchange ratio (RER), defined as $\dot{V}_{CO_2}/\dot{V}_{O_2}$, was monitored, and only metabolic results with RER values less than 1.1 were considered.

TABLE 3: Mechanical design parameters of the artificial gastrocnemius joint.

Parameter	Value
Maximum torque (Nm)	890
Torque sensor scale factor (Nm/V)	825
Number of teeth	90
Clutched-spring moment arm (mm)	32.1
Total orthosis mass (kg)	1.9

3. Results

3.1. Mechanical Design. The mechanical parameters of the design are summarized in Table 3. As in previous work [29], the teeth were manufactured using a contour milling process, and thus, the tooth spacing was limited by the 0.4 mm radius of the mills. Therefore, the tooth spacing was constrained to 4 degrees, or 90 teeth per revolution. The total mass of the orthosis, including battery and electronics, was 1.9 kg.

The aluminum teeth used in this design do not yield for applied knee flexion moments near to the expected values of 25 Nm [38] but start to yield at 35 times this expected load

Table 4: Net work per step by the powered ankle-foot prosthesis with the AG. The net work values were kept within the literature range of 0.045 to 0.16 J/kg.

	Baseline (J/kg)	Active (J/kg)
Subject 1	0.113 ± 0.028	0.130 ± 0.024
Subject 2	0.077 ± 0.016	0.082 ± 0.019

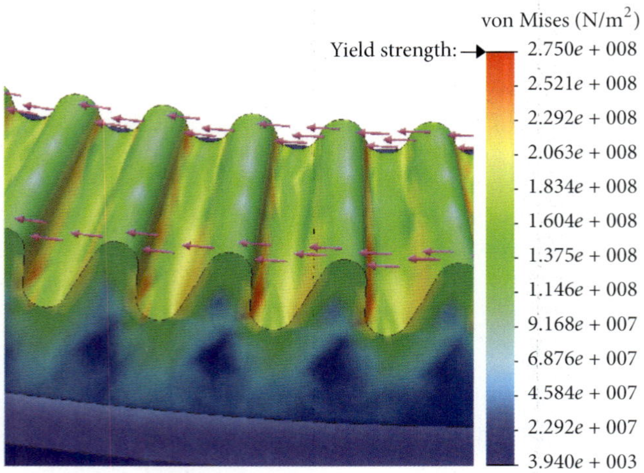

Figure 5: Results of finite element analysis with load evenly applied to all tooth tips tangentially. Results are shown for the yield condition, at which the applied torque to the clutch is 890 Nm.

(Table 3). The tooth stress at this failure load condition is shown in Figure 5.

3.2. Power Consumption. Given the solenoid specifications (22.2 V, 109 Ohms resistance), the power draw was 4.9 W. Assuming the maximum clutch-on time for each gait cycle of 0.4 ms and a typical cadence of 0.9 strides per second, the power consumption of the solenoid was 2.2 W. With the 1.1 W power draw of the Raspberry Pi, the average total power consumption of the device was 3.3 W during a typical walking gait.

3.3. Joint Spring Selection. The spring stiffness of the clutched-spring gastrocnemius element, derived from the modeling, was 92.9 Nm/rad [35]. To approximate this rotary stiffness about the knee joint, the two tangential linear compression springs were selected with spring constants of 51 N/mm. These springs acted on the joint with a moment arm of 32.1 mm, causing the equivalent joint rotary stiffness to be 105 Nm/rad.

3.4. Clinical Pilot Results

3.4.1. Acclimatization. In the time allotted for acclimatization, participants could walk with the AG and PAFP. Participants could walk comfortably with the AG, and the ankle prosthesis net work was appropriate for normal walking (see Table 4).

3.4.2. Ankle Prosthesis Net Work. As shown in Table 4, the net work produced by the powered ankle-foot prosthesis was within the desired range from literature [36].

3.4.3. Affected Knee Kinetics. Affected-side knee kinetics for both participants are shown in Figure 6 and summarized in Tables 5 and 6. Of particular note, the biological component of knee flexion moment impulse and the biological knee positive mechanical work during late stance knee flexion phase were both reduced by large percentage changes when subjects walked with the *Active* condition of the AG, compared to the *Baseline* condition.

3.4.4. Affected-Side Hip Kinetics. Hip flexion moment is shown in Figure 7, and hip flexion moment impulse values are summarized in Table 7. Hip net mechanical work values for both participants are shown in Table 8. As with the knee joint, both hip flexion moment impulse and positive mechanical work during late stance flexion were decreased slightly for both participants.

3.4.5. Metabolism. The net metabolic power (with the power required for standing subtracted out) is tabulated in Table 9. Both participants displayed small percentage reductions in metabolic cost when walking with the *Active* condition, as compared to the *Baseline* condition.

4. Discussion

Our novel design of a quasi-passive artificial gastrocnemius has been tested in our pilot study and found to be mechanically functional for the task of providing a flexion assist moment to the affected-side biological knee joint. The device exceeds the torque requirements while requiring only a small amount of electrical power. These properties make this a device a prime candidate for experiments for those with transtibial amputation, with the end goal of improving walking gaits for this population.

The preliminary findings from this study indicate that some measures of gait may be especially affected by the intervention. The pilot data support the hypothesis that this artificial gastrocnemius can reduce biological knee flexion moment and biological positive work of the affected-side knee, compared to a similarly weighted monoarticular prosthesis, during late stance knee flexion. Despite the short amount of time to get acclimated to the knee orthosis, the participants reduced their biological knee moment profiles, thereby allowing the AG to take over some of the kinetic loads. This behavior is consistent with other studies involving exoskeletal interventions at ankle [39] and hip joints [40]. The reduction of biological knee moment is beneficial from the standpoint of wearable device design, since it means that wearers of the AG were quickly able to replace biological function with that from the device.

Also as hypothesized, the affected-side hip flexion moment impulse and hip positive work of both participants were reduced during late stance knee flexion. These reductions are likely a result of the energy return of the spring, which served to help flex the hip. Although small, this effect

Table 5: Affected-side knee flexion moment impulse in late stance flexion with the AG compared between the two walking conditions with the quasi-passive artificial gastrocnemius: the Baseline condition, where the artificial gastrocnemius acted as a free-joint at the knee, and the Active condition, where the AG was appropriately engaged as per the controller. The values were computed over the late stance knee flexion phase of the gait cycle.

	Baseline moment impulse (Nm∗s/kg)	Active moment impulse (Nm∗s/kg)	Active Biological moment impulse (Nm∗s/kg)	Total % change from Baseline	Biological % change from Baseline
Subject 1	0.006 ± 0.004	0.009 ± 0.004	0.001 ± 0.002	+32	−78
Subject 2	0.055 ± 0.018	0.056 ± 0.013	0.039 ± 0.011	+2	−29

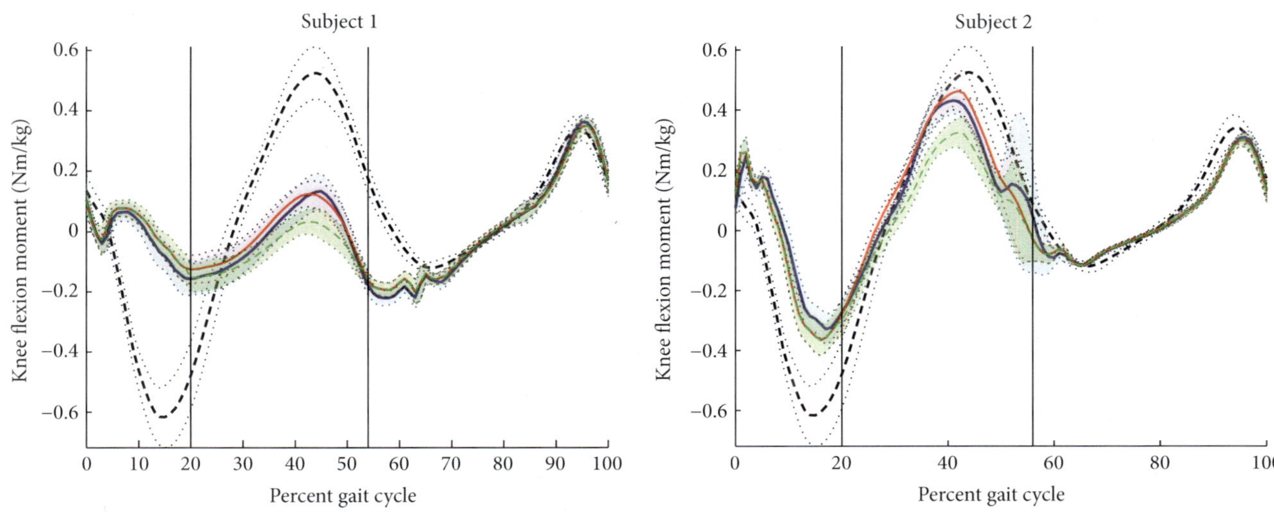

Figure 6: Affected-side knee flexion moment components with the AG shown for the *Baseline* condition where the clutch was disabled (thick solid blue line), the *Active* condition (thin solid red line), and the biological knee moment contribution during the *Active* condition (thin green dashed line). Shaded regions for these curves indicate ±1 standard deviation, bounded by light dotted lines. For reference, biological knee moment data are also shown for the nonamputee from which the clutch spring was tuned (thick black dashed line) ±1 standard deviation (black dotted lines). The vertical lines indicate the typical engagement and disengagement times for the clutch.

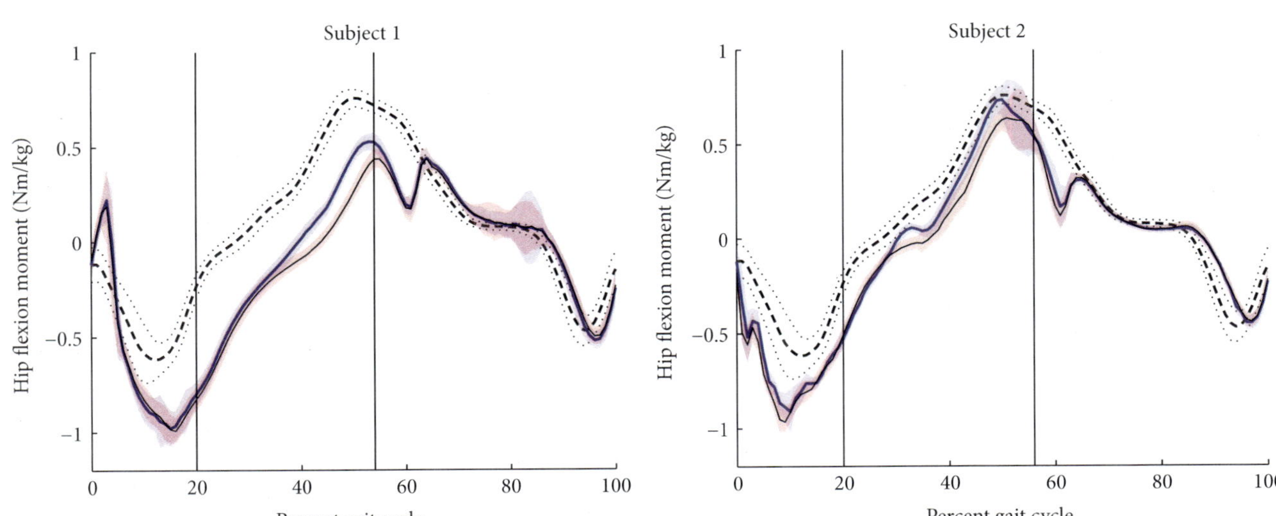

Figure 7: Affected-side hip flexion moment components with the AG shown for the *Baseline* condition where the clutch was disabled (thick solid blue line) and the *Active* condition (thin solid red line). Shaded regions for these curves indicate ±1 standard deviation. For reference, biological hip moment data are also shown for the nonamputee from which the clutch spring was tuned (thick black dashed line) ±1 standard deviation (black dotted lines). The vertical lines indicate the typical engagement and disengagement times for the clutch.

TABLE 6: Affected-side knee positive work in late stance flexion with the AG compared between the two walking conditions with the quasi-passive artificial gastrocnemius: the Baseline condition, where the artificial gastrocnemius acted as a free-joint at the knee, and the Active condition, where the AG was appropriately engaged as per the controller. The values were averaged over the late stance knee flexion phase of the gait cycle.

	Baseline total power (Nm∗s/kg)	Active total power (Nm∗s/kg)	Active biological power (Nm∗s/kg)	Total % change from Baseline	Biological % change from Baseline
Subject 1	0.006 ± 0.004	0.005 ± 0.003	0.002 ± 0.002	−6	−60
Subject 2	0.084 ± 0.064	0.062 ± 0.031	0.049 ± 0.030	−26	−41

TABLE 7: Hip flexion moment impulse in late stance flexion with the AG compared between the two walking conditions with the quasi-passive artificial gastrocnemius: the Baseline condition, where the artificial gastrocnemius acted as a free-joint at the knee, and the Active condition, where the AG was appropriately engaged as per the controller. The values were computed over the late stance knee flexion phase of the gait cycle.

	Baseline moment impulse (Nm∗s/kg)	Active moment impulse (Nm∗s/kg)	% change from Baseline
Subject 1	0.077 ± 0.006	0.057 ± 0.006	−26
Subject 2	0.134 ± 0.013	0.115 ± 0.012	−14

TABLE 8: Affected-side hip positive work in late stance flexion with the AG compared between the two walking conditions with the quasi-passive artificial gastrocnemius: the Baseline condition, where the artificial gastrocnemius acted as a free-joint at the knee, and the Active condition, where the AG was appropriately engaged as per the controller. The values were computed over the late stance knee flexion phase of the gait cycle.

	Baseline positive work (J/kg)	Active positive work (J/kg)	% change from Baseline
Subject 1	0.087 ± 0.011	0.063 ± 0.0090	−27
Subject 2	0.117 ± 0.016	0.095 ± 0.014	−19

TABLE 9: Metabolic power of the amputee participants with the AG.

	Subject 1	Subject 2
Baseline power (W/kg)	3.80 ± 0.19	3.04 ± 0.11
Active power (W/kg)	3.61 ± 0.19	2.94 ± 0.08
Percent change	−5%	−3%

may possibly compensate for the otherwise increased hip power exhibited by transtibial amputees [4].

Given the kinetic results, it is not surprising that slight metabolic improvements were demonstrated as well. Reductions in net positive biological mechanical work should correspond to reductions in concentric muscle work, since passive tissues, by definition, cannot generate net positive work. This reduction in muscle work, in turn, would be expected to reduce metabolic cost [41, 42].

5. Conclusions and Future Work

Despite the advances of today's monoarticular transtibial prostheses, their limitations still manifest in pathological gaits. This study builds on the hypothesis that dominant pathologies are caused by a lack of biarticular gastrocnemius function. Our approach of restoring this biarticular component via an artificial gastrocnemius has shown preliminary but promising results toward the improvement of transtibial prosthesis efficacy.

The quasi-passive nature of this novel artificial gastrocnemius means that the design could exclude heavy components like large batteries and motors. Yet still, this device has the capability to produce biological-levels of knee moments for assisting those with transtibial amputation. The low power draw also helps to make this device practical for daily use, as battery life would not likely be a problem over the course of a day.

The AG, despite lacking an ability to generate positive net mechanical work, did demonstrate an ability to reduce amputee metabolic cost of transport for the two amputees tested. This preliminary result suggests that biarticular devices may hold the potential to improve amputee quality of life beyond that with even the most advanced monoarticular devices. These metabolic benefits may stem from a combination of a reduction in knee and hip moments and powers in the affected-side leg.

However, given the lack of statistical power, more work is needed to verify this effect on more participants. In addition, the slight increase seen in the net mechanical work from the ankle-foot prosthesis during the ACTIVE conditions could account for a nonnegligible portion of this small metabolic benefit. If, however, the metabolic improvement is confirmed for other amputees when ensuring no increase in positive work from the prosthetic ankle, it would show that metabolic improvements are possible for transtibial amputees without the need for the injection of net positive mechanical work.

It would be beneficial to perform additional experiments that more broadly explore this device. Each individual has their own gait idiosyncrasies, and a matching to other human gait solely based on height and weight has its limitations. Therefore, it is possible that, although the joint stiffness of the device was chosen via optimization, this stiffness value may not have corresponded to the optimal values for the given amputee participants. Additionally, inherent compliance in the brace itself may change the effective device stiffness from the designed specification. Future trials could more systematically vary the joint stiffness to find the metabolically optimal value and then compare it to that from simulation.

In addition, it may be desirable to reach beyond the quasi-passive model of the gastrocnemius muscle. Seeing that the biological gastrocnemius generates several joules of net work per step [43, 44], greater benefits could possibly be achieved by providing this net work to amputees. By definition, this net mechanical work cannot be produced with a quasi-passive device, and, therefore, a different mechanism would be needed to test this hypothesis. In the design of either quasi-passive or active transtibial leg prostheses, we feel biarticular gastrocnemius actuation is an important consideration.

Conflicts of Interest

The authors declare that there are no conflicts of interest regarding the publication of this article.

Acknowledgments

This work was supported by departmental funding in the MIT Media Lab.

References

[1] P. A. Struyf, C. M. van Heugten, M. W. Hitters, and R. J. Smeets, "The Prevalence of Osteoarthritis of the Intact Hip and Knee Among Traumatic Leg Amputees," *Archives of Physical Medicine and Rehabilitation*, vol. 90, no. 3, pp. 440–446, 2009.

[2] R. Gailey, K. Allen, J. Castles, J. Kucharik, and M. Roeder, "Review of secondary physical conditions associated with lower-limb amputation and long-term prosthesis use," *Journal of Rehabilitation Research and Development*, vol. 45, no. 1, pp. 15–29, 2008.

[3] D. M. Ehde, J. M. Czerniecki, D. G. Smith et al., "Chronic phantom sensations, phantom pain, residual limb pain, and other regional pain after lower limb amputation," *Archives of Physical Medicine and Rehabilitation*, vol. 81, no. 8, pp. 1039–1044, 2000.

[4] P.-F. Su, S. A. Gard, R. D. Lipschutz, and T. A. Kuiken, "Gait characteristics of persons with bilateral transtibial amputations," *Journal of Rehabilitation Research and Development* , vol. 44, no. 4, pp. 491–501, 2007.

[5] D. A. Winter and S. E. Sienko, "Biomechanics of below-knee amputee gait," *Journal of Biomechanics*, vol. 21, no. 5, pp. 361–367, 1988.

[6] N. H. Molen, "Energy/speed relation of below-knee amputees walking on a motor-driven treadmill," *Internationale Zeitschrift für Angewandte Physiologie Einschließlich Arbeitsphysiologie*, vol. 31, no. 3, pp. 173–185, 1973.

[7] R. L. Waters, J. Perry, D. Antonelli, and H. Hislop, "Energy cost of walking of amputees: the influence of level of amputation," *The Journal of Bone & Joint Surgery*, vol. 58, no. 1, pp. 42–46, 1976.

[8] J. Hu and R. G. Gordon, "Textured aluminum-doped zinc oxide thin films from atmospheric pressure chemical-vapor deposition," *Journal of Applied Physics*, vol. 71, no. 2, p. 880, 1992.

[9] H. M. Herr and A. M. Grabowski, "Bionic ankle-foot prosthesis normalizes walking gait for persons with leg amputation," *Proceedings of the Royal Society B Biological Science*, vol. 279, no. 1728, pp. 457–464, 2012.

[10] D. Hill and H. Herr, "Effects of a powered ankle-foot prosthesis on kinetic loading of the contralateral limb: a case series," in *Proceedings of the IEEE International Conference on Rehabilitation Robotics (ICORR '13)*, vol. 10, pp. 1–6, IEEE, Seattle, WA, USA, June 2013.

[11] C. H. Soo and J. M. Donelan, "Mechanics and energetics of step-to-step transitions isolated from human walking," *Journal of Experimental Biology*, vol. 213, no. Pt 24, pp. 4265–4271, 2010.

[12] A. D. Kuo, J. M. Donelan, and A. Ruina, "Energetic consequences of walking like an inverted pendulum: step-to-step transitions," *Exercise and Sport Sciences Reviews*, vol. 33, no. 2, pp. 88–97, 2005.

[13] J. M. Donelan, R. Kram, and A. D. Kuo, "Mechanical work for step-to-step transitions is a major determinant of the metabolic cost of human walking," *Journal of Experimental Biology*, vol. 205, no. 23, pp. 3717–3727, 2002.

[14] A. D. Kuo, "Energetics of actively powered locomotion using the simplest walking model," *Journal of Biomechanical Engineering*, vol. 124, no. 1, pp. 113–120, 2002.

[15] M. Meinders, A. Gitter, and J. M. Czerniecki, "The role of ankle plantar flexor muscle work during walking," *Journal of rehabilitation medicine*, vol. 30, no. 1, pp. 39–46, 1998.

[16] R. R. Neptune, S. A. Kautz, and F. E. Zajac, "Contributions of the individual ankle plantar flexors to support, forward progression and swing initiation during walking," *Journal of Biomechanics*, vol. 34, no. 11, pp. 1387–1398, 2001.

[17] A. L. Hof, J. Nauta, E. R. van der Knaap, M. A. A. Schallig, and D. P. Struwe, "Calf muscle work and segment energy changes in human treadmill walking," *Journal of Electromyography & Kinesiology*, vol. 2, no. 4, pp. 203–216, 1993.

[18] J. M. Caputo and S. H. Collins, "Prosthetic ankle push-off work reduces metabolic rate but not collision work in non-amputee walking," *Scientific Reports*, vol. 4, article no. 7213, 2014.

[19] M. R. Williams, A. Grabowski, H. Herr, and S. D'Andrea, *Electromyographic Effects of Using a Powered Ankle-Foot Prosthesis*, American Society of Biomechanics, 2012.

[20] J. F. Veneman, R. Kruidhof, E. E. G. Hekman, R. Ekkelenkamp, E. H. F. Van Asseldonk, and H. van der Kooij, "Design and Evaluation of the Gait Rehabilitation Robot LOPES," *IEEE Transactions on Neural Systems and Rehabilitation Engineering*, vol. 15, no. 3, 2007.

[21] J. E. Pratt, B. T. Krupp, C. J. Morse, and S. H. Collins, "The RoboKnee: an exoskeleton for enhancing strength and endurance during walking," in *Proceedings of the IEEE International Conference on Robotics and Automation (ICRA '04)*, vol. 3, pp. 2430–2435, IEEE, May 2004.

[22] P. P. Pott, S. I. Wolf, J. Block et al., "Knee-ankle-foot orthosis with powered knee for support in the elderly," *Proceedings of the Institution of Mechanical Engineers, Part H: Journal of Engineering in Medicine*, vol. 231, no. 8, pp. 715–727, 2017.

[23] D. P. Ferris, J. M. Czerniecki, and B. Hannaford, "An ankle-foot orthosis powered by artificial pneumatic muscles," *Journal of Applied Biomechanics*, vol. 21, no. 2, pp. 189–197, 2005.

[24] G. S. Sawicki and D. P. Ferris, "A pneumatically powered knee-ankle-foot orthosis (KAFO) with myoelectric activation and inhibition," *Journal of NeuroEngineering and Rehabilitation*, vol. 6, no. 1, p. 23, 2009.

[25] D. P. Ferris, K. E. Gordon, G. S. Sawicki, and A. Peethambaran, "An improved powered ankle-foot orthosis using proportional myoelectric control," *Gait & Posture*, vol. 23, no. 4, pp. 425–428, 2006.

[26] M. Wehner, B. Quinlivan, P. M. Aubin et al., "A lightweight soft exosuit for gait assistance," in *Proceedings of the IEEE International Conference on Robotics and Automation (ICRA '13)*, pp. 3362–3369, May 2013.

[27] M. B. Wiggin, G. S. Sawicki, and S. H. Collins, "An exoskeleton using controlled energy storage and release to aid ankle propulsion," in *Proceedings of the IEEE International Conference on Rehabilitation Robotics (ICORR '11)*, pp. 1–5, Zurich, Switzerland, June 2011.

[28] C. J. Walsh, K. Endo, and H. Herr, "A quasi-passive leg exoskeleton for load-carrying augmentation," *International Journal of Humanoid Robotics*, vol. 4, no. 3, pp. 487–506, 2007.

[29] G. Elliott, A. Marecki, and H. Herr, "Design of a clutch-spring knee exoskeleton for running," *Journal of Medical Devices, Transactions of the ASME*, vol. 8, no. 3, Article ID 031002, 11 pages, 2014.

[30] M. S. Cherry, D. J. Choi, K. J. Deng, S. Kota, and D. P. Ferris, "Design and Fabrication of an Elastic Knee Orthosis: Preliminary Results," in *Proceedings of the ASME 2006 International Design Engineering Technical Conferences and Computers and Information in Engineering Conference*, pp. 565–573, Philadelphia, PA, USA, 2006.

[31] M. Ishikawa, P. V. Komi, M. J. Grey, V. Lepola, and G.-P. Bruggemann, "Muscle-tendon interaction and elastic energy usage in human walking," *Journal of Applied Physiology*, vol. 99, no. 2, pp. 603–608, 2005.

[32] K. Endo and H. Herr, "A model of muscle-tendon function in human walking," in *Proceedings of the 2009 IEEE International Conference on Robotics and Automation, ICRA '09*, pp. 1909–1915, IEEE, Kobe, Japan, May 2009.

[33] K. Endo and H. Herr, "Human walking model predicts joint mechanics, electromyography and mechanical economy," in *Proceedings of the 2009 IEEE/RSJ International Conference on Intelligent Robots and Systems, IROS 2009*, pp. 4663–4668, St. Louis, MO, USA, October 2009.

[34] K. Endo, D. Paluska, and H. Herr, "A quasi-passive model of human leg function in level-ground walking," in *Proceedings of the 2006 IEEE/RSJ International Conference on Intelligent Robots and Systems, IROS 2006*, pp. 4935–4939, Beijing, China, October 2006.

[35] K. Endo, E. Swart, and H. Herr, "An artificial gastrocnemius for a transtibial prosthesis," in *Proceedings of the 31st Annual International Conference of the IEEE Engineering in Medicine and Biology Society: Engineering the Future of Biomedicine, EMBC 2009*, pp. 5034–5037, Minneapolis, MN, USA, September 2009.

[36] M. Palmer, *Sagittal Plane Characterization of Normal Human Ankle Function Across a Range of Walking Gait Speeds , Masters Thesis [Masters, thesis]*, MIT, 2002.

[37] J. M. Brockway, "Derivation of formulae used to calculate energy expenditure in man," *Hum Nutr Clin Nutr*, vol. 41, no. 6, pp. 463–471, 1987.

[38] A. S. McIntosh, K. T. Beatty, L. N. Dwan, and D. R. Vickers, "Gait dynamics on an inclined walkway," *Journal of Biomechanics*, vol. 39, no. 13, pp. 2491–2502, 2006.

[39] P.-C. Kao, C. L. Lewis, and D. P. Ferris, "Invariant ankle moment patterns when walking with and without a robotic ankle exoskeleton," *Journal of Biomechanics*, vol. 43, no. 2, pp. 203–209, 2010.

[40] C. L. Lewis and D. P. Ferris, "Erratum to "Invariant hip moment pattern while walking with a robotic hip exoskeleton" [J. Biomech. 44 (5) (2011) 789-793]," *Journal of Biomechanics*, vol. 47, no. 7, p. 1748, 2014.

[41] A. Hill, "The heat of shortening and the dynamic constants of muscle," *Proceedings of the Royal Society B*, vol. 126, pp. 136–195, 1938.

[42] A. Hill, "The efficiency of mechanical power development during muscular shortening and its relation to load," *Proceedings of the Royal Society B Biological Science*, vol. 159, no. 975, pp. 319–324, 1964.

[43] R. J. Zmitrewicz, R. R. Neptune, and K. Sasaki, "Mechanical energetic contributions from individual muscles and elastic prosthetic feet during symmetric unilateral transtibial amputee walking: A theoretical study," *Journal of Biomechanics*, vol. 40, no. 8, pp. 1824–1831, 2007.

[44] R. R. Neptune, K. Sasaki, and S. A. Kautz, "The effect of walking speed on muscle function and mechanical energetics," *Gait & Posture*, vol. 28, no. 1, pp. 135–143, 2008.

12

Bioinspired Knee Joint for a Power-Assist Suit

Takehito Kikuchi, Kohei Sakai, and Isao Abe

Oita University, Dannoharu 700, Oita 870-1192, Japan

Correspondence should be addressed to Takehito Kikuchi; t-kikuchi@oita-u.ac.jp

Academic Editor: Liwei Shi

Movement of the knee joint of a human includes rolling and sliding. There also exist rotations in the frontal and horizontal planes. To assist the standing movement of a human, we developed a bioinspired knee joint and torque adjustment mechanism. We evaluated the motion, torque characteristics, and stress of the developed mechanism. This joint allows deep flexion of the knee with small resistance for both the user and the device. In addition, in spite of 33% error in deep flexion, the measured torque over less than 120 degrees fits the designed torque curve. We conducted evaluation tests for a human subject. The electromyogram (EMG) of musculus rectus femoris was measured during standing with or without the assistance. The result shows 30% and 63% reduction with the assistance from 100-degree and 80-degree knee angles, respectively. In addition, the proposed device reduced up to 80% of stress in the frontal plane during standing.

1. Introduction

The aging of society is rapidly progressing all over the world. According to the report of Department of Economic and Social Affairs, USA [1], the number of persons aged 60 years or more was 841 million in 2013 worldwide, and the number is expected to be 2 billion in 2050. Aging affects the overall physical functions. In particular, arthralgia (joint pain) is one of the most severe factors that prevent active lives of the aged. In Japan, the number of patients suffering from knee osteoarthritis is approximately 7 million, and the proportion continues to increase [2]. Knee joints are very important for many activities of daily living, for example, standing, walking, and climbing stairs; more than 0.5 Nm per body weight in torque is generated in the knee joint during walking [3]. The torque required in the standing action exceeds this value. The pain of knee osteoarthritis causes the losses of the range of motions, motivations of activities, and range of their activities in life.

The power-assist technology was originally developed as supporting technology for military persons [4]. However, the technology is now expected to promote the active life of the aged. Commercially available power-assist wearable devices already exist, for example, MuscleSuits (INNOPHYS Co., Ltd.) [5] and HAL (CYBERDYNE Inc.) [6]; these devices are used to reduce the loads of standing or the carrying burdens of persons. Such devices use several types of actuators to generate the assistive torque and can reduce the force that directly reacts on the user's body. However, due to the limitation of the mechanism, almost all of the devices do not support motions that require deep flexion of the knee. In addition, some devices that utilized electromagnetic motors tend to be quite heavy.

In recent years, pneumatic artificial muscles have been the focus for use in power-assist system because of their high force/weight ratio [7]. A pneumatic actuator contracts with the injection of air. The core element of a pneumatic actuator is a flexible membrane closed at both ends; this rubber bag is wrapped with the sleeves of hard plastic fibers. The rubber bag inflates with the injection of air, but the deformation toward radial direction is limited by the hard sleeves. As a result, the contraction toward axial direction occurs with the injection of air [8]. The above-mentioned MuscleSuits also utilize pneumatic artificial muscles.

The use of the pneumatic muscles in a power-assist system depends on the application and is mainly categorized into two types. In some devices, the pneumatic muscles are directly attached onto the surfaces of the human body, similar to clothes, for example, in the front and back of the target joint to assist its flexion and extension [9]. The other devices are combinations of the pneumatic actuators and mechanical joints, with a single-axis that is located in the outer side of the human joint [10]. The former types have an advantage that they do not require accurate alignment to the axis of the human joint. However, the inflated actuators give users the feeling of pressure. Alternatively, although the latter types of devices do not give such feelings of pressure, they must be accurately adjusted to the human joints. In addition, the above studies have not addressed the deep flexion of knee, although there are many such motions in daily life.

The complicated motion of knee joint results from the restraints of motion with the related ligaments and tendons. For example, the extension of the knee includes the roll-back motion, which is a combination of rotation and sliding [11]. At the same time, the rotational axis moves, depending on the knee angle. It is difficult to adequately fit the natural motion of human knee joint only using a mechanical joint with a single-axis. Some studies that utilize a bioinspired knee mechanism can be found in the field of humanoids [12, 13] and prosthetic knees [14]. Terada et al. [15] developed a knee-motion-assist mechanism for a wearable robot with a noncircular gear and grooved cams. In this paper, we also focused on the bioinspired knee joint mechanism but utilized a different mechanism.

Furthermore, during flexion of the knee joint, the femur normally has a 5–10 degrees' abduction against the tibia, reaching up to 30 degrees during extension in some cases [16]. From the 30-degree flexion to the terminal extension, the femur rotates 5–10 degrees in inner rotation. This motion is important to lock the knee motion in the terminal extension of the knee and prevent hyperextension [11]. There are few studies of the mechanical knee joint that considered the motions in the frontal and the horizontal planes.

A new mechanical knee joint that solves the above-mentioned problems is necessary for the pneumatic actuator-based power-assist suit to improve the usefulness and comfort of the suit. In this study, we propose a new bioinspired mechanism for mechanical knee joint and flexible elements that can reduce the stress from the motion with the misaligned joint.

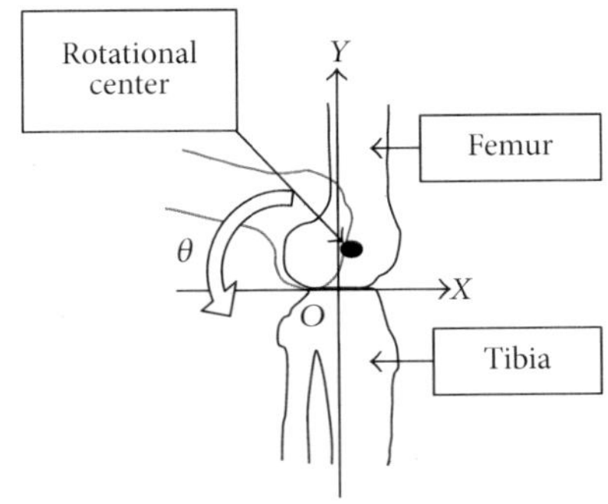

FIGURE 1: Coordination system of the tibia and femur [17].

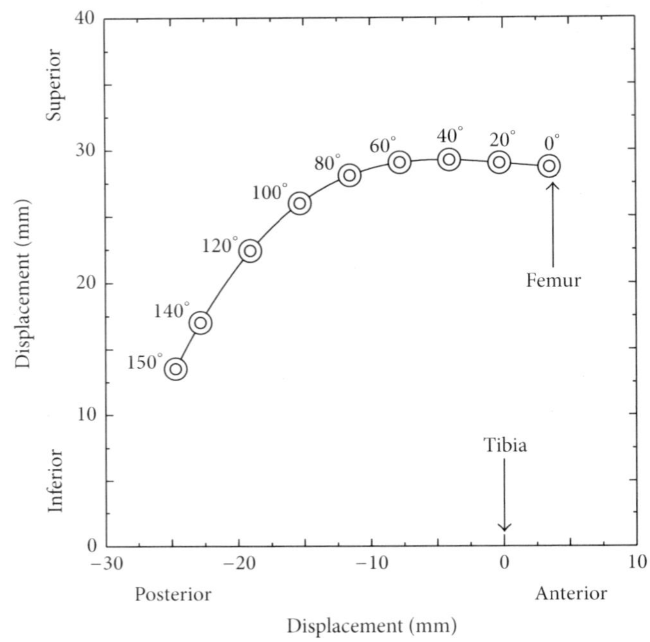

FIGURE 2: Trajectory of the center of the femur in the sagittal plane for a Japanese male (original data are obtained from [16]).

2. Target Setting

2.1. Trajectory of the Axis of the Knee Joint. The motion of the axis of knee joints is a combination of rotation and sliding [11]. Therefore, the rotational axis is not fixed in a point. According to the definition of the coordination system in [17] (Figure 1), the origin of the femur's rotation in the sagittal plane with respect to the tibia's origin moves as shown in Figure 2 during the standing motion [16]. The vertical axis of this figure shows the displacement of the superior-inferior motion of the rotational axis (positive values denote the superior positions). The horizontal axis shows the anteroposterior motion (positive values denote anterior positions). The number indicated at each plot denotes the knee angles, and the maximum extension and flexion are 0 and 150 degrees, respectively.

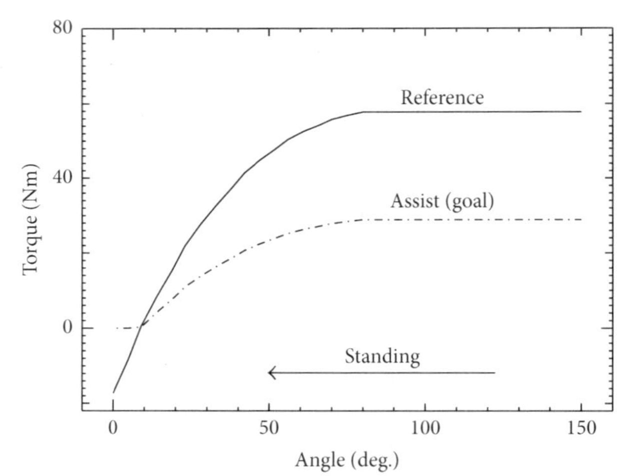

FIGURE 3: Natural standing torque and assisted torque around the knee joint (original data of solid line are obtained from [18]).

2.2. Goal of Assistive Torque. The solid line in Figure 3 shows the average torque profile during standing for a Japanese male

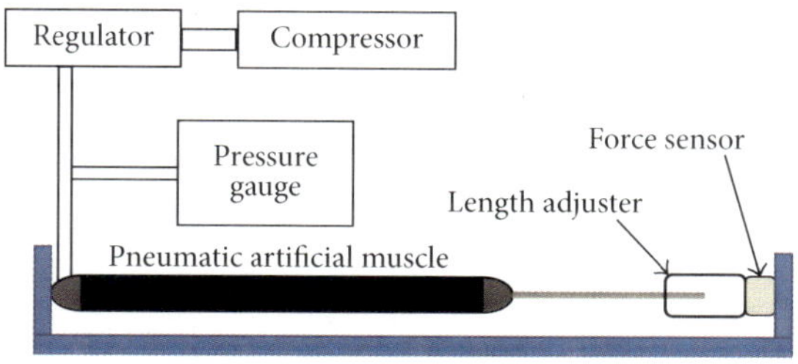

FIGURE 4: Measurement method for determining the force characteristics of the pneumatic actuator.

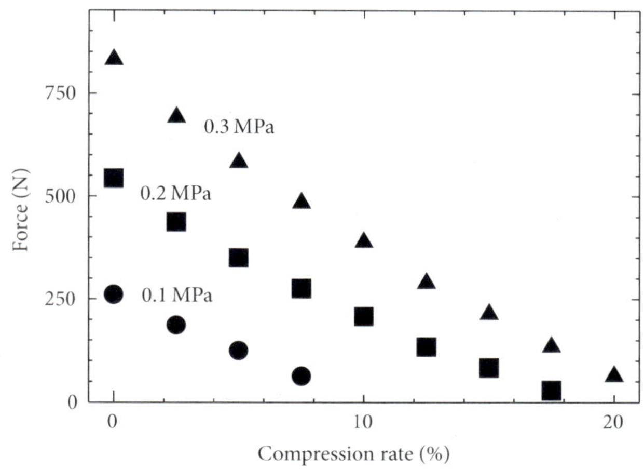

FIGURE 5: Compression rate versus compression force of the pneumatic actuator.

(40 years old, weight: 65.3 kg, and height: 171 cm) obtained from [18]. The maximum torque is 57.8 Nm at the moment of standing. The maximum torque remains flat within 80–150 degrees' knee angle. This line is defined as a reference in this paper. In this study, we set a half of the reference as the goal of the assistive torque (Figure 5). For simplification of the structure, only extension of the knee was assisted with the device. As a result, the negative torque in the reference was translated to zero in the goal.

2.3. Force Characteristics of the Pneumatic Actuator. In this study, we used AIR-MUSCLE (Kanda Tsushin Kogyo Co., Ltd., A300B20C20D, nominal length: 800 mm, and max. pressure: 0.6 MPa) as a power source. In addition, the maximum air pressure was defined as 0.3 MPa to allow for use of a compact compressor. This actuator is a McKibben-type artificial muscle, which can generate a compression force with air input. The force characteristics of the actuator were measured with the setup shown in Figure 4 because the actuator has individual difference in the force characteristics.

Figure 5 shows the experimental results of the static force characteristics of this actuator. The horizontal axis shows the contraction ratio with respect to the nominal length. The vertical axis shows the compression force with each value of inner pressure and length. This actuator generates more than 750 N with 0.3 MPa at the nominal length. The force decreases with the compression rate, and it becomes zero at approximately 20% compression.

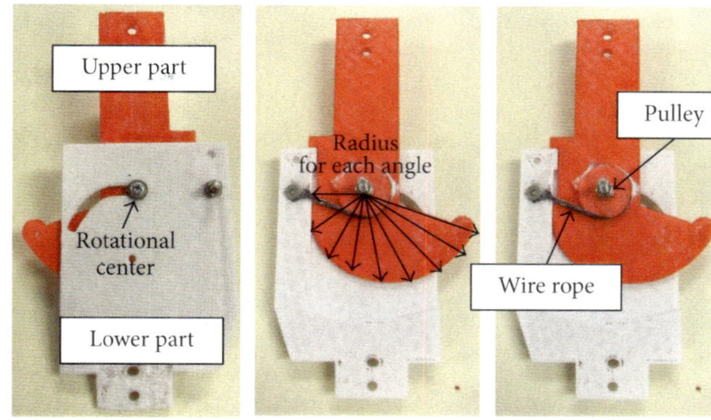

FIGURE 6: Knee mechanism of the device.

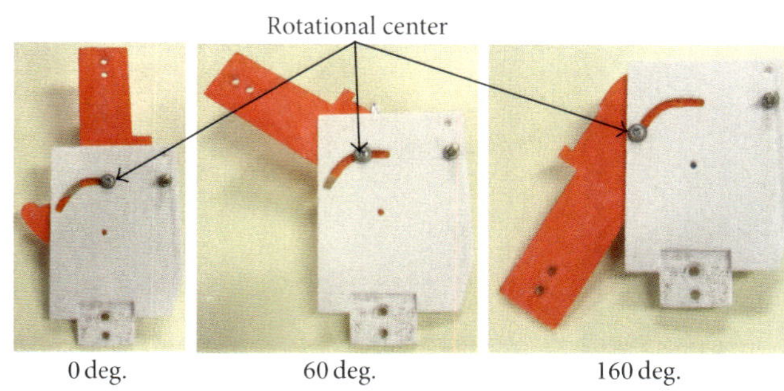

FIGURE 7: Motion of the knee mechanism.

3. Materials

3.1. Bioinspired Knee Joint. We developed a prototype of the joint mechanism by referring to the trajectory of the rotational center [19], as shown in Figure 2. Figure 6 shows the structure of the prototype, which consists of an upper part, a lower part, a wire rope, and a pulley. The upper part and the lower part move relatively with sliding and rolling. The rotational center of the upper part has a shaft that can move in a curved hole. A circular pulley is fixed with the upper part and is joined with a wire rope, one edge of which is fixed on a point of the lower part. The anteroposterior motion of the rotational axis depends on the knee angle because the shape of the sliding surface is noncircular (see the middle figure of Figure 6). The wire-pulley mechanism is also used; this mechanism defines the length between the rotational center and the fixed point (see the right figure of Figure 6). The actual motion of the mechanism is shown in Figure 7. This joint allows for deep flexion of knee with small resistance for both the user and the device (Figure 8).

3.2. Lever Arm for the Pneumatic Actuator. Figure 9 shows the basic structure of the proposed power-assist suit. The joint mechanism mentioned above is activated with a McKibben-type artificial muscle. For simplification of the mechanism, there are no reduction gears. In this structure, 40 mm of

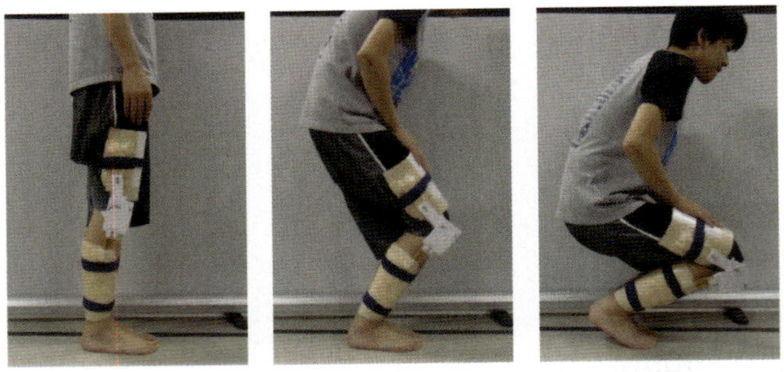

Figure 8: Squat with the knee mechanism.

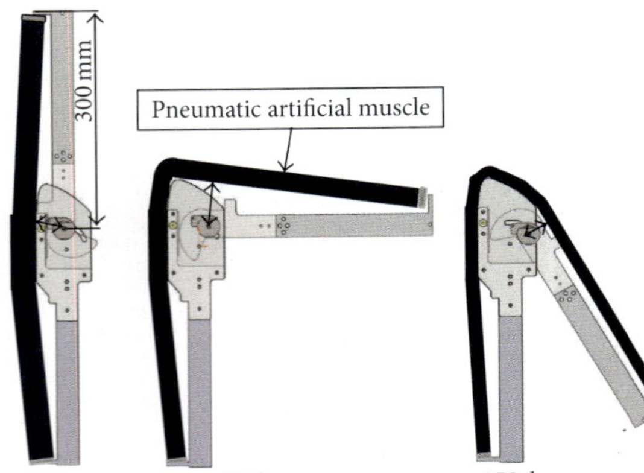

Figure 9: Structure and motion of the device.

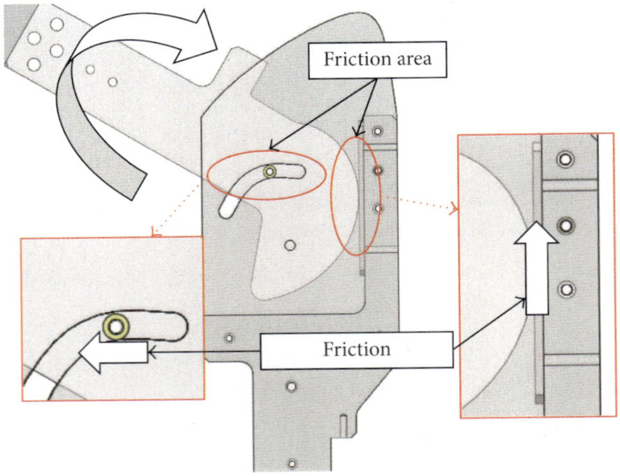

Figure 10: Motion of the torque adjustment mechanism.

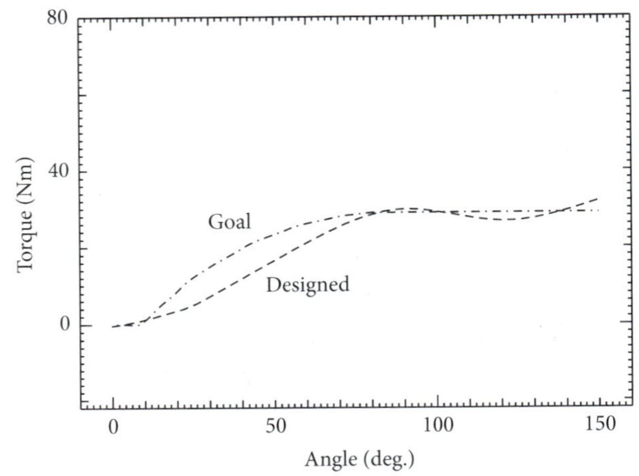

Figure 11: Designed torque.

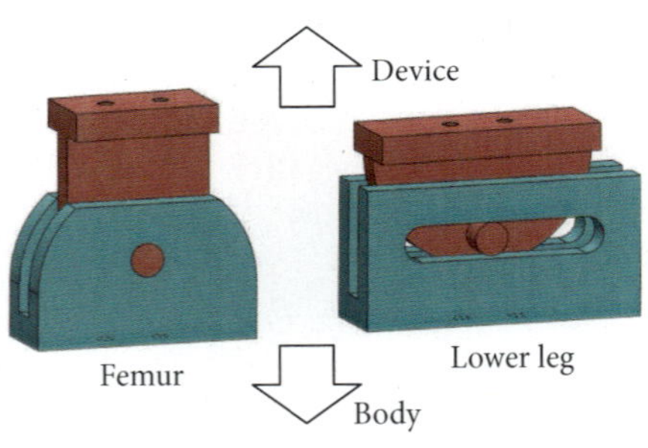

Figure 12: Hinge parts for flexibility.

compression is required to achieve 150 degrees of flexion. Thus, we selected the AIR-MUSCLE as the actuator.

To meet the goal of Figure 3 with the force characteristics of Figure 5, the adjustment of the lever arm is required. Hence, we designed the shape of the outer surface as the patella of the device (Figure 9). As shown in Figure 5, the actuator can generate a large force over a low range of the contraction rate. The force gradually decreases with the contraction and reaches zero at a specific contraction rate. However, the required torque remains constant within knee angles 80–150 degrees (Figure 3). To meet this requirement, the shape was designed to have a large lever arm at 90-degree flexion of the knee.

To simulate the torque generation with this structure, we assumed two friction areas, as shown in Figure 10. We attached fluororesin tapes onto the surface of the lower part, and then the friction coefficient between the upper and lower parts was defined as 0.3. The friction coefficient between the slit (aluminum) and rotational shaft (stainless steel) was defined as 0.4. The calculation result is shown in Figure 11. The designed curve almost achieves the goal.

3.3. Hinge Mechanism for Flexibility. To enhance the compatibility of the wearable device with the human body, the mechanism should have suitable flexibility. However, rigidity is required for the control of direction. To satisfy these conflicting requirements, we use a pair of hinge joints, as shown in Figure 12, between the human body and the machine. One of the joints has a simple rotational joint and is attached to the femur (upper) part of the device. The other has a pin and a slot to allow for rotation and sliding and is attached to the lower part.

The hinge parts were installed between the knee mechanism and the plastic cuffs. The plastic cuffs were tightened on the user's thigh and shin by using wide and stretchable supporters. Figure 13 shows the motion of the knee

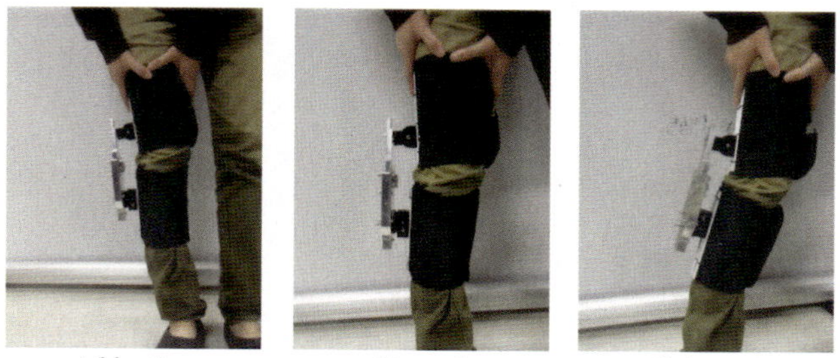

FIGURE 13: Installation of hinge parts and their motion.

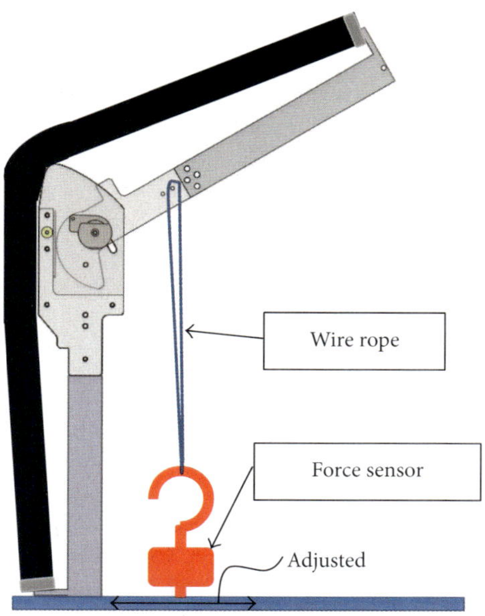

FIGURE 14: Setup for the torque measurement.

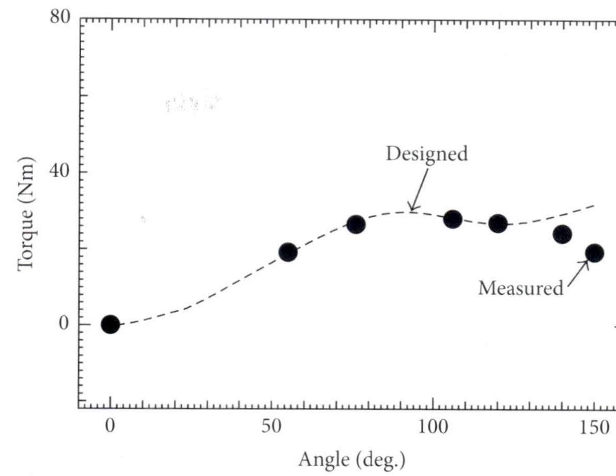

FIGURE 15: Comparison of the measured and design torque.

FIGURE 16: Flattened tube at deep flexion of the knee.

mechanism with the hinge parts. The upper and lower cuffs fit on the legs during abduction and adduction of the knee. The smooth flexion and extension of knee were achieved.

4. Evaluation of the Torque Output

4.1. Method. To evaluate the torque output, we prepared an experimental setup shown in Figure 14. The basic structure of the assist device was fixed on a rigid base. A wire rope was hooked at a specific point on the device with one end. The other end was hooked with a force sensor and the force sensor was manually fixed onto the base such that the wire rope is perpendicular to the base. The static torque measurement was conducted on each knee angle (7 points between 0 and 150 degrees). A constant air pressure of 0.3 MPa was applied to the actuator for each angle of knee (0–150 degrees). The torque around the axis of the knee joint was calculated with the torque and lever arm.

4.2. Result. Figure 15 shows the experimental results (plots) and the design torque curve. The measured torque for angles less than 120 degrees fit the designed curve. The friction model mentioned in the Section 3.2 adequately estimated the static assistive torque of the device for angles less than 120 degrees. However, an error exists between the measured and design torque in deep flexion for angles of over 140 degrees. At the maximum flexion, the measured torque was 33% less than the design torque.

4.3. Discussion. To explain the cause of the error in deep flexion, Figure 16 shows the condition of the pneumatic actuator at a knee angle of 150 degrees. As shown in the picture, the main tube of the pneumatic actuator is flattened in the deep flexion; this separates compressible volume in this state. This flattened state possibly corresponds to a 33% loss from the maximum output of the actuator. This problem can be solved by using a longer actuator; however, the required margin of air volume results in the delayed response at the initial moment of standing motion. We will attempt to solve this problem in the future.

5. Evaluation of Assistance

5.1. Method. Despite the shortness of output in the deep flexion condition, we conducted evaluation tests using a human

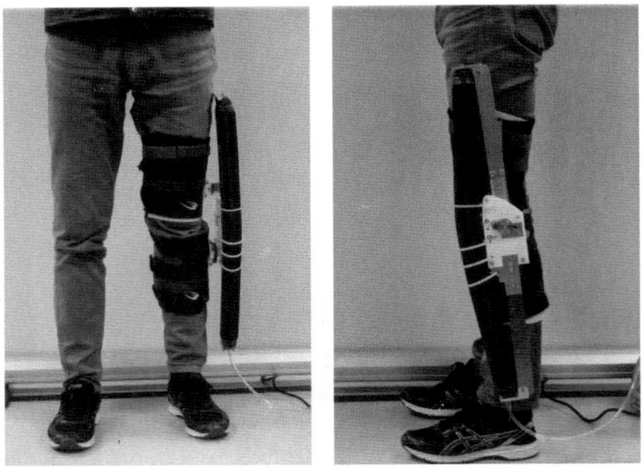

Figure 17: Device attached on the body.

subject. A male subject (22 years old, weight: 65 kg, and height: 177 cm) without any physical or mental disabilities was recruited. He wore the assistive device on his left leg (Figure 17). The plastic cuffs were tightened on his thigh and shin by using wide and stretchable supporters. He sat on seats of two different heights (210 mm and 420 mm) at the initial state before standing. A wireless electromyogram (EMG) sensor (Logical Product, LP-WS1221) was attached on the musculus rectus femoris of his left thigh, and the EMG was measured during standing. A Flexible Goniometer System (Biometrics Ltd., K800 and SG110) was used to measure the knee angle. We just opened a valve for the air tube of the actuator at the instant of standing with assistance from the device. The air pressure was maintained at 0.3 MPa during assistance. The start switch was pushed by the subject. The experiment was repeated five times for each condition.

5.2. Result. The knee angles at the initial state were 100 degrees for the 210 mm chair and 80 degrees for the 420 mm chair. Figure 18 shows the experimental results for the four conditions described as follows:

(1) EMG without assistance from 100 degrees.

(2) EMG with assistance from 100 degrees.

(3) EMG without assistance from 80 degrees.

(4) EMG with assistance from 80 degrees.

The horizontal axis shows the knee angle; the angle starts from 100 degrees for conditions (1) and (2) and 80 degrees for conditions (3) and (4).

The root mean square (RMS) with 100 ms interval was calculated from the raw EMG signals. The results of repeating the test five times are averaged for the same angles. The maximum for condition (1) was defined as the maximum voluntary contraction (MVC).

5.3. Discussion. The maximum outputs appeared just after standing, followed by a gradual decrease. This trend is consistent with the characteristics of knee torque during

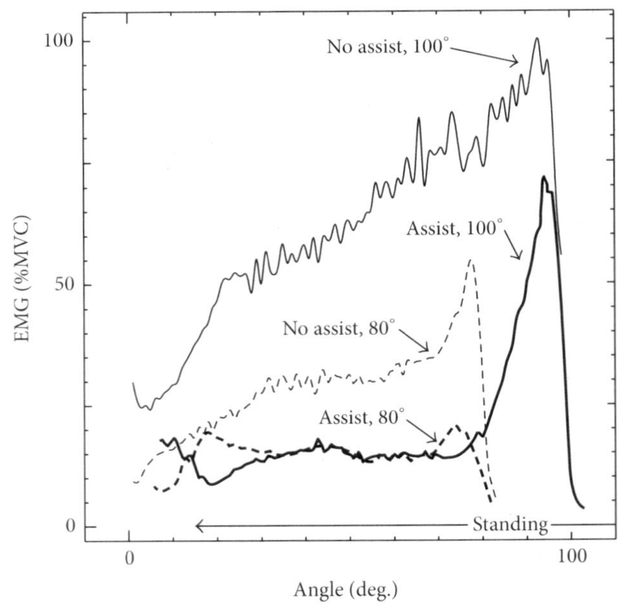

Figure 18: Muscle activation with and without assistance.

standing (Figure 3). The peak EMGs with the assistance show 30% and 63% reduction from that without the assistance for 100 degrees' standing and 80 degrees' standing, respectively. After the peaks, the EMGs with the assistance were significantly smaller than those without the assistance. From these results, the developed device was found to successfully help standing only with the user's effort at the initial moment of standing. Although the designed assistive torque is just half of the reference torque (see Figures 3 and 15), the device significantly reduces the muscle activation without the initial period of standing. The reason for this result remains unclear; however, there is a possibility that the assist device also helps to reduce the muscle activation for maintaining posture. In this study, we conducted the experiments for a healthy subject. However, the final goal of this device is to assist the standing of disabled people. To achieve this goal, we must establish the biofeedback control using the EMG signals. One of the merits of this device is the simplicity of the control. A feedback control system was not used in the proposed device. We only input a constant pressure of air in the pneumatic actuator because the output characteristics are mechanically adjusted for the standing motion. This merit is helpful for developing a rehabilitative assistance suit for standing.

6. Evaluation of Stress

6.1. Method. Misalignments of assist suits generate unexpected stress on users' joints and skins. In order to reduce such stress, we proposed the bioinspired knee joint and flexible hinge parts. In this section, we measured the stress applied to the device in three conditions:

(1) A single-axis knee joint (its rotational center is fixed at the position of knee extension) without the flexible parts.

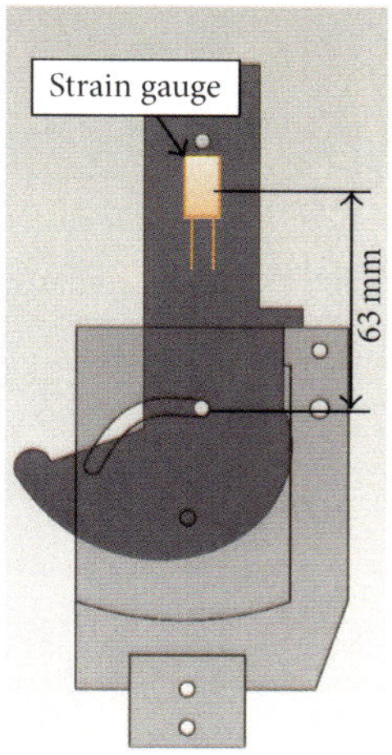

FIGURE 19: Position of strain gauge.

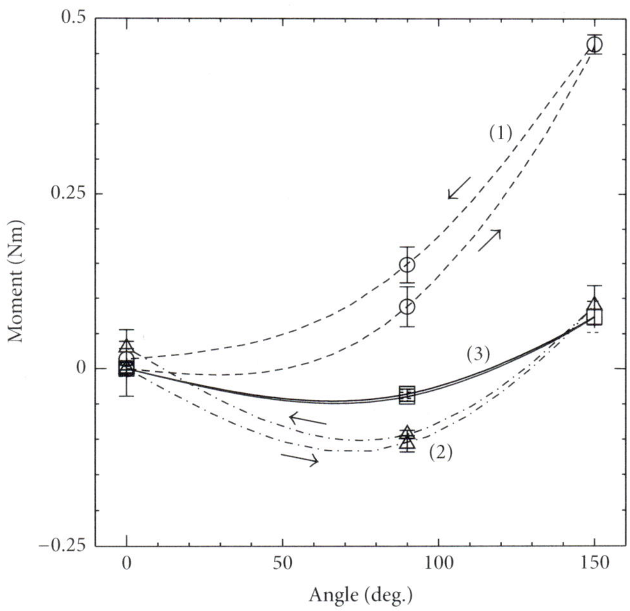

FIGURE 20: Bending moment on knee joint.

(2) The single-axis knee joint with the flexible parts.

(3) The bioinspired knee joint with the flexible parts.

A strain gauge was attached on the upper part of the knee joint (Figure 19). A healthy subject wore the knee mechanism and flexed (0–150 degrees) and extended (150–0 degrees) his knee joints 5 times without the assistance. The bending moment in the frontal plane applied to the frame of the device was calculated from the measured stress.

6.2. Result. The experimental results are shown in Figure 20. The vertical axis is the bending moment on the upper part, and the horizontal axis is the angle of knee joint. The negative direction of the bending moment represents the deformation away from the user. Each mark denotes the average value in each condition. The error bars denote the standard deviations.

6.3. Discussion. The bending moments at the maximum flexion are 0.46 Nm, 0.09 Nm, and 0.07 Nm for conditions (1), (2), and (3), respectively. But, for condition (2), the maximum absolute value reached 0.11 Nm in negative direction. During flexion of the knee joint, the femur normally has a 5–10 degrees' abduction against the tibia [11]. This motion generates undesired stress on both users and assist suits. The flexible parts reduced the stress 80% at the maximum flexion. In addition, the maximum variations are 0.20 Nm and 0.11 Nm, for conditions (2) and (3), respectively. The bioinspired knee joint reduced 45% of the variation of stress during standing.

7. Conclusions

In this study, we developed a new knee joint mechanism that includes a torque adjustment mechanism. We evaluated the motion, torque characteristics, and stress of the device. The combination of the sliding mechanism and wire-pulley mechanism fitted the subject's flexion-extension motions. The result of the EMG measurement during standing shows 30% and 63% reduction with the assistance of motion over 100-degree and 80-degree knee angles, respectively. In addition, the proposed device reduced up to 80% of stress in the frontal plane during standing.

Competing Interests

The authors declare that they have no competing interests.

Acknowledgments

This research was supported by Fluid Power Technology Promotion Foundation.

References

[1] United Nations Department of Economic and Social Affairs/Population Division, "World population ageing 2013," Tech. Rep. ST/ESA/SER.A/348, 2013.

[2] H. Kurosawa, "Rehabilitation as a treatment for osteoarthritis of the knee: the effectiveness of home exercise," *The Japanese Journal of Rehabilitation Medicine*, vol. 42, pp. 124–130, 2005 (Japanese).

[3] J. Perry and J. Burnfield, *Gait Analysis: Normal and Pathological Function*, Slack, 2010.

[4] H. Kazerooni, "Exoskeletons for human power augmentation," in *Proceedings of the IEEE/RSJ International Conference on Intelligent Robots and Systems (IROS '05)*, pp. 3120–3125, August 2005.

[5] INNOPHYS, June 2015, https://innophys.jp/.

[6] Cyberdyne Inc, 2015 http://www.cyberdyne.jp/.

[7] X. Li, T. Noritsugu, M. Takaiwa, and D. Sasaki, "Design of wearable power assist wear for low back support using pneumatic actuators," *International Journal of Automation Technology*, vol. 7, no. 2, pp. 228–236, 2013.

[8] F. Daerden and D. Lefeber, "Pneumatic Artificial Muscles: actuators for robotics and automation," in *Proceedings of the IEEE/ASME International Conference on Advanced Intelligent Mechatronics*, vol. 2, pp. 738–743, Como, Italy, July 2001.

[9] T. Noritsugu, D. Sasaki, M. Kameda, A. Fukunaga, and M. Takaiwa, "Wearable power assist device for standing up motion using pneumatic rubber artificial muscles," *Journal of Robotics and Mechatronics*, vol. 19, no. 6, pp. 619–628, 2007.

[10] K. Tadano, H. Araya, K. Kawashima, C. Youn, and T. Kagawa, "Development of jump assist system using pneumatic rubber muscle," in *Proceedings of the 8th JFPS International Symposium on Fluid Power*, pp. 677–682, October 2011.

[11] V. H. Frankel and M. Nordin, *Basic Biomechanics of the Skeletal System*, Lea & Febiger, Philadelphia, Pa, USA, 1980.

[12] Y. Asano, H. Mizoguchi, M. Osada et al., "Biomimetic design of musculoskeletal humanoid knee joint with patella and screw-home mechanism," in *Proceedings of the IEEE International Conference on Robotics and Biomimetics (ROBIO '11)*, pp. 1813–1818, Phuket, Thailand, December 2011.

[13] Y. Morita, Y. Kawai, H. Ukai, K. Sanaka, H. Nakamuta, and K. Takao, "Development of leg robot for physical therapy training—proposal of knee joint mechanism with rolling, sliding and coming off," in *Proceedings of the International Conference on Mechatronics and Information Technology*, pp. 333–334, 2009.

[14] A. C. Etoundi, R. J. Lock, R. Vaidyanathan, and S. C. Burgess, "A bio-inspired condylar knee joint for knee prosthetics," *International Journal of Design and Nature & Ecodynamics*, vol. 8, no. 3, pp. 213–225, 2013.

[15] H. Terada, Y. Zhu, M. Suzuki, C. Cheng, and R. Takahashi, "Developments of a knee motion assist mechanism for wearable robot with a non-circular gear and grooved cams," in *Mechanisms, Transmissions and Applications*, vol. 3 of *Mechanisms and Machine Science*, pp. 69–76, Springer, Berlin, Germany, 2012.

[16] Y. Shiraishi, T. Shimoto, H. Higaki et al., "Functional assessment for the natural knee joints in squat activity by simulation of 2D X-ray images based on 3D CT images," *Transactions of the Japan Society of Mechanical Engineers, Part C*, vol. 77, no. 782, pp. 3761–3769, 2011 (Japanese).

[17] T. P. Andriacchi, E. J. Alexander, M. K. Toney, C. Dyrby, and J. Sum, "A point cluster method for in vivo motion analysis: applied to a study of knee kinematics," *Journal of Biomechanical Engineering*, vol. 120, no. 6, pp. 743–749, 1998.

[18] M. Ohashi, Y. Ehara, K. Shimada, J. Maeda, Y. Katano, and F. Sato, "Hip, knee, and ankle torque measurement during standing-up and sitting-down motion by link model method," *The Japanese Journal of Rehabilitation Medicine*, vol. 27, no. 2, pp. 107–113, 1990 (Japanese).

[19] K. Sakai, T. Kikuchi, and I. Abe, "Development of bio-inspired knee joint for power assist suit," in *Proceedings of the IEEE International Conference on Robotics and Biomimetics (ROBIO '15)*, pp. 523–528, Zhuhai, China, December 2015.

An Alignment Method of Human-Robot Collaboration based on the Six-Dimensional Force/Torque Dynamic Measurement for Large-Scale Components

Ke Wen,[1] **Fuzhou Du**,[2] **Jiabo Zhang**,[1] **and JiZhi Yang**[1]

[1] *Beijing Spacecrafts, China Academy of Space Technology, Beijing 100194, China*
[2] *School of Mechanical Engineering and Automation, Beihang University, Beijing 100191, China*

Correspondence should be addressed to Fuzhou Du; du_fuzhou@163.com

Academic Editor: Keigo Watanabe

A Stewart parallel robot (SPR) is a promising choice for alignment or assembly of components that are large or heavy. This paper presents a method for human-robot collaboration, for positioning and orientation of large components. Use of interactive force measurements is important for human-robot collaboration. It is based on six-dimensional force/torque (F/T) measurements. First, the six-dimensional F/T data are calculated based on the six-actuator SPR geometry and screw theory. Second, the effects of gravity forces (dynamic gravity compensation) are considered, and a method to offset their effects is explained. Third, force estimation experiments were performed using an S-type force sensor and known applied test forces. Finally, the F/T-driven feedback was tested for the alignment of a large-scale component. The experimental results show that the calculated six-dimensional F/T can accurately track the force applied to a large and/or heavy component by a human worker. It can also accurately predict the F/T required to compensate for inertial forces and components' weight. Thus, the alignment method of human-robot collaboration based on the six-dimensional force/torque dynamic measurements for large-scale components is correct and effective.

1. Introduction

With the constant increase in the difficulty of assembly and the numbers of large-scale satellite cabins, rocket cabins, ship cabins, and other large and/or heavy assemblies, the assembly requirements for large-size and heavy-weight components present obstacles. These can include safety risks for workers. Currently, assembly methods based on manual operations have many problems such as physical difficulty for workers, high collision risks, inadequate precision, and low productivity.

Modern industrial practice is moving towards intelligence, diversification, and personalization. Human-robot collaborations have become hot spots in the field of robotics. Such robots allow people to enter their work spaces and collaborate to complete operations. Robots and humans no longer need to work separately. This type of collaborative process not only takes advantage of the large load capacity, high stability, and high-precision position and orientation (P&O) capabilities of robots but also complements the capabilities of people for on-site observation and task flexibility [1, 2]. The team of Mitsantisuk introduced a new force sensor less haptic human-robot collaboration system based on 3D parallel mechanism delta robot, and the controller of delta robot was designed with considering of gravity compensation [3]. The team of Geravand introduced two examples of reaction schemes for collaboration, with the user pushing/pulling the robot at any point of its structure (e.g., for manual guidance) or with a compliant-like robot behaviour in response to forces applied by the human [4].

An industrial robot's characteristics of high load capacity, high-precision P&O control, and good stability mean it can offer stable and accurate positioning of heavy components. In recent years, parallel robots have been widely used for P&O of the large components and assemblies, due to these advantages as well as high stiffness and rapid motion [5, 6]. Stewart parallel robots (SPRs) are the most frequently used type of parallel robot structure. They are suitable for

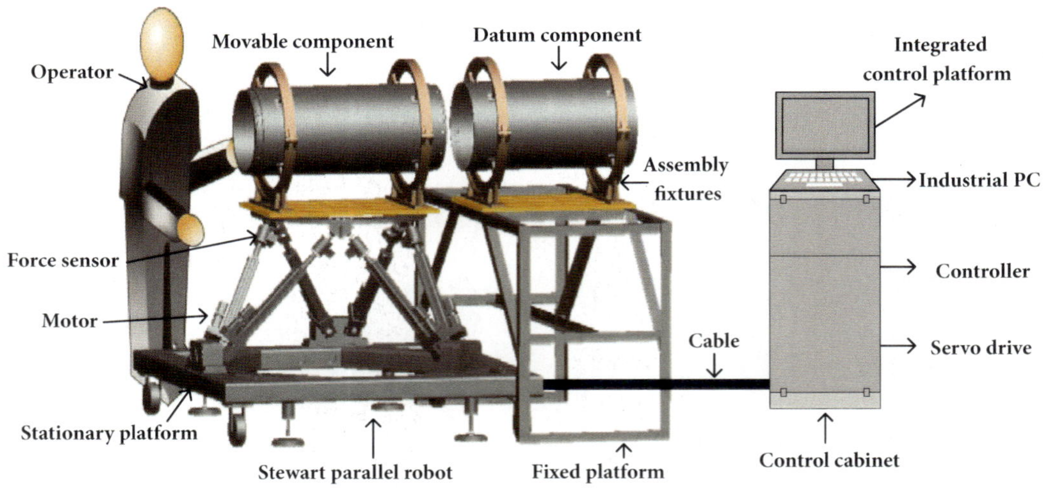

Figure 1: The large-scale alignment system with an SPR for large components.

machining and manufacturing [7, 8], surgical operations [9], flight simulators [10], flexible and precise assembly of aircraft sections [11], P&O of spacecraft on the ground [12], and low-impact alignments [13]. An SPR is composed of a moving platform and a base platform, which are connected by six extensible links with spherical/universal joints. In its operational range, 6-Degrees-of-Freedom (DOF) motion of the moving platform can be achieved by the motions of the six limbs as a group [14, 15]. The SPR's limbs' position and their force measurements are also very important [16]. In 2016, the team of Du from Beihang University presented an analytical algorithm of six-dimensional F/T by measuring limbs' position and their force based on a SPR [17]. It laid the foundation for the research of this paper.

In the process of human-robot collaboration, methods of compensating for the weight of an unknown load are a research hot spot. In general, the mass and center of gravity of the load are not known in advance, so it is necessary to estimate them online. In 2015, the team of Liu from South China University of Technology analyzed the center of gravity of the load according to the relationship between force and torque and then used this calculation to compensate for the effect of load center of gravity on the required screw torque [18]. In 2016, Mitsantisuk of Thai Agricultural University proposed a solution for human-robot collaboration using a delta parallel robot. A disturbance observer was used at the drive joint of the delta robot, to assist in judging the load characteristics by observing additional torque at each drive joint beyond the torque required for motion of the unloaded robot. They also compensated for gravity [3]. In 2017, Boisclair of Laval University in Canada studied a method combining of gravity compensation and the six-dimensional force/torque (F/T) sensor with a cylindrical Halbach array. The cylindrical Halbach array was installed at the joint of an industrial robot and generated joint torque through a magnetic field to compensate the additional torque of the load precisely [19]. In 2017, using six-dimensional F/T sensors and a set of calibration and calculation methods, the team of Zhang from Beijing Institute of Spacecraft Environment Engineering eliminated the effect of sensor zero and load gravity on force perception and accurately obtained the external force and torque applied at the load end of the robot [20, 21].

As mentioned above, human-robot collaboration is becoming more and more important in the assembly of large and heavy components, and SPR's are widely used in this type of process. In this paper, an SPR design is described which can adjust the P&O of the component and dynamically measure interactive forces. Then, the dynamic gravity deviations are considered and accurately compensated for. Lastly, the large-scale alignment experiments were carried out, using an SPR incorporating the six-dimensional F/T dynamic measurements.

2. Large-Scale Alignment System Using an SPR and a Human

The large-scale alignment system with a Stewart parallel robot (SPR) for large components is shown in Figure 1. It includes an SPR, fixed platform, control cabinet, assembly fixtures, movable component, datum component, and an operator. The SPR is designed based on a 6-SPS (six limbs, each with an actuator, spherical hinge-prismatic pair-spherical hinge) Stewart parallel robot and six force sensors, which can adjust the P&O of the moving platform and dynamically measure six-dimensional F/T from the interactive force. It contains a moving platform, stationary platform (base platform), force sensors, AC servo motors, and branched chains. One of the SPR's six force sensors is placed in each limb to measure the limbs' actuation force to calculate the six-dimensional F/T data.

The principles for positioning and aligning components are described in this paper. They are based on the measured six-dimensional F/T data and the input forces from a human operator on a component. These principles are shown by the flowchart in Figure 2.

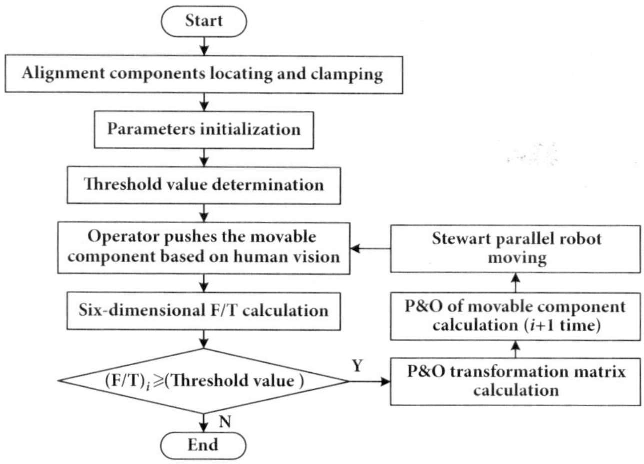

FIGURE 2: The human-SPR component alignment flowchart based on the measured six-dimensional F/T data.

3. Six-Dimensional Force/Torque Dynamic Measurement of the SPR

3.1. Method of Measuring Six-Dimensional F/T Based on the SPR. As presented in Figure 3, the SPR consists of a moving platform and a stationary platform, which are connected to each other with six limbs, adjustable in length through sliding joints. In the operation range, the 6-DOF motion of the moving platform could be achieved by the motions of the six limbs. Force sensors are placed in each limb to measure the limbs' actuation force f_i. Encoders are used to measure the limbs' length l_i ($i = 1, \ldots, 6$). The Cartesian coordinate system of o_0-$x_0y_0z_0$ is located in the center of the top surface of the base platform, while the Cartesian coordinate system of o_1-$x_1y_1z_1$ is located in the center of the bottom surface of the moving platform. The centers of the spherical joints are denoted as A_i and B_i. $[\boldsymbol{F}_s\ \boldsymbol{M}_s]^T = [F_x, F_y, F_z, M_x, M_y, M_z]^{\mathrm{T}}$ is expressed the external load which is decomposed in o_1-$x_1y_1z_1$. $\boldsymbol{A}_i$ is the coordinate in o_1-$x_1y_1z_1$. $\boldsymbol{B}'_i$ is the coordinate in o_1-$x_1y_1z_1$ and $\boldsymbol{B}_i$ is the coordinate in o_0-$x_0y_0z_0$. The two vectors are different descriptions of the same point; hence they can be related using a linear transformation as follows:

$$B'_i = R_{3\times3}{}^{-1} \cdot \left((B_i)_{3\times1} - M_{3\times1}\right) \tag{1}$$

where $\boldsymbol{R}_{3\times3}$ represents a rotation matrix and $\boldsymbol{M}_{3\times3}$ represents a translation matrix. Once the distance between A_i and B_i (limb length l_i) is set, the P&O parameters $\{x\ y\ z\ \alpha\ \beta\ \gamma\}$ between o_1-$x_1y_1z_1$ and o_0-$x_0y_0z_0$ could be solved by the newton iteration method. Among the P&O parameters, $\{x\ y\ z\}$ is the position vector of o_1-$x_1y_1z_1$, with respect to o_0-$x_0y_0z_0$, and $\{\alpha\ \beta\ \gamma\}$ is the orientation angle of o_1-$x_1y_1z_1$, with respect to o_0-$x_0y_0z_0$. Thus, $\boldsymbol{R}_{3\times3}$ and $\boldsymbol{M}_{3\times3}$ can be expressed as

$$R = \begin{bmatrix} \cos\alpha\cos\beta & \cos\alpha\sin\beta\sin\gamma - \sin\alpha\cos\gamma & \cos\alpha\sin\beta\cos\gamma + \sin\alpha\sin\gamma \\ \sin\alpha\cos\beta & \sin\alpha\sin\beta\sin\gamma + \cos\alpha\cos\beta & \sin\alpha\sin\beta\cos\gamma - \cos\alpha\sin\gamma \\ -\sin\beta & \cos\beta\sin\gamma & \cos\beta\cos\gamma \end{bmatrix} \tag{2}$$

$$M = [x\ \ y\ \ z]^T \tag{3}$$

The force equilibrium equation could be defined in the o_1-$x_1y_1z_1$ using the screw theory as

$$F_s + \epsilon M_s = \sum_{i=1}^{6} f_i \$_i \tag{4}$$

where $\boldsymbol{\$}_i$ is the unit screw along the ith leg and could be obtained by $\boldsymbol{\$}_i = [\boldsymbol{S}_i, \boldsymbol{S}_{0i}]^T$. $\boldsymbol{S}_i$ is the unit vector of f_i with respect to o_1-$x_1y_1z_1$. $\boldsymbol{S}_{0i}$ is the torque vector of $\boldsymbol{S}_i$ with respect to o_1-$x_1y_1z_1$. According to (1), $\boldsymbol{S}_i$ and $\boldsymbol{S}_{0i}$ could be calculated as

$$S_i = \frac{A_i - B_i}{\|A_i - B_i\|} \qquad S_{0i} = A_i \times S_i \tag{5}$$

Equation (4) can be rewritten in the form of matrix equation as

$$F = [G]\, f \tag{6}$$

where

$$F = [F_s\ \ M_s]^T = [F_x\ \ F_y\ \ F_z\ \ M_x\ \ M_y\ \ M_z]^T \tag{7}$$

$$f = [f_1\ \ f_2\ \ f_3\ \ f_4\ \ f_5\ \ f_6]^T \tag{8}$$

$$[\mathbf{G}] = \begin{bmatrix} \mathbf{S}_1 & \mathbf{S}_2 & \mathbf{S}_3 & \mathbf{S}_4 & \mathbf{S}_5 & \mathbf{S}_6 \\ \mathbf{S}_{01} & \mathbf{S}_{02} & \mathbf{S}_{03} & \mathbf{S}_{04} & \mathbf{S}_{05} & \mathbf{S}_{06} \end{bmatrix} \tag{9}$$

Hence, the external load $[\boldsymbol{F}_s\ \boldsymbol{M}_s]^T$ can be calculated by (6).

3.2. Dynamic Gravity Compensation. During the assembly process, the moving platform, assembly fixtures, and components are relatively heavy and bulky, so their barycenter and gravity deviations, which are caused by manufacturing errors and installation errors, will lead to wrong calculation results of the six-dimensional F/T. The dynamic gravity compensation is needed.

The influential factors of the calculation results, which are the barycenter and gravity of the moving platform, assembly fixtures, and other components, cannot be ignored. This

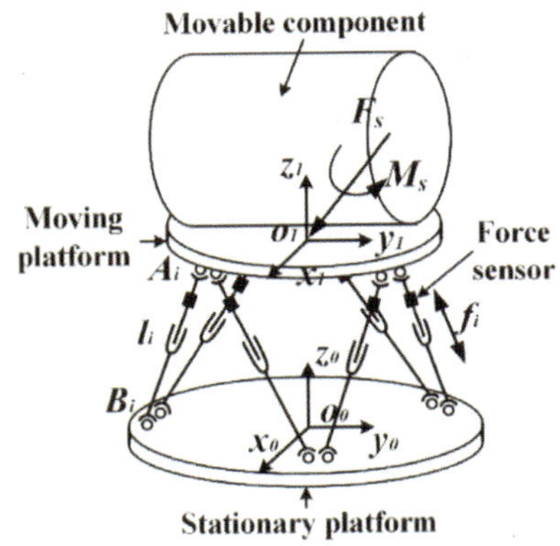

FIGURE 3: The theoretical model for a six-dimensional F/T based on the SPR.

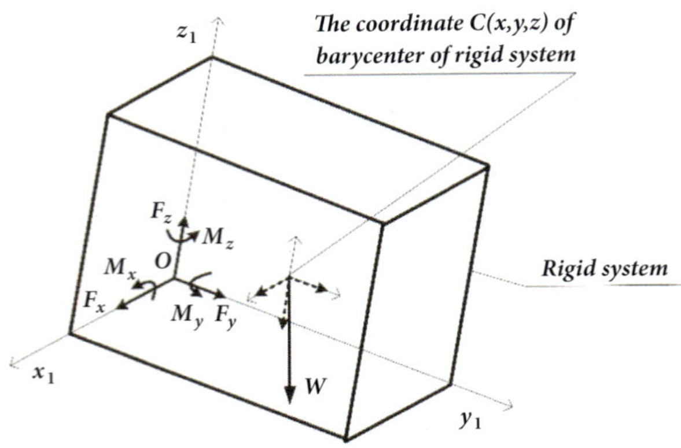

FIGURE 4: Schematic diagram of gravity of the rigid system in o_1-$x_1y_1z_1$.

paper considers them as a rigid system. Equation (6) can be thus rewritten as

$$F + \begin{bmatrix} S_G \\ S_{0G} \end{bmatrix} W = [G] f \tag{10}$$

where W is the dimensionless value of the gravity, so it is not a vector. $\mathbf{S}_G$ is the gravity unit vector of the said rigid system, so it is a 3-column vector. $\mathbf{S}_{0G}$ is the torque vector of $\mathbf{S}_G$ with respect to o_1-$x_1y_1z_1$, so $\mathbf{S}_{0G}$ is also a 3-column vector.

When the external load is zero, the six-dimensional F/T is caused by the gravity of the rigid system (as in Figure 4). The coordinate $\mathbf{C}=[x\ y\ z]^{\mathrm{T}}$ indicates the barycenter of the rigid system in o_1-$x_1y_1z_1$. According to the least square principle, $\mathbf{C}$ and W can be solved by the six-dimensional F/T under three different P&O sets. As an example, four measurements are performed here:

$$\begin{aligned} &\begin{bmatrix} M_{x1} & M_{x2} & M_{x3} & M_{x4} \\ M_{y1} & M_{y2} & M_{y3} & M_{y4} \\ M_{z1} & M_{z2} & M_{z3} & M_{z4} \end{bmatrix} \\ &= \begin{bmatrix} 0 & -z & y \\ z & 0 & -x \\ -y & x & 0 \end{bmatrix} \begin{bmatrix} F_{x1} & F_{x2} & F_{x3} & F_{x4} \\ F_{y1} & F_{y2} & F_{y3} & F_{y4} \\ F_{z1} & F_{z2} & F_{z3} & F_{z4} \end{bmatrix} \end{aligned} \tag{11}$$

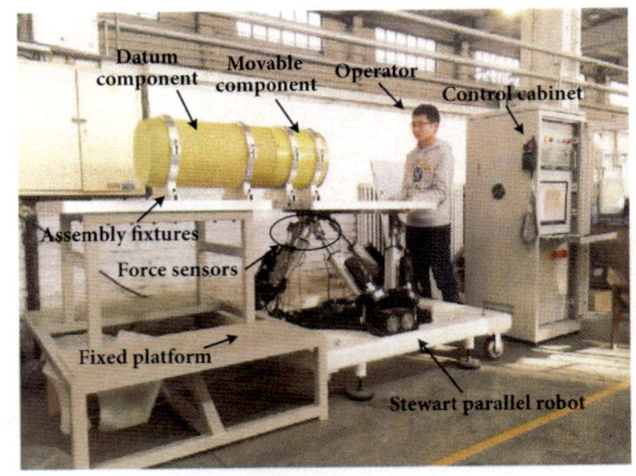

FIGURE 5: The large-scale alignment system with a SPR supporting a large component.

The resolving process of $\mathbf{C}$ and W from (11) is similar to that of the generalized inverse matrix of

$$\begin{bmatrix} F_{x1} & F_{x2} & F_{x3} & F_{x4} \\ F_{y1} & F_{y2} & F_{y3} & F_{y4} \\ F_{z1} & F_{z2} & F_{z3} & F_{z4} \end{bmatrix}. \tag{12}$$

When P&O changes, $\mathbf{C}$ and W of the rigid system do not differ in o_1-$x_1y_1z_1$, and the direction $\mathbf{S} = [0\ 0\ -1]^{\mathrm{T}}$ of gravity does not vary in o_0-$x_0y_0z_0$, meaning

$$\begin{aligned} S_G &= \frac{R^{-1} \cdot S}{\left\| R^{-1} \cdot S \right\|} \\ S_{0G} &= C \times S_G \end{aligned} \tag{13}$$

which could serve for the solution of (10).

For the preparation of a six-dimensional F/T measurement, experiments without external loads were carried out first, and the measured F/Ts could be used for the calculations of $\mathbf{C}$ and W, using (11), after which $\mathbf{C}$ is substituted into (13) for the vector $[\mathbf{S}_G\ \mathbf{S}_{0G}]^T$. Then, for an arbitrary external load, the six-dimensional F/T in o_1-$x_1y_1z_1$ could be obtained using (10) and the gravity of the rigid system can be dynamically compensated.

4. Experimental Results and Discussion

The component alignment system using an SPR to align large components is designed as shown in Figure 5. It includes the SPR, fixed platform, control cabinet, assembly fixtures, movable component, datum component, and operator. The parameters of the SPR are listed in Table 1.

4.1. Measurement of Known Applied Force. An S-type force sensor (with an accuracy of ±0.03 N) is used to measure the magnitude of applied force in real time. The operator applies the force to the SP or component to be moved, and the force is transmitted through the SPR to the S-type force sensor. The value of the six-dimensional F/T data can then be calculated and compared with the measurement value of the S-type force sensor. Datapoints were collected every 0.05 s during the experiment. The results are presented in Figure 6.

TABLE 1: The parameters of the SPR.

Parameter name	Value
Moving range along x direction	±50 mm
Moving range along y direction	±50 mm
Moving range along z direction	±50 mm
Rotation range along x direction	±5°
Rotation range along y direction	±5°
Rotation range along z direction	±5°
Maximum load	800 kg
Accuracy	0.05 mm
Overall size	1.5m×1.5m×1m

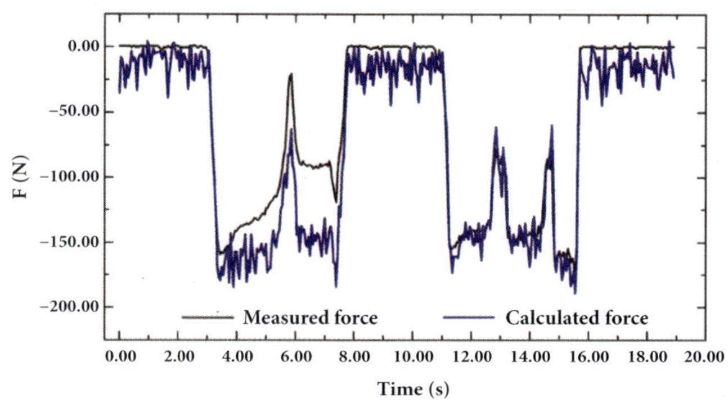

FIGURE 6: Comparison between the measured forces and calculated forces.

To measure an applied force, three steps were followed in this experiment. The first step is to compensate for the weight of the SPR and component, and the dynamic forces due to motion. When no external forces are applied to the SPR, any four groups of P&O parameters are selected as inputs to the SPR. Then, the four groups of the six-dimensional F/T can be calculated. The weight and center of gravity of the rigid body (moving platform, assembly fixture, and movable component) are inputs to the dynamic gravity compensation method. The weight is 1969.1017 N. The barycentric coordinate of the rigid body is $[-36.7749, -6.2296, -208.3247]^T$ in o_1-$x_1y_1z_1$. Second, the mean values of the six-dimensional F/T data are calculated. Due to measurement errors that caused the six-dimensional F/T data to fluctuate, the mean values of 100 groups of the six-dimensional T/F data are taken as the reference zero values. Finally, the six-dimensional F/T data are calculated. Figure 6 compares actual force data, as applied by the operator, to the input (interactive) force calculated by the system. These F/T data allows the system to estimate the interactive force between a person and the SPR or load component.

The results compared in Figure 6 show that the SPR system can accurately track the actual (measured) interactive force. The measured force and the calculated force are in good agreement, with an average error magnitude of 14.70 N (the difference between measured force and calculated force is taken as the absolute value, and then the average error magnitude is taken). Thus, these experimental results verify that this method of using six-dimensional SPR F/T data can track the operator's actual force inputs.

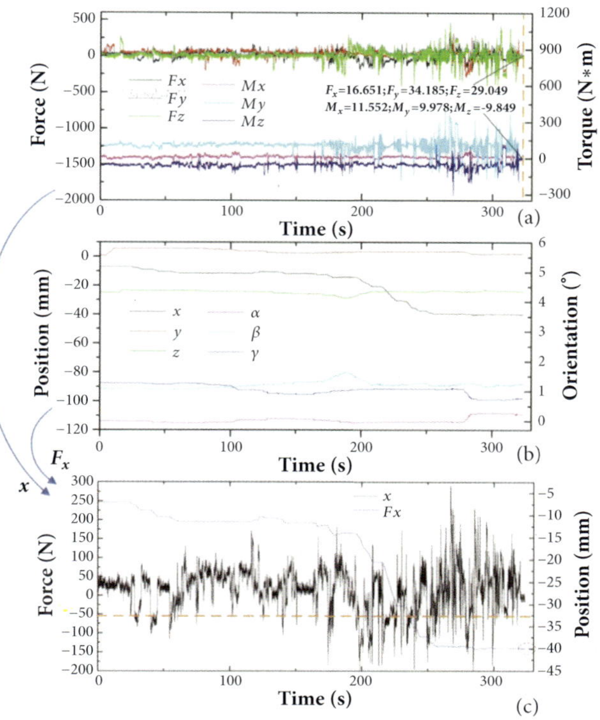

FIGURE 7: P&O and F/T curves over time.

4.2. F/T-Driven Alignment of Large Components. The operator applies the force to the SPR, and the SPR adjusts the P&O of large component to follow the intentions of the operator. Thus, the F/T feedback-driven alignment of large components can be implemented as in Figure 2. It is described as follows. First, some parameters are initialized, and the threshold values of six-dimensional F/T are obtained by experimental optimization. The applied threshold value of the force magnitude was 60 N, and the applied threshold value of the torque magnitude was 25 N∗m. Second, the operator visually judged the direction of the desired motion, and the external force was applied to the SPR by the operator. Third, the six-dimensional F/T tensor was calculated in real time by the dynamic force and weight compensation. This interactive force is taken as an input to the SPR. Fourth, the process of the F/T-driven alignment of large components was then executed though the direction and magnitude of interactive force and torque inputs. At last, the SPR adjusts the P&O of large component to follow the intentions of the operator. The experiment and the result are presented in Figure 7.

In Figure 7(a), when the calculated force $|F_x|$ is greater than the threshold value 60N, the SPR moved along the -x directions, until it stops at about -40 mm. At this location, the alignment process was successfully completed. Small changes in other degrees of freedom indicate that the components have high assembly accuracy and the adjustment range is small aside from the -x movement (as shown in Figure 7(b)). After the alignment process is completed, the maximum force value in the six-dimensional F/T is 35 N, and the maximum torque value is 12 N∗m (as shown in Figure 7(a)). These experimental results demonstrate that the calculation method of the six-dimensional F/T based on the SPR is correct, and the F/T-driven alignment of large components is effective.

5. Conclusions

The SPR can adjust the P&O of the component and dynamically measure forces from interaction between a robot and a human. The experimental results show that the forces due to weight can be reliably compensated for. The contributions of the paper are summarized as follows:

(1) A method of measuring six-dimensional F/T, based on the SPR characteristics and screw theory, is proposed for the accurate estimation of the external forces during alignment of large-scale components. Forces due to gravity are taken into consideration, and a precise weight compensation method is provided.

(2) Known applied force measurement experiments are performed on the SPR using an S-type force sensor. The six-dimensional F/T can accurately track the change of the measurement force and the average error is about 14.70 N. The experiment of F/T-driven alignment for large component was successfully completed.

(3) Interactive force measurements are important for human-robot collaboration. The proposed method fulfills the requirement for accurate force estimation. It provides satisfactory compensation for forces due to gravity, in order to accurately separate forces due to the interaction of a human with a robot. The measurements from the six-dimensional F/T based on the SPR could be further improved with more accurate measurements of the forces and the limb lengths' variation with motion.

Conflicts of Interest

The authors declare that there are no conflicts of interest regarding the publication of this paper.

Acknowledgments

This work was supported by the National Key R&D Program of China.

References

[1] B. Yao, Z. Zhou, L. Wang, W. Xu, Q. Liu, and A. Liu, "Sensorless and adaptive admittance control of industrial robot in physical human–robot interaction," *Robotics and Computer-Integrated Manufacturing*, vol. 51, pp. 158–168, 2018.

[2] J. Krüger, T. K. Lien, and A. Verl, "Cooperation of human and machines in assembly lines," *CIRP Annals - Manufacturing Technology*, vol. 58, no. 2, pp. 628–646, 2009.

[3] C. Mitsantisuk and K. Ohishi, "Haptic human-robot collaboration system based on delta robot with gravity compensation," in *Proceedings of the 42nd Conference of the Industrial Electronics Society, IECON 2016*, pp. 5796–5801, Italy, October 2016.

[4] M. Geravand, F. Flacco, and A. De Luca, "Human-robot physical interaction and collaboration using an industrial robot with a closed control architecture," in *Proceedings of the 2013 IEEE International Conference on Robotics and Automation, ICRA 2013*, pp. 4000–4007, Germany, May 2013.

[5] A. Seibel, S. Schulz, and J. Schlattmann, "On the Direct Kinematics Problem of Parallel Mechanisms," *Journal of Robotics*, vol. 2018, Article ID 2412608, 9 pages, 2018.

[6] R. Yao, W. Zhu, and P. Huang, "Accuracy analysis of Stewart platform based on interval analysis method," *Chinese Journal of Mechanical Engineering*, vol. 26, no. 1, pp. 29–34, 2013.

[7] B. Denkena, C. Holz, and H. Abdellatif, "Model-based control of a hexapod with linear direct drives," *International Journal of Computer Integrated Manufacturing*, vol. 19, no. 5, pp. 463–472, 2006.

[8] S. Pedrammehr, M. Mahboubkhah, and N. Khani, "A study on vibration of Stewart platform-based machine tool table," *The International Journal of Advanced Manufacturing Technology*, vol. 65, no. 5–8, pp. 991–1007, 2013.

[9] M. Moradi Dalvand and B. Shirinzadeh, "Motion control analysis of a parallel robot assisted minimally invasive surgery/microsurgery system (PRAMiSS)," *Robotics and Computer-Integrated Manufacturing*, vol. 29, no. 2, pp. 318–327, 2013.

[10] A. Pisla, T. Itul, D. Pisla, and A. Szilaghyi, "Considerations upon the influence of manufacturing and assembly errors on the kinematic and dynamic behavior in a flight simulator stewart-gough platform," *Mechanisms and Machine Science*, vol. 3, pp. 215–223, 2012.

[11] C. Löchte, F. Dietrich, and A. Raatz, "A Parallel Kinematic Concept Targeting at More Accurate Assembly of Aircraft Sections," in *Intelligent Robotics and Applications*, vol. 7101 of *Lecture Notes in Computer Science*, pp. 142–151, Springer Berlin Heidelberg, Berlin, Heidelberg, 2011.

[12] Y. Xu, J. Yuan, J. Zhao, and Y. Zhao, "Robust attitude control and simulation of a Stewart spacecraft," in *Proceedings of the 2015 27th Chinese Control and Decision Conference (CCDC)*, pp. 477–482, Qingdao, China, May 2015.

[13] H. Zhao, S. Zhang Y, and D. Chen X, "Compliant force control in space docking," in *Proceedings of the 2007 IEEE International conference on mechatronics and automation*, Harbin, China, Aug, 2007.

[14] G. Zhang, J. Du, and S. To, "Calibration of a small size hexapod machine tool using coordinate measuring machine," *Proceedings of the Institution of Mechanical Engineers, Part E: Journal of Process Mechanical Engineering*, vol. 230, no. 3, pp. 183–197, 2014.

[15] W. Zhou, W. Chen, H. Liu, and X. Li, "A new forward kinematic algorithm for a general Stewart platform," *Mechanism and Machine Theory*, vol. 87, pp. 177–190, 2015.

[16] S. A. Maged, A. M. R. Fath El Bab, and A. A. Abouelsoud, "An adaptive observer for a Stewart platform manipulator using leg position and force measurements," *International Journal of Modelling, Identification and Control*, vol. 24, no. 1, pp. 62–74, 2015.

[17] K. Wen, F. Du, and X. Zhang, "Algorithm and experiments of six-dimensional force/torque dynamic measurements based on a Stewart platform," *Chinese Journal of Aeronautics*, vol. 29, no. 6, pp. 1840–1851, 2016.

[18] B. Liu W, *Research on Industrial Robot Grinding based on Force Control*, South China University of Technology, 2014.

[19] J. Boisclair, P.-L. Richard, T. Laliberté, and C. Gosselin, "Gravity Compensation of Robotic Manipulators Using Cylindrical Halbach Arrays," *IEEE/ASME Transactions on Mechatronics*, vol. 22, no. 1, pp. 457–464, 2017.

[20] L.-J. Zhang, R.-Q. Hu, and W.-M. Yi, "Research on Force Sensing for the End-load of Industrial Robot Based on a 6-axis Force/Torque Sensor," *Zidonghua Xuebao/Acta Automatica Sinica*, vol. 43, no. 3, pp. 439–447, 2017.

[21] L. Zhang J, R. Hu Q, and W. Yi M, "A Study of Flexible Force Control Method on Robotic Assembly for Spacecraft," *Applied Mechanics Materials*, vol. 681, pp. 79–85, 2014.

14

Development and Evaluation of a Powered Artificial Gastrocnemius for Transtibial Amputee Gait

Michael F. Eilenberg, Jiun-Yih Kuan, and Hugh Herr

Biomechatronics Group, MIT Media Lab, Massachusetts Institute of Technology, Cambridge, MA, USA

Correspondence should be addressed to Hugh Herr; hherr@media.mit.edu

Academic Editor: Kazuo Kiguchi

Existing robotic transtibial prostheses provide only ankle joint actuation and do not restore biarticular function of the gastrocnemius muscle. This paper presents the first powered biarticular transtibial prosthesis, which is a combination of a commercial powered ankle-foot prosthesis and a motorized robotic knee orthosis. The orthosis is controlled to emulate the human gastrocnemius based on neuromuscular models of matched nonamputees. Together with the ankle-foot prosthesis, the devices provide biarticular actuation. We evaluate differences between this biarticular condition and a monoarticular condition with the orthosis behaving as a free-joint. Six participants with transtibial amputation walk with the prosthesis on a treadmill while motion, force, and metabolic data are collected and analyzed for differences between conditions. The biarticular prosthesis reduces affected-side biological knee flexion moment impulse and hip positive work during late-stance knee flexion, compared to the monoarticular condition. The data do not support our hypothesis that metabolism decreases for all participants, but some participants demonstrate large metabolic reductions with the biarticular condition. These preliminary results suggest that a powered artificial gastrocnemius may be capable of providing large metabolic reductions compared to a monoarticular prosthesis, but further study is warranted to determine an appropriate controller for achieving more consistent metabolic benefits.

1. Introduction

For those living with transtibial (below-the-knee) amputation, prosthetic technology has yet to fully restore the functionality of the missing biological structures. As evidence of this technological limitation, transtibial amputees wearing conventional prostheses naturally select slower walking speeds than those with biological limbs, and, when walking at the same pace as nonamputees, they require more metabolic power than their nonamputee counterparts [1–4]. Further, transtibial amputees exhibit kinetic pathologies, particularly during the late-stance knee flexion phase of gait, where the knee flexes to prepare for the swing phase. Transtibial amputees walking with conventional prostheses increase their affected-side hip positive and negative power during this phase, compared to nonamputees, possibly in response to the lack of power from the calf muscles [4].

Unlike conventional prostheses, new robotic powered ankle-foot prostheses (PAFP) can output net mechanical work similar to that of the human ankle-foot complex. Consequently, amputees using powered prostheses have preferred walking speed, metabolic cost at a given speed, and contralateral limb impacts that are not significantly different from those of nonamputees [5, 6].

These robotic prostheses, however, still act only at the ankle joint and hence can only provide monoarticular ankle function. In contrast, the gastrocnemius calf muscle both plantar flexes the ankle and flexes the knee. Hence, amputees walking with monoarticular prostheses must compensate for this lack of function by increasing activation of the remaining knee flexors in the hamstring group [7–9]. This increase in activation is still seen with robotic ankle-foot prostheses [7]. Therefore, it is likely that biarticular interventions are required to provide a more complete restoration of the missing musculature.

Only few studies have been done to evaluate how a biarticular robotic device could improve metrics of gait for transtibial amputees beyond monoarticular devices. In the first such study, researchers approximated the gastrocnemius muscle as isometric during walking and built a clutched-spring joint into a knee orthosis, called an artificial gastrocnemius (AG), to replicate the modeled function of an isometric muscle in series with a compliant tendon [10]. This AG-PAFP combination was shown to reduce metabolic cost of a bilateral transtibial amputee, as compared to a conventional passive transtibial prosthesis.

However, this previous study was limited in that it did not evaluate the individual contributions of the PAFP and AG on metabolism. Hence, it is not known whether the metabolic effects stemmed from the actions of the PAFP, AG, or both devices. Researchers have since attempted to isolate the incremental influence of the AG on metabolism and kinetic measures by comparing an AG-PAFP combination to a PAFP-only condition with identical mass distribution. This study provided preliminary evidence that the inclusion of the AG may, in fact, improve metabolism and influence several kinetic measures, compared to the PAFP-only condition [11].

Still, these two aforementioned studies only considered isometric behavior of the biological gastrocnemius muscle. In contrast, the human gastrocnemius muscle is responsible for 3–5 Joules of positive net mechanical work per step [12, 13]. This positive net work is impossible to reproduce with a quasi-passive device, yet its presence may be critical to the advancement of active transtibial prostheses of greater efficacy.

For the present study, we develop a novel AG that can provide positive net mechanical work. We develop a controller, based on a neuromuscular model, for said device that emulates the functionality of the missing gastrocnemius. We also develop a torque controller to enforce this functionality. We anticipate that the applied moments and positive net work from the AG will replace affected-side biological knee positive work and flexion moments. Further, we expect that the AG's improvement of calf muscle function will reduce the late-stance pathologies typically seen in the affected-side hip [4]. Therefore, we hypothesize that the use of this AG, in conjunction with a PAFP, will reduce both biological flexion moment impulse and positive mechanical work of the affected-side knee and hip joints during late-stance knee flexion, as compared to the same conditions with only the PAFP. We further hypothesize that these changes will provide a corresponding reduction in metabolic cost of walking via the reduction in metabolically expensive positive muscle work.

2. Powered Biarticular Transtibial Prosthesis Design

It has been shown that adding mass to the lower extremities can have a detrimental effect on walking metabolism [14]. Hence, the goal in this project was to provide interventions to a participant walking on a treadmill by remote actuation, so to minimize mass affixed to the wearer. To this end, we developed a cable-driven tethered exoskeleton system, comprising a motor drive module, a cable transmission linkage, a robotic knee joint, and a custom knee orthosis. We first describe the mechatronic design of these four elements and then describe the torque control of the system in the next section.

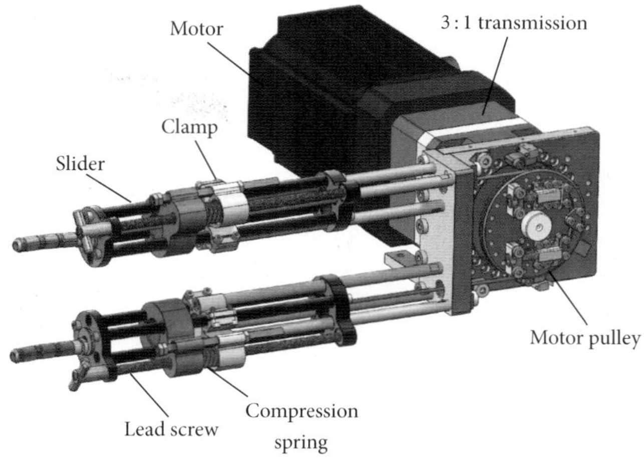

Figure 1: Motor drive module: actuation source. The tensioner attached to the motor drive unit and provided the ability to preload the cables to a desired tension without adding series compliance to the system.

2.1. Motor Drive Module. The core of the motor module was a powerful offboard actuation source and real-time sensing and control system. The actuation source included a 3 kW AC servo motor (SGMSV-30A3A61, YASKAWA), a 3 : 1 gear reducer (042PLX0030-LB-04027, CGI INC), and a custom cable tensioner. The sensing and control system, designed for real-time control and system safety, included a host personal computer, an EtherCAT® master controller (SPiiPlusNTM, ACS Motion Control), local servo controllers (V200AE1A002000200, Yaskawa), and an EtherCAT® slave I/O system (750-534 EtherCAT Coupler, WAGO). This system had multiple analog and digital inputs that were used for gathering feedback signals in real-time from the device for the purposes of data collection and feedback control. A 90 mm diameter pulley was mounted to the transmission output shaft to actuate the drive cable.

A spring-based tensioner system was used to preload the cable. This tensioner was a bracket with an internal slider, driven by a lead screw (Figure 1). This joint enabled the bracket to expand to take up cable slack. A spring was placed in series between the sliding element of the bracket and the stationary element, serving as an indicator of cable preload. As the lead screw was manually rotated, the bracket expanded and provided tension on the cable. The reaction force from the cable conduit was transmitted through the bracket and loaded the spring. Therefore, by measuring the spring deflection during the cable preloading, the cable tension could be approximated. Once the cable was pretensioned, screws were tightened to lock the slider to make the tensioner a rigid structure. Therefore, the tensioning springs did not introduce series compliance to the system during operation of the cable drive.

2.2. Cable Linkage. The cable linkage consisted of an inner cable, flexible conduits, rigid tubing, and a pulley system. A 3 mm diameter 7 × 19 braided steel inner cable was used to transmit power from the motor to the knee joint pulley

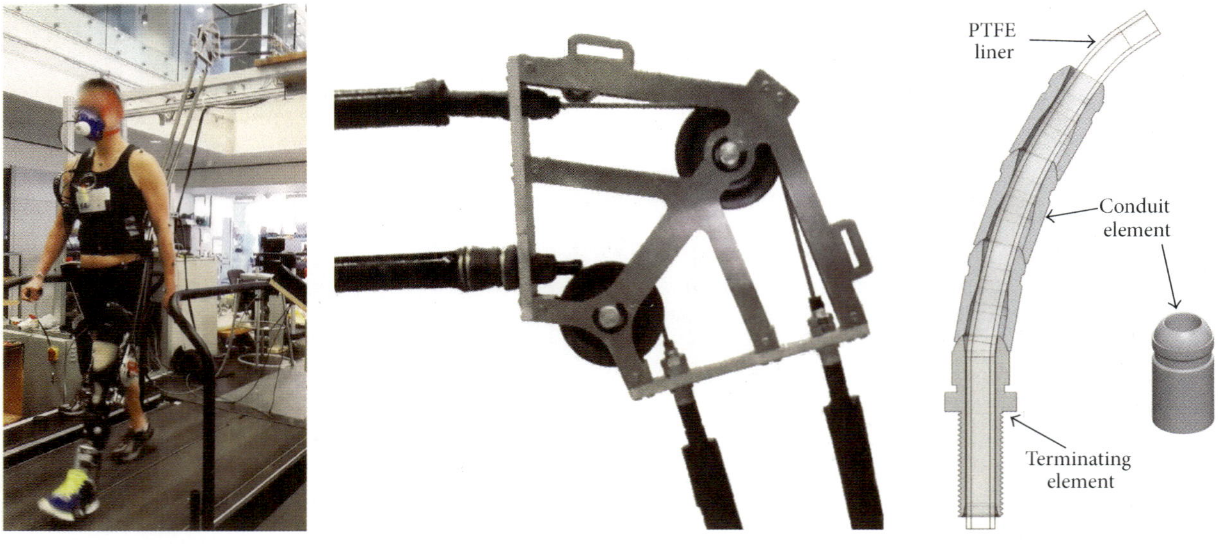

FIGURE 2: Conduit linkage as it interfaces with the artificial gastrocnemius and participant. The complete biarticular prosthesis system is shown (a) along with close-up views of the conduit linkage pulley system (b) and components of the flexible conduit sections (c).

in a similar way as a Bowden cable drive. However, instead of a flexible conduit throughout the entire tether, sections of rigid tubing and a pulley system were used (Figure 2). The tubing was made of 0/90 carbon fiber, with 22.2 mm outer diameter and 19.1 mm inner diameter. This system had a distinct advantage over a traditional, flexible Bowden cable drive: the losses due to friction between the cable and conduit were near-zero through the tubing sections, as the cable did not contact the sides of the conduit.

Short, flexible conduit sections were used to link the straight rigid tubing together in order to provide articulation. Typical Bowden cable conduits have the tendency to straighten when tension is applied to the cable. Rather than using such a conduit, short, interlocking aluminum conduit elements were designed (Figure 2(c)). These elements were designed in case the length of the inner bore of the conduit did not change upon bending. Each element had a male and female end, that, when mated, produced a ball-and-socket joint, through which the cable could pass. These elements were strung together in series to produce flexible conduit sections of the desired lengths. Polytetrafluoroethylene (PTFE) liners (3 mm inner diameter, 6 mm outer diameter) provided low-friction interface within these sections of conduit elements.

The conduit linkage was arranged behind and above the treadmill, so to provide minimal interference to a wearer of the knee joint (Figure 2(a)). The conduit linkage allowed two degrees of freedom in the sagittal plane, via the vertical and horizontal hinging of the conduit sections. Some limited motion in the coronal and transverse planes was also possible with this system. The remaining flexibility needed near the waist of a human subject was provided by a section of several conduit elements linked in series, forming a flexible section of conduit.

The added mass from this conduit system was expected to provide some downward force to the wearer during walking. This force was estimated by using a force gauge to hold up the AG with the conduit linkage at the location where a typical participant's knee would be during walking.

2.3. Robotic Knee Joint. The actuated robotic knee consisted of a pulley, driven by the tether cable, which sat lateral to the wearer's affected-side biological knee joint (Figure 3). This pulley was free to rotate on ball bearings in the sagittal plane about the knee center with respect to the joint housing. The pulley itself was connected to the output link through a torsional spring (stiffness = 1200 Nm/rad), which acted as a series compliance between the cable drive and the output link.

A rotary potentiometer (8 kOhm), mounted between the joint housing and the output link, was used to measure joint angle. This potentiometer was mounted on the lateral side of the joint, with the potentiometer shaft going through the center of the joint and attaching to the output link on the medial side (Figure 3). Torque sensing was accomplished via the use of a full-bridge strain gauge (SGT-2/350-FB13, Omega Engineering, Inc.), mounted on the output link so that bending of the link in the sagittal plane would be detected by the strain gauge. The strain gauge was connected to a custom preamplifier that included a common-mode input filter to reject electromagnetic interference and a second-order low-pass filter with a cutoff frequency of 585 Hz.

The designed robotic joint had physical parameters within the desired values (see Table 1). Although the robotic joint itself had a mass of only 0.6 kg, the orthosis for attaching it to the body required the addition of 0.8 kg and 1.3 kg, respectively, for the amputee and nonamputee subjects. In addition, the suspended cable linkage provided approximately a 0.6 kgf downwards force at the knee. Regarding cable pretension, it was found that 30–50 N of cable pretension kept cable friction low without introducing bandwidth-limiting cable slack.

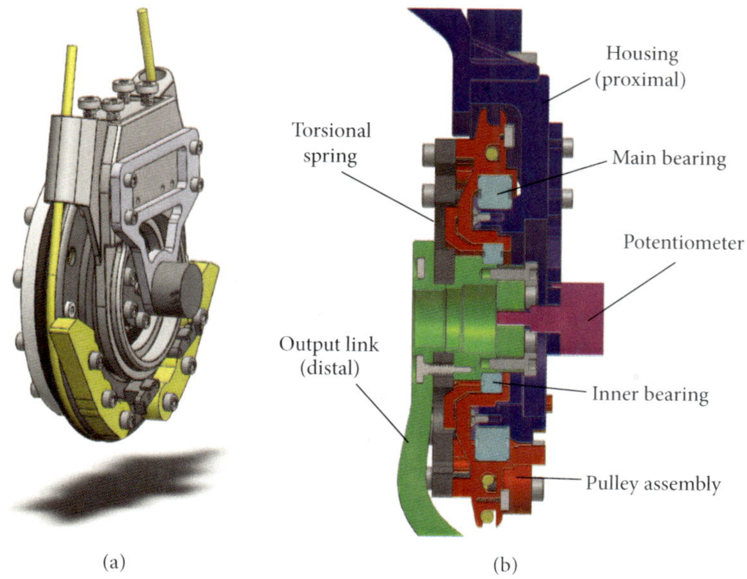

FIGURE 3: Powered artificial gastrocnemius joint. The assembled joint is shown without proximal and distal attachments (a). A color-coded coronal plane section view of the same mechanism is also shown (b), where parts that connect as a rigid body are shaded with the same color. The housing (blue) attaches to the proximal mount that is affixed to the thigh cuff. The pulley (red) rotates with respect to the housing via the main bearing (teal). The output link (green) is affixed to the distal mount that connects to the socket. This link rotates with respect to the pulley via the inner bearing, with the torsional spring connecting the output link to the pulley. The potentiometer measures the relative rotation of the housing and the output link. Hard stops (not shown) are mounted to the pulley assembly and engage with the housing at the limits of joint rotation.

TABLE 1: Design parameters of the powered artificial gastrocnemius.

Parameter	Desired value	Achieved value
Joint range of motion (radians)	1.3	2.1
Size	(Minimal)	50 mm radius, 53 mm wide
Weight (kg)	(Minimal)	1.4–1.9
Center-of-rotation range (mm)	20	20
Maximum output torque (Nm)	60	90
Torque sensing resolution (Nm)	0.01	0.01

2.4. Custom Knee Orthosis and Robotic Ankle-Foot Prosthesis. A custom knee orthosis was built for each amputee participant to use the AG (Figure 4). Each amputee participant was brought to a certified prosthetist to get a cast made of their residual limb while their normal prosthetic socket was being worn. The prosthetist then created carbon fiber orthoses that were form-fitted for each individual. Consequently, an orthosis could be strapped to the affected-side leg while the prosthetic socket was being worn. The orthoses each comprised a carbon fiber thigh cuff and a carbon fiber shank cuff that fitted over the existing prosthetic socket. These cuffs were joined by a pair of polycentric knee orthosis hinges.

A set of bearings was designed to transmit only tangential and sagittal plane forces from the output link of the robotic joint to the shank cuff of the custom orthosis (Figure 4(c)). This bearing set consisted of (1) a linear bearing to allow for distal-proximal freedom between the robotic joint and orthosis, (2) a ball joint to provide 3-axis rotational freedom, and (3) A linear brass sleeve bearing on a 6 mm steel shaft to allow medial-lateral freedom.

The orthosis was designed to accommodate a robotic ankle-foot prosthesis (PAFP) (EmPower, BionX Medical Technologies, Inc., Bedford, MA) attached to a given amputee's socket with a standard prosthetic pylon (Figure 4). This prosthesis was a commercial product capable of large amounts of positive net work during walking.

3. System Dynamics and Torque Control Design

Despite the large offboard power source, the AG still required system dynamics analyses and control strategies to compensate for parasitic effects like cable friction and other dynamics between the motor and output joint. The goal was to make the system behave as though these dynamics did not exist, so that, from a high-level control standpoint, the AG could be considered close to an ideal torque source. In this section, we describe the characterization of the AG system dynamics. We also describe the development of a low-level torque controller

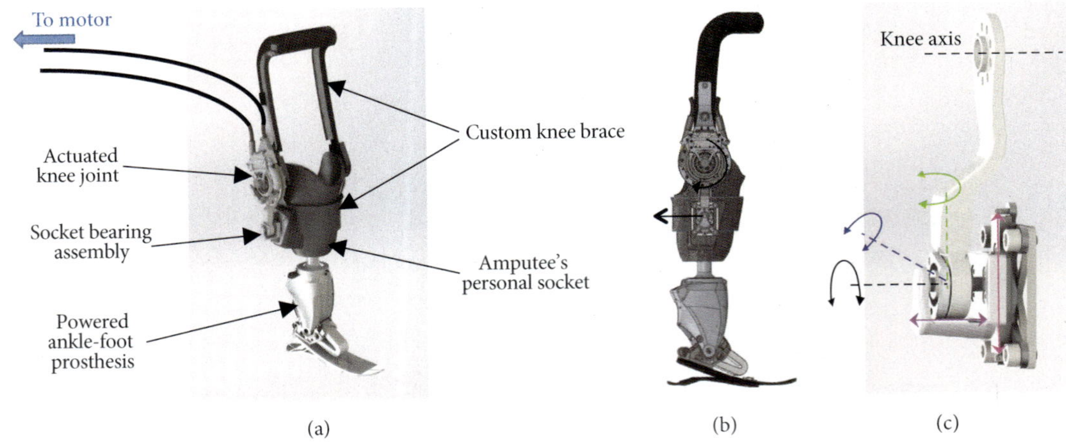

FIGURE 4: Knee orthosis and socket attachment. Rendering of the powered AG knee orthosis is shown with the robotic knee joint and powered ankle-foot prosthesis (a). A side-view of the same is shown, with arrows indicating the intended applied torque from the knee actuator and the resulting tangential force on the socket from the actuator (b). Also shown is the bearing assembly with resulting degrees of freedom (c) that is mounted on the socket attachment of the orthosis. This assembly transmits the tangential force from the actuator while allowing for motion from the polycentric knee joints as well as misalignment in the orthosis.

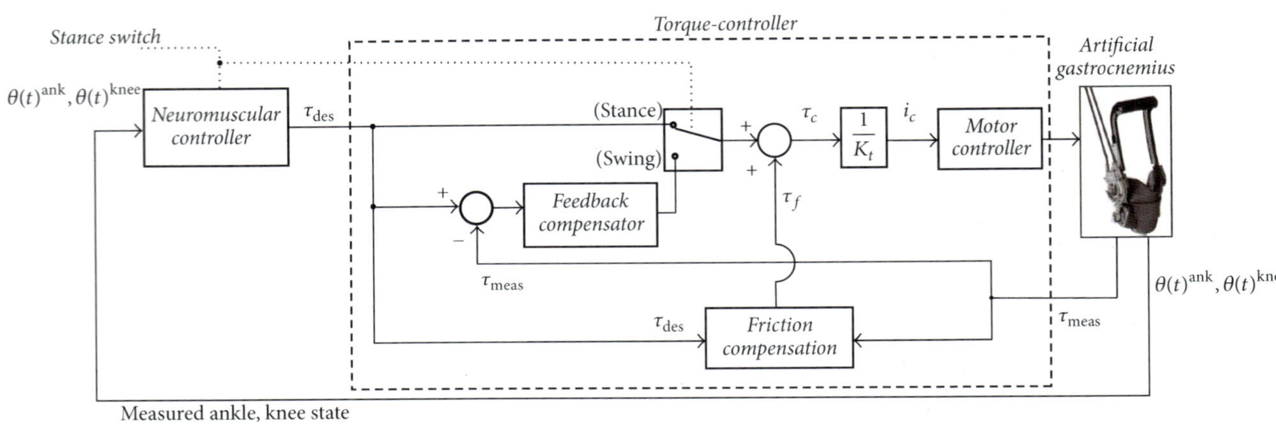

FIGURE 5: Torque control schematic. The control system for the AG is composed of an inner torque-controller loop that enforces desired torque at the joint and an outer loop involving a neuromuscular model as a controller that receives joint kinematics as input. The controller switches between open-loop control (stance) and closed-loop torque control (swing) as appropriate given the gait cycle phase. Motor torque control is performed at the lowest level by controlling motor current, scaled by the torque constant, K_t.

in terms of torque bandwidth and also performance during walking.

3.1. Open-Loop System Characterization. System identification tests for the AG were performed using a knee orthosis built for nonamputee participants. The orthosis was worn by a healthy adult male participant so to include the full dynamics of the human-machine system. The participant stood still, keeping his knee bent slightly, but resisting motion from the device. A swept-sine torque signal of varying amplitudes from 2 to 10 Nm was commanded to the motor, with frequencies ranging from 0.1 to 200 rad/s in increments of 10 rad/s. Additional data were collected with the flexible conduit section in the cable conduit linkage bent approximately 15 degrees, so as to capture the range of friction in this section. The magnitude of the command signal was adjusted for each frequency to maximize torque while preventing excessive discomfort to the wearer. Joint moment data were collected from the AG strain gauge. When attached to the human leg as in the system identification experiments, the robotic joint had a −3 dB open-loop torque bandwidth of 100 rad/s.

3.2. Artificial Gastrocnemius Torque Control Scheme. The AG was a torque-controlled system. A high-level neuromuscular controller (NMC) took as input affected-side ankle angle $\theta(t)^{\text{ank}}$ and knee angle $\theta(t)^{\text{knee}}$ as functions of time and produced a desired knee joint torque, τ_{des}. A low-level torque controller enforced this desired torque at the physical joint. For the purposes of this torque control, the gait cycle was divided into two phases: stance and swing (Figure 5). Stance phase began at heel-contact and ended at ToeOff. Swing phase began at ToeOff and ended at subsequent heel-contact. A footswitch (model: FSW, B&L Engineering, Santa Ana, CA), worn as an insole in the shoe on the leg wearing the knee orthosis, was used to detect stance and swing phases.

TABLE 2: Amputee matching to nonamputee participants.

Amputee number	Nonamputee match	Amputee height (cm)	Matched height (cm)	Amputee weight (kg)	Matched weight (kg)
S1	NA2	180	180	92	90
S2	NA1	175	175	74	73
S3	NA3	193	188	89	95
S4	NA3	188	188	98	95
S5	NA4	175	180	110	103
S6	NA4	180	180	104	103

3.2.1. Stance Phase Torque Control. An open-loop control methodology was used for the stance phase of the gait cycle, wherein the control input was τ_{des}, and the output was motor torque, τ_c (Figure 5). A feed-forward friction compensation term, τ_f with iterative learning, was used to reduce the torque errors between the desired torque, τ_{des} and the measured torque, τ_{meas}, resulting from friction in the cable drive. This friction compensation was implemented as a torque profile as a function of time within a gait cycle. A running counter was used to estimate the elapsed fraction of the current gait cycle by comparing the time spent in the current gait cycle to the total time of the previous gait cycle: gc = *counter*/*lastsamples* where gc is the estimated elapsed fraction of the current gait cycle, *counter* is the number of control cycles elapsed in the current gait cycle, and *lastsamples* is the total number of control cycles of the previous gait cycle.

The function was updated at each gait step by smoothly incorporating the observed torque error in two steps:

$$\tau_{\text{learn}}(\text{gc}, n+1) = \tau_{\text{err}}(\text{gc}, n) \cdot \alpha + \tau_f(\text{gc}, n) \cdot (1-\alpha), \quad (1)$$

$$\tau_f(\text{gc}, n+1) = \tau_{\text{learn}}(\text{gc}, n), \quad (2)$$

where gc is the elapsed fraction of the gait cycle, n is the gait cycle number, and α is the learning rate. Once a new learned profile τ_{learn} was computed for an entire gait cycle (1), it was subsequently used in the feed-forward term τ_f on the subsequent gait cycle (2). Thus, the error recorded during a gait cycle would be incorporated in the controller two gait cycles later. A value of α of 0.5 was found to provide a reasonable learning rate, for which the learned friction compensation profile was updated such that a new learned behavior was fully incorporated in approximately 10 seconds. Similar iterative learning methods have been successful in wearable robotics in the presence of cable frictional losses [15].

3.2.2. Swing Phase Torque Control. To reduce resistance during swing phase, a feedback torque controller was used along with the iterative friction compensation. This controller took the form

$$\tau_{\text{comm}} = K_i \int (\tau_{\text{des}} - \tau_{\text{meas}})\, dt, \quad (3)$$

where τ_{comm} is the motor torque command from the feedback controller. Standard loop-shaping techniques were performed in MATLAB using the open-loop data to determine this control law and the parameter K_i. In order to provide a fast response during state changes, this torque controller was run throughout the gait cycle, but its output was applied to the robotic joint only during the swing phase.

3.2.3. Closed-Loop System Characterization. System characterization tests were performed for the swing phase feedback torque controller (without the friction compensation), with a swept-sine desired joint torque as input and measured strain gauge joint torque as output, in a similar manner as the open-loop system characterization tests. The closed-loop bandwidth (−3 dB) and the phase margin were estimated from the data using a fast Fourier transforms of both inputs and outputs. When applying the swing phase closed-loop torque controller, the closed-loop system bandwidth was 14 Hz (87 rad/s), with a gain margin of 1.6 and a phase margin of 69 degrees.

3.3. Torque Control Performance during Walking. In addition to the transparency requirement, it was necessary to ensure that the nonzero desired torque values could be tracked by the AG within a reasonable range. The root mean squared (RMS) stance phase torque error, between the desired torque from the NMC and the measured joint torque from the AG strain gauge, was 1.45 ± 0.72 Nm (6 percent of the average maximum torque of 23 Nm) on average across the amputee participants listed in Table 2. A representative plot of this performance is shown in Figure 6.

3.4. Torque Control Metabolic Validation. Clinical metabolic tests were performed to test the transparency of the robotic knee joint; it was required that the described torque control techniques could provide sufficiently low resistance to wearers as not to adversely affect their metabolism. Success was defined as a nonsignificant difference in metabolism between two conditions during treadmill walking: (1) torque-controlled AG knee orthosis worn on one leg with zero torque commanded and (2) AG knee orthosis worn, acting as a free-swinging joint (with the cable drive disconnected). For the purposes of this metabolic validation trials with nonamputees, a custom-built, carbon fiber orthosis, molded by a prosthetist, was used to mount the robotic joint on the leg of nonamputee participants.

Six healthy participants (age = 27 ± 5.7 years, body mass = 71 ± 10 kg, and height = 181 ± 4 cm) were asked to walk on a treadmill while metabolic data were collected. The clinical evaluation was conducted at MIT (Cambridge, MA) and was approved by MIT's Committee on the Use of Humans

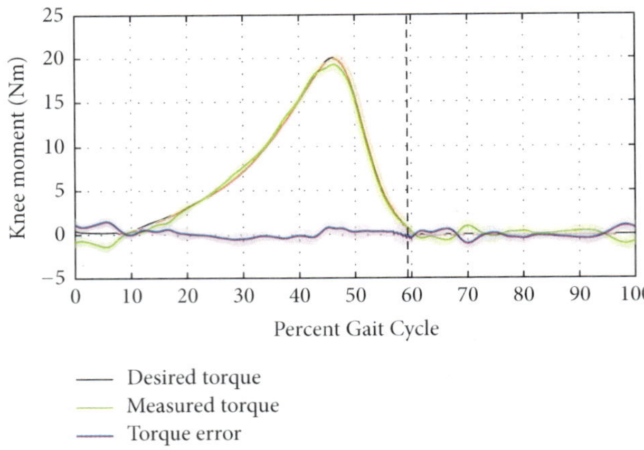

FIGURE 6: Representative closed-loop system response during walking trials. Data were collected with an amputee participant. The torque error (blue) represents the difference in joint moment between the desired AG device torque (red) and the measured strain-gauge device torque (green). The vertical dashed line indicates ToeOff.

as Experimental Subjects (COUHES). Each participant provided written, informed consent which was obtained before data collection was initiated. Each participant was asked to walk at a typical walking speed of 1.25 m/s under three conditions, in random order: (1) walking normally without the AG, (2) wearing the orthosis with the cable drive disconnected, and (3) wearing the orthosis with torque controller set to zero torque. For all conditions, participants wore a portable oxygen consumption mask attached to a standard open-circuit gas analysis system (model K4B^2, COSMED, Rome, Italy). This system estimated metabolic energy consumption based on measurements of oxygen inspired and expired carbon dioxide.

Metabolic cost for each walking speed was computed by taking average oxygen and carbon dioxide data over a two-minute window at the end of each six-minute trial. The metabolic power was computed using the equation

$$P = K_{O_2}\dot{V}_{O_2} + K_{CO_2}\dot{V}_{CO_2}, \quad (4)$$

where P is the metabolic power in Watts, $\dot{V}_{O_2}$ is the volume flow rate of oxygen inhaled, $\dot{V}_{CO_2}$ is the volume flow rate of carbon dioxide exhaled, and K_{O_2} and K_{CO_2} are constants with values from literature [16], given as $K_{O_2} = 16{,}580$ W/L and $K_{CO_2} = 4{,}510$ W/L. It was confirmed that the average ratio of $\dot{V}_{CO_2}/\dot{V}_{O_2}$ was less than 1.1 for all trials.

The metabolic consumption of the participants walking with the zero-torque control condition (3.40 ± 0.40 W/kg) was not significantly different from that of the disconnected condition (3.55 ± 0.45 W/kg), $p = 0.18$, although significantly lower metabolism was found for the normal-walking condition without the AG (2.60 ± 0.44 W/kg), $p < 0.001$, as compared to the zero-torque condition.

The results of these system dynamics and metabolic tests show two important performance characteristics. First, the AG can accurately provide high levels of torque to the wearer; the average observed RMS error of 1.45 Nm is well within biological variation; in comparison, the average step-to-step standard deviation in peak knee flexion moment during mid-stance phase for the nonamputee participants is 4.8 Nm. Second, the AG can behave sufficiently close to a free-hinging joint so that the metabolism of the wearer reflects the use of the latter. These two performance characteristics allow the AG to be treated as a near-ideal torque source at both low and high knee moment values. Hence, a higher-level control strategy, such as a neuromuscular controller, may be used without needing to reconsider the system dynamics of the hardware.

4. Neuromuscular Controller

A neuromuscular controller (NMC) was used to provide torque commands to the AG so to emulate the behavior of the biological gastrocnemius, similar to earlier work in robotic prosthesis control [17, 18]. The NMC was based on a musculoskeletal model of a human gastrocnemius. This model took joint kinematics and neural stimulation signals as input and produced joint moments as output. For control purposes, a simulated spinal reflex produced the neural stimulation signals by feeding back the internal state of the musculoskeletal model. Together, the musculoskeletal model and neural reflex took only joint kinematics as input and produced joint moments as output and formed the NMC for controlling the AG.

4.1. Musculoskeletal Model. Nonamputee kinematic, kinetic, and muscle activation data (Appendices A and B) were used to inform the parameter selection of a musculoskeletal model of the human legs. This model comprised a hill-type muscle model for nine effective muscles and attachment geometry of the human leg as described in [19], including optimizer settings and parameter bounds.

The free parameters of the musculoskeletal model were chosen via optimization. The goal was to have the output moments from the optimization closely resemble those from the biological data while keeping simulated metabolic cost close to the measured values. The optimization scheme in [19] was used. For each amputee participant, a corresponding nonamputee was chosen whose height and weight most closely matched that of the given amputee (Table 2). The parameter values of the musculoskeletal model were then optimized based on the walking data at the preferred walking speed for the given nonamputee participant. This preferred speed was that which had the lowest metabolic energy cost per unit distance. This method produced a model that was individually tuned for each amputee.

Several variations of these cost functions have been utilized for determining the parameters of similar models [19, 20]. In the current work, two methods of evaluating the optimizations were utilized (Appendix C). For both methods, the cost included both kinetic and metabolic similarities of the model to the recorded human walking data. The metabolic cost was represented as cost of transport (COT), which was the metabolic energy cost per unit distance traveled. Upper and lower bounds for the model parameters were

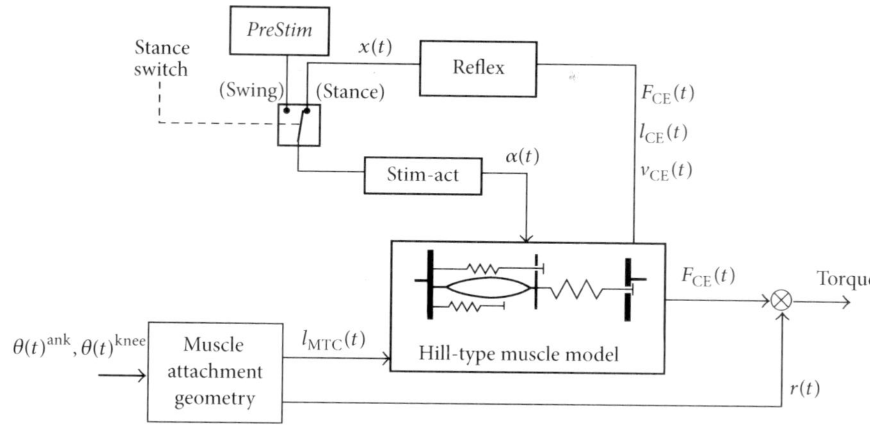

FIGURE 7: Neuromuscular controller including spinal reflex loop. The input to the neuromuscular controller was the kinematics of the ankle and knee joints $\theta(t)^{\text{ank}}$ and $\theta(t)^{\text{knee}}$. The internal reflex loop fed back muscle force, length, and velocity, and produced an internal activation signal to drive the simulated hill-type muscle model. Lookup tables were used to convert from joint kinematics to muscle-tendon length $l_{\text{MTC}}(t)$ and moment arm $r(t)$.

taken from [19]. Optimization settings were also identical to those from [19], except that the present study limited the number of generations to 50, as convergence was observed. See Appendix D for details on the optimization results.

4.2. Spinal Reflex Model. For the purposes of device control, the muscle model required a neural driving signal $x(t)$ to produce the activation $\alpha(t)$. Force, $F_{\text{CE}}(t)$, length, $l_{\text{CE}}(t)$, and velocity, $v_{\text{CE}}(t)$, feedback was used to model the spinal reflex of the gastrocnemius muscle, in the form described in [18] (Figure 7), with a lower bound of the reflex set to PreStim = 0.001. The stimulation-activation dynamics described in [21, 22] were used to obtain $\alpha(t)$ from the spinal reflex signal, $x(t)$, in the same manner as [19]. Additional details can be found in Appendix E.

The reflex gains and threshold values were optimized for the gastrocnemius muscle by simulating the effects of the reflex loop with the musculoskeletal model in the same manner as a previous study in ankle-foot prosthesis control [18]. These parameters were optimized to minimize the mean squared error between the reflex-based activation signals and the muscle EMG patterns from EMG data, given kinematic data. The optimization was performed using a genetic algorithm followed by a gradient descent method. For each data set corresponding to a nonamputee participant, the gastrocnemius morphological parameters of the musculoskeletal model were taken from the results of the corresponding morphological optimization.

4.3. Neuromuscular Controller Implementation. For the purpose of prosthesis control, the musculoskeletal model took as input walking state information and produced desired torque commands to the robotic joint during the stance phase. The ankle joint angle was read from an electrogoniometer (Delsys, Boston, MA) which measured the affected-side ankle angle, and the affected-side knee joint angle was read from the AG onboard joint angle sensor. The NMC was disabled during the swing phase by setting the neural stimulation to zero using the stance phase switch and ignoring any residual commanded torque. The muscle-tendon length and moment arm geometry were encoded in lookup tables to allow real-time computation of these values from joint kinematics. For additional tuning, the force gain of the reflex controller was made adjustable in real-time. This gain was chosen for adjustment since the controller's output power in simulation, based on nonamputee data, was found to be the most sensitive to changes in this parameter. The NMC was implemented in MATLAB Simulink and compiled to a C executable that communicated at 1 kHz to the master controller that, in turn, controlled the motor.

For each amputee participant, a corresponding model was used, which was based on a height-weight matched nonamputee participant's walking data (Table 2). The COT values for all nonamputee participants across walking speed are shown in Table 3. The participants all had minimal COT at either 1.25 or 1.5 m/s.

With this neuromuscular controller, the AG emulated the behavior of a biological gastrocnemius muscle by taking only the sensor-based ankle and knee joint kinematics as input. Thus, the AG was a self-contained unit that acted as an externally worn gastrocnemius muscle with neural and musculoskeletal dynamics encoded in its control.

5. Clinical Trials

In the final clinical trials, six participants with transtibial amputation walked on an instrumental treadmill while wearing a PAFP-AG combination while biomechanical and metabolic data were collected (Figure 2(a)).

5.1. Experimental Setup. The walking speed for a given amputee was set as the preferred speed determined from the matched nonamputee walking data (Table 3). This speed was set for all walking conditions. The clinical evaluation was conducted at MIT (Cambridge, MA) and was approved by MIT's Committee on the Use of Humans as Experimental Subjects (COUHES). Each participant provided written, informed consent before data collection was initiated.

Table 3: Metabolic cost of transport for nonamputee participants.

Participant	0.75 m/s	1.0 m/s	1.25 m/s	1.5 m/s	1.75 m/s	2.0 m/s	Optimal speed (m/s)
NA1	0.419	0.364	0.326	0.339	0.362	0.380	1.25
NA2	0.462	0.386	0.371	0.375	0.404	0.467	1.25
NA3	0.481	0.389	0.347	0.341	0.368	0.400	1.5
NA4	0.395	0.339	0.305	0.315	0.348	0.408	1.25

Each amputee was first asked to don only the PAFP, and the electrogoniometer was installed on the PAFP for subsequent use with the AG. The participant was then instructed to walk on the treadmill while the prosthesis power was adjusted using the commercial tuning app so to achieve average prosthesis power within two standard deviations from the average biological ankle power at the given walking speed [23]. Prosthesis tuning was subsequently adjusted slightly during subsequent trials as needed in an attempt to maintain prosthesis power within these limits.

Next the amputee donned the AG orthosis along with the PAFP. A short time period was given to ensure that both devices were functioning properly. Then the amputee was asked to walk on an instrumented treadmill at the given walking speed with the same data collection methods used for developing the NMC, including the instrumented treadmill, and reflective markers. Each trial consisted of 90 seconds of data collection.

In addition to using the optimized parameters of the NMC, additional conditions were tested wherein the force feedback reflex gain of the NMC was adjusted manually. This adjustment allowed some level of controller tuning based on the actual participant feedback. While a participant walked on the treadmill with the AG and PAFP, this gain was incrementally increased from its optimized, Default value (the *Default* gain setting) until either further increases in gain were said to be undesirable by the participant or the average power from the device was no longer increasing with higher controller gain (the *Max* gain setting). The gain was then decreased until the average power coming from the device was in the range of the biological gastrocnemius, of 3–5 Joules (the *BioGain* setting) [12, 13]. If the *Default* setting provided less than or equal to 3–5 Joules of net mechanical work per step, an additional gain was set to mid-way between the *Default* gain and the *Max* gain. This additional gain setting was called *Half*.

Once the aforementioned conditions were recorded with the motion capture system, metabolic data were collected using the same gas analysis system as the other data collection sessions (see Section 3.4). The participants were first asked to perform one standing trial to measure standing metabolism. Next, the walking conditions described above were repeated, in random order. Each metabolic trial lasted 6 minutes.

5.2. Data Processing. The same data processing methods as in the NMC development (see Appendix B) were used to obtain gait events, sagittal plane joint moments and angles, and metabolic cost from the recorded data, except that a point-mass of 2 kg was added at the affected-side knee joint of the SIMM model to represent the dynamical effects of the robotic joint and conduit mass. Also affected-side biological knee moment contribution was computed by subtracting the measured AG knee moment from the total knee moment estimated from the inverse dynamics.

The walking metabolism was compared for each amputee participant across trials, where each trial involved a different value for the force feedback reflex gain of the NMC. The optimal gain was found as the one that resulted in the lowest metabolic power for a given amputee. The walking condition (called the OPT condition) corresponding to this optimal gain was then compared to the walking condition where the AG was controlled to behave as a free-joint (called the ZERO condition).

Joint powers were computed as the product of joint velocity from joint kinematics and joint moments derived from the motion capture data, where positive power was defined as that produced by the joint on the environment. Where available, mechanical power from the powered prosthesis was taken from PAFP telemetry data, computed using the prosthesis onboard torque and angle sensors. For two subjects (S2 and S3), these data were not available, so PAFP mechanical power was computed in the same manner as the other joints, using SIMM-derived ankle angular velocities and moments from the motion capture data. AG applied mechanical power was computed from motion capture-based knee joint angular velocity and internal AG torque from strain gauge data.

Positive and negative joint work contributions were defined as the time integral of the positive and negative components of joint power, respectively, integrated over the late stance knee flexion phase (from mid-stance maximum knee extension to ToeOff), and divided by the time in that phase. Flexion and extension moment impulse values were computed in the same way, but using joint moments instead of powers.

Theoretical reductions in metabolic power were computed by assuming the average mechanical power provided by the combined AG knee actuator and ankle prosthesis in the OPT condition, as compared to the ZERO condition (with zero net work assumed for the AG knee in the ZERO condition), could perfectly replace muscle fiber mechanical power. With the efficiency of skeletal muscle assumed to be 25% [24], the reduction in muscle metabolic power was computed as 4 times the observed mechanical power provided by the AG knee actuator and ankle-foot prosthesis in the OPT condition. The resulting change in metabolic power was then compared to the measured metabolic power from the ZERO condition to determine theoretical percent reduction.

Statistics were run on metabolic powers and also bilateral ankle, knee, and hip joint features (flexion and extension

TABLE 4: Walking conditions for minimal metabolic power (OPT condition). For all but the last row, numbers represent the settings and results during the OPT condition. In the last row, positive values indicate that the given robotic joint produced greater net work in the OPT condition than in the ZERO condition.

Participant	S1	S2	S3	S4	S5	S6
Condition name	Bio	Max	Default	Default	Default	Half
Force gain scaling	0.1	8	1	1	1	1.5
Force gain (1/N)	$3.69e-5$	$4.99e-3$	$3.99e-4$	$6.19e-4$	$5.11e-4$	$7.66e-4$
AG mean power (W/kg)	0.054	0.118	0.028	0.027	0.105	0.031
Change in AG mean power from ZERO (W/kg)	−0.001	+0.020	+0.056	+0.077	+0.011	+0.009

TABLE 5: Metabolic cost of transport using the artificial gastrocnemius. Values given represent the net metabolic power, above that of basal metabolic rate. Positive values for percent change indicate that the participant required less metabolic power with the neuromuscular controller, compared to the control condition of zero-torque at the knee. Theoretical metabolic reduction, based on the change in knee orthosis and ankle-foot prosthesis applied average power from the ZERO condition to the OPT condition, and assuming 25% muscle efficiency, is also shown for comparison.

	S1	S2	S3	S4	S5	S6
ZERO Condition Net Metabolic Power (W/kg)	2.95	3.51	3.60	5.06	3.34	3.52
OPT Condition Net Metabolic Power (W/kg)	3.13	3.08	3.87	4.95	2.97	3.38
Percent change in metabolic power from ZERO condition to OPT condition	5.8	−12.1	7.4	−2.3	−11.1	−4.0
Theoretical percent change in metabolic power	−7.2	−15.7	−9.3	−8.3	−14.1	−4.6

moment impulses and positive and negative joint work contributions) by performing a two-sided, paired t-test between these features during the OPT condition and ZERO condition. Statistically significant differences were defined as those with p values less than or equal to 0.05.

5.3. Clinical Results

5.3.1. AG Overall Device Performance. In the final amputee trials with the AG-PAFP combination, the metabolically optimal (OPT) conditions were found for each participant, based on the adjusted parameter of the force-reflex gain. The gain settings and average output power from the AG joints are summarized in Table 4.

5.3.2. Metabolic Power. Walking metabolism is shown in Table 5 for the walking conditions in which the metabolic cost was found to be lowest, compared to the theoretically calculated values using device average power. Large percentage reductions in metabolism were found in two of the six participants. Two participants showed small metabolic reductions, and the two remaining participants had metabolism values higher with the AG intervention, compared to the zero-torque condition. Within the participants with a metabolic reduction in the OPT condition, these metabolic reductions were well correlated with the theoretical values based on device mechanical power ($r = 0.9$), as shown in Table 5. However, Subject 1 and Subject 3 displayed increases in metabolism in the OPT condition, as compared to the ZERO condition, which was not predicted with the theoretical calculations. Overall, the metabolic power between the ZERO and OPT conditions was not found to be statistically significant ($p = 0.43$).

5.3.3. Joint Kinetics. Comparing the ZERO and OPT conditions, statistically significant differences were found in the biological knee flexion moment impulse and biological hip positive work of the affected leg during the late-stance knee flexion phase of gait. The knee flexion moment impulse was reduced by an average of 0.0223 Nm∗s/kg ($p = 0.03$), from the ZERO condition (0.0413 ± 0.0262 Nm∗s/kg) to the OPT condition (0.0190 ± 0.0252 Nm∗s/kg). Figure 8 illustrates the AG and biological knee moment contributions over the gait cycle. The affected-side hip showed a decrease in positive

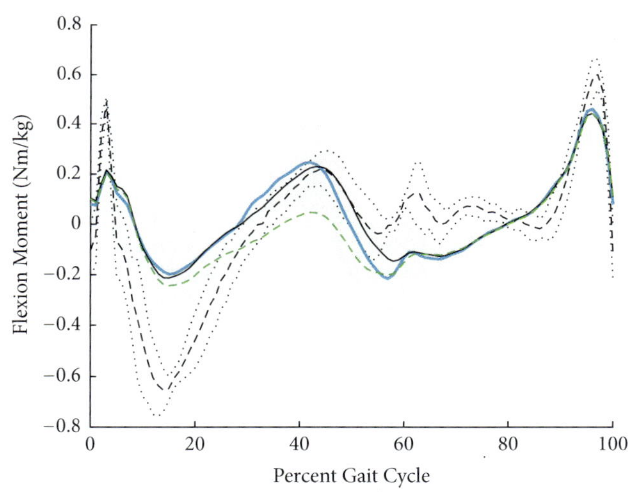

FIGURE 8: Components of affected-side knee moment of amputee participants with the powered AG, averaged across participants. The net value, which is the sum of biological knee and AG values, is shown for two walking conditions: the ZERO condition (thick solid blue line) and the OPT condition (thin solid red line). The contribution from the biological knee joint is also shown for the OPT condition (dashed thin green line). Nonamputee data, averaged across participants, are shown for reference, with the mean (dashed black line) ± one standard deviation (dotted black lines).

work of 0.0739 J/kg ($p = 0.004$) from the ZERO condition (0.2621 ± 0.0848 J/kg) to the OPT condition (0.1882 ± 0.0734 J/kg). All other tested kinetic features did not indicate significant differences between conditions.

5.4. Discussion. Although a statistically significant metabolic reduction across participants was not found in this study, the large percentage reductions in metabolism for some participants, as in Table 5, indicate that it is likely that a metabolic advantage may be achieved with a powered artificial gastrocnemius, provided appropriate conditions are satisfied. It is yet unclear what exact conditions must be met to provide metabolic reductions for all participants. However, within the participants with a metabolic reduction in the OPT condition, these reductions were fairly well correlated with the theoretical values based on device mechanical power ($r = 0.9$), as shown in Table 5. Therefore, it is possible that higher levels of mechanical power from the AG would increase the chance of producing a statistically significant metabolic reduction across participants.

The kinetic changes, as also illustrated in Figure 8, support the hypothesis that the AG causes a reduction in affected biological knee flexion moment impulse and hip positive net work, but not knee positive work or hip moment impulse. Considering the short trial times, this result indicates that participants were capable of rapid kinetic responses to the interventions, on the order of several minutes or less. The knee moment impulse reduction may stem from an attempt by the participants to prevent the total knee moment from changing in the presence of the external disturbance. This result is consistent with other studies, as humans have been observed to maintain invariant joint moments at both the ankle [25] and hip joints [26] when an external torque from an exoskeleton is applied. This finding has a potential consequence: if the applied joint moment from the AG exceeded the affected knee moment from the ZERO condition, the participant might attempt to maintain total joint kinetics by providing an active extension moment with their quadriceps, so to prevent the total knee flexion moment from increasing. This behavior may put an upper-limit on the flexion torque that can be provided by the AG without incurring a cost from the quadriceps in the form of active knee extension efforts.

5.4.1. Potential Mechanisms for Metabolic Reductions. First, it is possible that this reduction in knee moment itself may play a role in the metabolic reduction in some of the participants, as the production of muscle force exacts a metabolic cost [27]. However, the generation of isometric muscle force alone is comparatively less costly compared to the generation of mechanical work. Therefore, other factors are likely at play.

The similarity of the observed metabolic reductions to the theoretical values (Table 5) suggests that much of the power provided by the AG goes to replacing muscle fiber positive work, as modeled in the theoretical cost values. The reduction in affected-side hip positive work with AG assistance indicates that energy may be transferred from the AG up to the hip from the knee. These hip work reductions seen in late-stance knee flexion coincide with the initiation of hip flexion during preswing, indicating that the AG probably helps to initiate leg swing. This result is consistent with what is known about the normal function of the human gastrocnemius muscle during walking [28, 29].

6. Conclusions and Future Work

To create the first fully powered artificial gastrocnemius for transtibial amputees, we develop a versatile knee orthosis that is capable of kinetic behavior comparable to the biological gastrocnemius. The control strategies provide low errors between desired and measured joint torque across a wide dynamic range. Good torque tracking of zero-torque commands allow for metabolic transparency, while the torque tracking at higher torque levels enables the system to replicate desired behavior. This large dynamic range is attributed to the tethered system that enabled large power capability without the commensurate cost of added actuator mass. Others have made use of this type of platform for this reason [30, 31]. Unlike other researchers, however, our use of a cable linkage mitigates transmission losses in the cable drive compared to a pure Bowden cable transmission, since only part of the cable is in contact with the conduit surfaces. This system is well suited for simulating a gastrocnemius muscle at the knee for amputee participants.

The clinical study presented here represents an effort to improve upon earlier biarticular transtibial prostheses by using an active element for the affected knee joint. Such an active device can provide positive net work, unlike previous quasi-passive counterparts [10, 11]. If this net robotic work replaces positive muscle work, we expect to see large reductions in metabolism with the biarticular conditions, as compared to the monoarticular conditions. Indeed, with the application of sufficient positive work, the results suggest that it may be possible to provide metabolic and kinetic assistance to transtibial amputee walking gaits beyond that provided by a powered ankle-foot prosthesis alone. This study also provides suggestive evidence that metabolic improvements may stem from not only reductions in muscle activity at the assisted joint, but other joints as well, such as the hip. This finding is consistent with other work in lower-extremity exoskeletons [32].

However, this study also highlights the difficulty of reducing metabolic demand in humans, even in the presence of an existing pathology. Even the smallest unwanted perturbations can severely hinder an intervention that would otherwise cause a successful metabolic improvement. Thus, in the development of assistive wearable devices, it is of utmost importance to consider the biomechanical consequences of applying torques and powers to the human body. Future work could utilize a torque control strategy to more systematically vary output power levels of the AG to determine if higher net positive work could further improve metabolic cost of the wearer, as researchers have performed with an ankle exoskeleton [33]. An additional next step would be to use a modified NMC to command both the knee and ankle joints of the powered artificial gastrocnemius with one, unified model. Indeed, such a control strategy would provide better synchronization between the knee and ankle actuators.

TABLE 6: Nonamputee body parameters.

Participant number	Height (cm)	Weight (kg)
NA1	175	73.4
NA2	180	90.3
NA3	188	95.4
NA4	180	103

In the advancement of active transtibial prostheses, we feel a powered artificial gastrocnemius is of critical importance. It is our hope that this study will underscore the potential of this novel mechatronic approach to the biomechanical and physiological gait performance of persons with transtibial leg amputation.

Appendix

A. Neuromuscular Controller Data Collection

The parameters of the neuromuscular controller (NMC) were selected in two stages. In the first stage, the musculoskeletal model parameters were optimized for a full leg musculoskeletal model so to best match biological joint moments and metabolism, given biological kinetics and EMG data. In the second stage, the morphological model parameters for the gastrocnemius were fixed at their optimized values, and the reflex parameters were optimized for this muscle.

The gait data of four healthy, nonamputees were used to inform the parameters of the NMC. These nonamputee participants were chosen so to match the amputees as closely as possible (see Table 6). The clinical evaluation was conducted at MIT (Cambridge, MA) and was approved by MIT's Committee on the Use of Humans as Experimental Subjects (COUHES). Each participant provided written, informed consent which was obtained before data collection was initiated. These data were collected in two phases.

In the first phase, kinematic, kinetic, and electromyographic (EMG) data were collected for 90 seconds at each of six speeds (0.75 m/s, 1.00 m/s, 1.25 m/s, 1.50 m/s, 1.75 m/s, and 2.00 m/s) in random order. An infrared camera system (12 cameras, model: T40s, Vicon Motion Systems Ltd., Oxford, UK) was used to track the motion of subjects as they walked in the capture volume. Reflective markers were placed at 47 locations on each participant's body, based on the Helen Hayes marker model, and their three-dimensional trajectories were recorded at 100 Hz. The marker locations were chosen specifically to track joint motion. The ground reaction forces and contact centers of pressure were measured using an instrumented force plate treadmill (Bertec Corporation, Columbus, OH) with a sampling rate of 1 kHz. Electromyographic signals were collected at 2 kHz using a wireless surface system (Trigno, Delsys Inc., Natick MA). Twelve muscles (tibialis anterior, soleus, medial gastrocnemius, lateral gastrocnemius, rectus femoris, vastus lateralis, vastus medialis, semimembranosus, biceps femoris long head, adductor magnus, iliacus, and gluteus maximus) on the right leg of each subject were recorded, with symmetry being assumed for the other leg. Prior to the walking trials, the muscle maximum voluntary contraction (MVC) tests were performed for each of the muscles of interest, in which the participants were instructed to push as hard as possible, three times in succession for 3–5 seconds each with the given muscle group, while electromyographic data were collected.

Once the first data collection phase was completed, the markers and electrodes were removed, and the participants were equipped with an open-circuit gas exchange measurement system for estimating metabolic cost (model: K4b2, COSMED, Rome, Italy). The participants were asked to perform one standing trial to measure standing metabolism, and then walking trials were performed on the treadmill for the same six walking speeds. Each walking trial lasted six minutes.

B. Neuromuscular Controller Data Processing

Fourth-order Butterworth filters were used to filter the marker position and ground reaction force data with 6 Hz and 25 Hz cutoff frequencies, respectively. The marker and force data were postprocessed through the inverse dynamics module of SIMM (MusculoGraphics Inc., Evanston, IL), to produce joint moments and angles for ankle, knee, and hip joints, as well as muscle-tendon lengths and moment arms, all as a function of time. Only sagittal plane dynamics were considered.

Gait events were determined using vertical ground reaction force data from the embedded force plates. Approximate event timing was found by determining the times when the force increased beyond a 40 N threshold. Exact heel strike and ToeOff times where found by progressing backward and forward in time, respectively, until the force value dropped to zero. Data were then cut to gait cycles based on heel strike times and resampled to 101 points. Gait cycles were discarded for the beginning and end of each trial, during the speed transients of the treadmill. Gait cycles in which the stride times were below 0.7 seconds or above 1.3 seconds or in which a foot crossed the midline of the treadmill were also discarded. Metabolic cost for each walking speed was computed in the same manner as in the nonamputee validation trials (Section 3.4).

Several processing steps were taken to prepare the EMG signal to be processed for the estimation of muscle activation. EMG signals were first processed internally by the EMG system by applying a 4th-order bandpass filter with a pass band from 20 Hz to 450 Hz. A 60-Hz notch filter was also applied to remove mains hum. Any DC offset was removed, and the remaining signal was saturated at 5 standard deviations from zero. The signal was then normalized to the new maximum value of 5 standard deviations and rectified.

The processed EMG data were further processed with the method described in [19] to achieve estimates of muscle activation from the measured EMG signals. The activation was initially scaled by those resulting from the MVC trials, but the resulting scaled EMG signals were unreasonably scaled when compared to literature values. Therefore, the final magnitudes of the estimated activation signals were scaled so that the average values of the activation signals

TABLE 7: Optimized morphological parameters.

Parameter	S1	S2	S3	S4	S5	S6
F_{max}	1094	1052	1094	1936	3049	3049
l_{sl}	0.4565	0.4115	0.4565	0.4568	0.4217	0.4217
l_{opt}	0.0504	0.0454	0.0504	0.0504	0.0465	0.0465
K_{sh}	2.58	2.78	2.58	2.722	2.84	2.84
λ_{ref}	0.0593	0.0684	0.0593	0.0599	0.0846	0.0846

matched those from [34] during the range of activation for the literature data.

As was found in a previous study [19], many of the muscles spanning the hip were found to have signals too weak to be reliably measured (rectus femoris, iliacus, biceps femoris short head, hip abductors, adductor longus, and adductor magnus). For these muscles, neural excitation profiles from wire electrode experiments [34] were delayed 40 ms to simulate the delay between the EMG signals and the development of force [35] and then passed through the same activation dynamics used for the collected electromyographic data. The resulting activation signals were then used as input to the musculoskeletal model along with those derived from the electromyography.

C. Morphological Optimization Cost Functions

In the current work, two methods of evaluating the morphological model parameters were utilized. For the first cost function (used for Subjects 2 and 3), the musculoskeletal model parameters were optimized using the method described by [19]. However, using this method, it was found that the measured electromyographic data failed to produce satisfactory values for both kinetic fits and metabolic cost estimates; either kinetic fit R^2 values were low (less than a cutoff threshold of 0.7) or the metabolic cost estimates of the model were more than one standard deviation higher than the mean of the empirical metabolic cost values (from the nonamputee participants walking at their preferred speeds). As an alternative, the activation profiles derived from wire electrode experiments [34] provided both kinetic and metabolic results closer to that of biology. Therefore, for this optimization method, all muscle activation profiles were derived from the wire electrode literature [34], processed through the same activation dynamics used for the EMG [19]. Selection of the final parameter set was also performed as in [19]. For Subject 2, the HAB muscle group was used for selecting the final parameter set. For Subject 3, the VAS muscle group was used. These muscles were selected based on their dominance in the simulated fractional metabolic cost.

Some biological walking data did not produce acceptable kinetic and metabolic cost values (as described above) with the first optimization method (those for Subjects 1, 5, and 6), or the resulting model produced low average mechanical power when tested in simulation with walking kinematics and EMG as input (Subject 4). For these data sets, a different cost function was used. This metabolic cost estimation function was based on empirical measures of muscle metabolic energy consumption as a function of fiber contractile velocity and activation [36, 37]. The simulated metabolic cost was found by integrating muscle power for each muscle in the leg over the gait cycle, and summing up all muscles for the leg, assuming bilateral symmetry. The measured basal energy consumption was added to the resulting value to determine the full-body metabolic energy cost. As with the first cost function, both the simulated metabolic cost and the average kinetic fit were considered for the evaluation of parameter solutions. However, it was found that the kinetic fit was largely invariant in the range of solutions where the simulated COT was near that of the measured values. Therefore, in selecting a single solution, the parameter set with the best kinetic fit was chosen, provided that the simulated COT was within the standard error of the measured COT value. A similar parameter selection method had been used to select optimal parameter sets in previous work [38].

TABLE 8: Final cost values for the morphological parameter optimization. The COT values were the simulated metabolic cost of transport values, as computed by the musculoskeletal model, when given the final optimized parameter set and biological nonamputee data as input.

Cost value	S1	S2	S3	S4	S5	S6
R^2	0.71	0.76	0.80	0.85	0.80	0.80
COT	0.360	0.324	0.246	0.330	0.301	0.301

D. Morphological Parameter Optimization Results

The optimized morphological parameters for the gastrocnemius muscle are shown in Table 7. The cost values are shown in Table 8. These parameters correspond to those in the model described in previous work [19].

E. Spinal Reflex Model

The muscle model required a neural driving signal $x(t)$ to produce the activation $\alpha(t)$. Herein, the neural driving signal can be represented as [18]

$$\begin{aligned} x(t) = {} & G_F \cdot (F(t - \Delta t_F) - F_0) \cdot u(F - F_0) + G_l \\ & \cdot (l(t - \Delta t_{lv}) - l_0) \cdot u(l - l_0) + G_v \\ & \cdot (v(t - \Delta t_{lv}) - v_0) \cdot u(|v| - v_0). \end{aligned} \tag{E.1}$$

Here, G_F, G_l, and G_v are the gains; Δt_F, Δt_l, and Δt_v are the time delays for the force, muscle length, and muscle velocity

TABLE 9: Optimized neuromuscular model reflex parameters. Each column represents the settings for an amputee participant, with the corresponding nonamputee data set in parentheses.

	S1 (NA2)	S2 (NA1)	S3 (NA3)	S4 (NA3)	S5 (NA4)	S6 (NA4)
G_F (1/N)	$3.687e-4$	$6.238e-4$	$3.994e-4$	$6.185e-4$	$5.11e-4$	$5.11e-4$
F_0 (N)	0.1786	0.3935	0.3624	0.291	0.532	0.532
G_l (1/m)	6.88	5.22	1.3465	6.32	8.19	8.19
l_0 (m)	0.0507	0.0490	0.0528	0.0448	0.0382	0.0382
G_v (s/m)	0.9606	1.34	2.00	0.0193	2.00	2.00
v_0 (m/s)	0.1230	−0.1175	−0.1168	−0.0897	0.0340	0.0340

values, respectively. The function $u(\cdots)$ is a step function, indicating that each reflex only activates for values greater than the corresponding threshold value (F_0, l_0, and v_0 for force, length, and velocity, resp.). The time delays model the neural delay for a signal to traverse the reflex arc from muscle to spinal cord and back to the muscle. All three delays were fixed at 20 ms [39–41]. The reflex signal $x(t)$ represents the neural stimulation. The optimized parameter values for the reflex loop are shown in Table 9. The stimulation-activation dynamics described in previous work [19] were used to obtain the muscle activation, α.

Conflicts of Interest

The authors declare no conflicts of interest regarding the publication of this article.

Acknowledgments

This work was supported by NASA [NNX12AM16G] as part of the National Robotics Initiative and departmental funding from the MIT Media Lab. The authors thank Bob Emerson and Jared Markowitz for sharing their valuable expertise.

References

[1] N. H. Molen, "Energy/speed relation of below-knee amputees walking on a motor-driven treadmill," *Internationale Zeitschrift für Angewandte Physiologie Einschließlich Arbeitsphysiologie*, vol. 31, no. 3, pp. 173–185, 1973.

[2] R. L. Waters, J. Perry, D. Antonelli, and H. Hislop, "Energy cost of walking of amputees: the influence of level of amputation," *The Journal of Bone & Joint Surgery*, vol. 58, no. 1, pp. 42–46, 1976.

[3] G. Colborne, S. Naumann, P. Longmuir, and D. Berbrayer, *Am J Phys Med Rehabil*, vol. 71, no. 5, pp. 272–278, 1992.

[4] P.-F. Su, S. A. Gard, R. D. Lipschutz, and T. A. Kuiken, "Gait characteristics of persons with bilateral transtibial amputations," *Journal of Rehabilitation Research and Development*, vol. 44, no. 4, pp. 491–501, 2007.

[5] H. M. Herr and A. M. Grabowski, "Bionic ankle-foot prosthesis normalizes walking gait for persons with leg amputation," *Proceedings of the Royal Society B Biological Science*, vol. 279, no. 1728, pp. 457–464, 2012.

[6] D. Hill and H. Herr, "Effects of a powered ankle-foot prosthesis on kinetic loading of the contralateral limb: a case series," in *Proceedings of the IEEE International Conference on Rehabilitation Robotics (ICORR '13)*, pp. 1–6, Seattle, Wash, USA, June 2013.

[7] M. R. Williams, A. Grabowski, H. Herr, and S. D'Andrea, "Electromyographic effects of using a powered ankle-foot prosthesis," *American Society of Biomechanics*, 2012.

[8] E. Isakov, O. Keren, and N. Benjuya, "Trans-tibial amputee gait: Time-distance parameters and EMG activity," *Prosthetics and Orthotics International*, vol. 24, no. 3, pp. 216–220, 2000.

[9] E. Isakov, H. Burger, J. Krajnik, M. Gregoric, and C. Marincek, "Knee muscle activity during ambulation of trans-tibial amputees," *Journal of Rehabilitation Medicine*, vol. 33, no. 5, pp. 196–199, 2001.

[10] K. Endo, E. Swart, and H. Herr, "An artificial gastrocnemius for a transtibial prosthesis," in *Proceedings of the 31st Annual International Conference of the IEEE Engineering in Medicine and Biology Society: Engineering the Future of Biomedicine, EMBC 2009*, pp. 5034–5037, USA, September 2009.

[11] M. F. Eilenberg, K. Endo, and H. Herr, "Biomechanic and Energetic Effects of a Quasi-Passive Artificial Gastrocnemius on Transtibial Amputee Gait," *Journal of Robotics*, In press.

[12] R. J. Zmitrewicz, R. R. Neptune, and K. Sasaki, "Mechanical energetic contributions from individual muscles and elastic prosthetic feet during symmetric unilateral transtibial amputee walking: A theoretical study," *Journal of Biomechanics*, vol. 40, no. 8, pp. 1824–1831, 2007.

[13] R. R. Neptune, K. Sasaki, and S. A. Kautz, "The effect of walking speed on muscle function and mechanical energetics," *Gait & Posture*, vol. 28, no. 1, pp. 135–143, 2008.

[14] R. C. Browning, J. R. Modica, R. Kram, and A. Goswami, "The effects of adding mass to the legs on the energetics and biomechanics of walking," *Medicine & Science in Sports & Exercise*, vol. 39, no. 3, pp. 515–525, 2007.

[15] J. Zhang, C. C. Cheah, and S. H. Collins, "Experimental comparison of torque control methods on an ankle exoskeleton during human walking," in *Proceedings of the 2015 IEEE International Conference on Robotics and Automation, ICRA 2015*, pp. 5584–5589, USA, May 2015.

[16] J. M. Brockway, "Derivation of formulae used to calculate energy expenditure in man," *Hum Nutr Clin Nutr*, vol. 41, no. 6, pp. 463–471, 1987.

[17] M. F. Eilenberg, H. Geyer, and H. Herr, "Control of a powered ankle-foot prosthesis based on a neuromuscular model," *IEEE Transactions on Neural Systems and Rehabilitation Engineering*, vol. 18, no. 2, pp. 164–173, 2010.

[18] J. Markowitz, P. Krishnaswamy, M. F. Eilenberg, K. Endo, C. B. Chris, and H. Herr, "Speed adaptation in a powered transtibial prosthesis controlled with a neuromuscular model," *Philosophical Transactions of the Royal Society B: Biological Sciences*, vol. 366, no. 1570, pp. 1621–1631, 2011.

[19] J. Markowitz and H. Herr, "Human Leg Model Predicts Muscle Forces, States, and Energetics during Walking," *PLoS Computational Biology*, vol. 12, no. 5, Article ID e1004912, 2016.

[20] P. Krishnaswamy, E. N. Brown, and H. M. Herr, "Human leg model predicts ankle muscle-tendon morphology, state, roles and energetics in walking," *PLoS Computational Biology*, vol. 7, no. 3, Article ID e1001107, 2011.

[21] J. He, W. S. Levine, and G. E. Loeb, "Feedback Gains for Correcting Small Perturbations to Standing Posture," *IEEE Transactions on Automatic Control*, vol. 36, no. 3, pp. 322–332, 1991.

[22] C. C. Raasch, F. E. Zajac, B. Ma, and W. S. Levine, "Muscle coordination of maximum-speed pedaling," *Journal of Biomechanics*, vol. 30, no. 6, pp. 595–602, 1997.

[23] M. Palmer, *Sagittal Plane Characterization of Normal Human Ankle Function Across a Range of Walking Gait Speeds , Masters Thesis [Masters, thesis]*, MIT, 2002.

[24] R. Margaria, *Biomechanics and Energetics of Muscular Exercise*, Clarendon Press, Oxford, 1976.

[25] P.-C. Kao, C. L. Lewis, and D. P. Ferris, "Invariant ankle moment patterns when walking with and without a robotic ankle exoskeleton," *Journal of Biomechanics*, vol. 43, no. 2, pp. 203–209, 2010.

[26] C. L. Lewis and D. P. Ferris, "Erratum to "Invariant hip moment pattern while walking with a robotic hip exoskeleton" [J. Biomech. 44 (5) (2011) 789-793]," *Journal of Biomechanics*, vol. 47, no. 7, p. 1748, 2014.

[27] T. M. Griffin, T. J. Roberts, and R. Kram, "Metabolic cost of generating muscular force in human walking: insights from load-carrying and speed experiments," *Journal of Applied Physiology*, vol. 95, no. 1, pp. 172–183, 2003.

[28] R. R. Neptune, S. A. Kautz, and F. E. Zajac, "Contributions of the individual ankle plantar flexors to support, forward progression and swing initiation during walking," *Journal of Biomechanics*, vol. 34, no. 11, pp. 1387–1398, 2001.

[29] R. R. Neptune, F. E. Zajac, and S. A. Kautz, "Muscle force redistributes segmental power for body progression during walking," *Gait & Posture*, vol. 19, no. 2, pp. 194–205, 2004.

[30] J. M. Caputo and S. H. Collins, "An experimental robotic testbed for accelerated development of ankle prostheses," in *Proceedings of the IEEE International Conference on Robotics and Automation (ICRA '13)*, pp. 2645–2650, Karlsruhe, Germany, May 2013.

[31] Y. Ding, I. Galiana, A. T. Asbeck et al., "Biomechanical and physiological evaluation of multi-joint assistance with soft exosuits," *IEEE Transactions on Neural Systems and Rehabilitation Engineering*, vol. 25, no. 2, pp. 119–130, 2017.

[32] L. M. Mooney and H. M. Herr, "Biomechanical walking mechanisms underlying the metabolic reduction caused by an autonomous exoskeleton," *Journal of NeuroEngineering and Rehabilitation*, vol. 13, no. 1, article no. 111, 2016.

[33] R. W. Jackson and S. H. Collins, "An experimental comparison of the relative benefits of work and torque assistance in ankle exoskeletons," *Journal of Applied Physiology*, vol. 119, no. 5, pp. 541–557, 2015.

[34] J. Perry, *Gait Analysis: Normal and Pathological Function*, SLACK, Inc, New Jersey, NJ, USA, 1992.

[35] E. M. Arnold, S. R. Hamner, A. Seth, M. Millard, and S. L. Delp, "How muscle fiber lengths and velocities affect muscle force generation as humans walk and run at different speeds," *Journal of Experimental Biology*, vol. 216, no. 11, pp. 2150–2160, 2013.

[36] A. V. Hill, "The heat of shortening and the dynamic constants of muscle," *Proceedings of the Royal Society B*, vol. 126, no. 843, pp. 136–195, 1938.

[37] A. V. Hill, "The efficiency of mechanical power development during muscular shortening and its relation to load," *Proceedings of the Royal Society B Biological Science*, vol. 159, no. 975, pp. 319–324, 1964.

[38] K. Endo and H. Herr, "A model of muscle-tendon function in human walking," in *Proceedings of the 2009 IEEE International Conference on Robotics and Automation, ICRA '09*, pp. 1909–1915, Japan, May 2009.

[39] H. Geyer, A. Seyfarth, and R. Blickhan, "Positive force feedback in bouncing gaits?" *Proceedings of the Royal Society B Biological Science*, vol. 270, no. 1529, pp. 2173–2183, 2003.

[40] R. B. Stein and C. Capaday, "The modulation of human reflexes during functional motor tasks," *Trends in Neurosciences*, vol. 11, no. 7, pp. 328–332, 1988.

[41] M. Knikou and W. Rymer, "Erratum: effects of changes in hip joint angle on H-reflex excitability in humans," *Experimental Brain Research*, vol. 143, no. 4, pp. 149–159, 2002.

15

A Wearable Robotic Device based on Twisted String Actuation for Rehabilitation and Assistive Applications

Mohssen Hosseini, Roberto Meattini, Gianluca Palli, and Claudio Melchiorri

Università di Bologna (DEI), Viale Risorgimento 2, Bologna, Italy

Correspondence should be addressed to Gianluca Palli; gianluca.palli@unibo.it

Academic Editor: Shahram Payandeh

The preliminary experimental study toward the implementation of an arm rehabilitation device based on a twisted string actuation module is presented. The actuation module is characterized by an integrated force sensor based on optoelectronic components. The adopted actuation system can be used for a wide set of robotic applications and is particularly suited for very compact, light-weight, and wearable robotic devices, such as wearable rehabilitation systems and exoskeletons. Thorough presentation and description of the proposed actuation module as well as the basic force sensor working principle are illustrated and discussed. A conceptual design of a wearable arm assistive system based on the proposed actuation module is presented. Moreover, the actuation module has been used in a simple assistive application, in which surface-electromyography signals are used to detect muscle activity of the user wearing the system and to regulate the support action provided to the user to reduce his effort, showing in this way the effectiveness of the approach.

1. Introduction

Research laboratories worldwide are concurrently designing a novel era of robots much more advanced than their predecessors in terms of cognitive capabilities and are able to adapt to constantly changing environments. These new robots have been intentionally designed for physical interaction with unstructured environments and humans such as servoactuated prostheses and exoskeletons. An exoskeleton is a wearable robotic system usually composed of an external structural mechanism with joints and links corresponding to those of the human body and provided with a suitable actuation system to support the user's movements. In these devices, the force and torque measurements are used to support and fulfill the human goals. Exoskeletons can be applied to a number of fields, and in particular in rehabilitation and haptic applications, ultimately benefiting all populations, both disabled and healthy ones.

In this paper, the design of a novel twisted string actuation (TSA) module is presented. The main purpose of this device is the development of wearable assistive and rehabilitation systems. Due to its particular structure, TSA is well suited in case the force is directly applied to the user's limbs, without external rigid structures supporting the limb and/or the actuation, as schematized in Figure 1. The design of the proposed wearable assistive and rehabilitation system aims to (i) remove all rigid joints to have a lighter and more comfortable system, which can be easily adapted to any user; (ii) decrease the weight, size, and mechanical complexity of the exoskeleton, avoiding complex regulation mechanisms, to reduce the costs and improve its reliability and affordability; (iii) design a modular actuation system that can be reused for implementing different assistive movements. The proposed TSA module is characterized by an integrated force sensor and embedded acquisition and control electronics, and its main components and structure are conceptually depicted in Figure 2.

The paper presents a complete overview of the actuation module structure and the integrated force sensor based on optoelectronic components and the sensor working principle. The motor module structure is manufactured in ABS plastic by 3D printing, and a commercial optoelectronic component, called light fork because of its forked structure, has been adopted for the force sensor implementation. The light fork

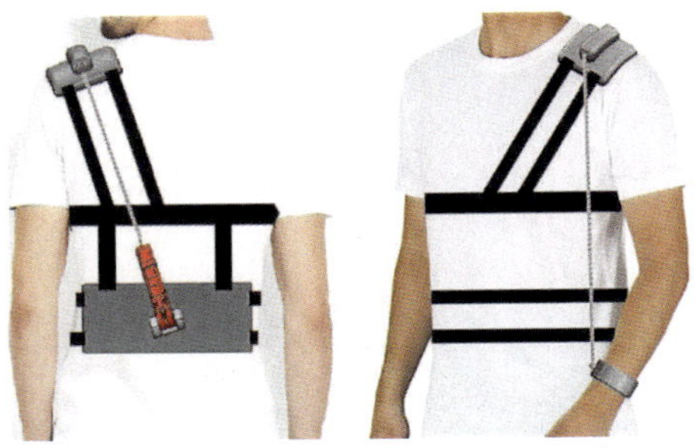

FIGURE 1: Conceptual view of an elbow assistive device based on the twisted string actuation module.

hosts in its package both the Light Emitting Diode (LED) and the Photodetector (PD) required for measuring the compliant frame deformation. This solution presents several advantages with respect to alternative strain-gauge based or optoelectronic based solutions, such as a simpler assembly, because of the well-defined package, high sensitivity, and low cost. The module is driven by a low-cost DC motor, and a combined axial-radial bearing is adopted to support the output shaft at the location of the twisted string connection. Moreover, the actuation module is also provided with an optical encoder for output shaft position measurement. The Finite Element Analysis (FEA) together with experimental measures has been executed to validate the actuation module design, and the sensor calibration process has been performed. To validate the properties of the proposed TSA module, two experiments have been performed in dynamic conditions, and the results have been validated by means of a reference strain-gauge based force sensor. Finally, as an application example, the TSA module has been used together with a surface-ElectroMyoGraphic (sEMG) signal acquisition system to implement an assistive device able to support the user during a lifting task involving his/her forearm, and preliminary results are reported.

The paper is organized as follows. Section 2 reviews some related works, whereas Section 3 describes the mechanical design of the TSA modules and the working principle of the integrated force sensor. In Section 4 the force sensor calibration and the experimental evaluation of the TSA module are presented, while in Section 5 the results about the use of the proposed device for the assistive application is reported. Finally, Section 6 draws the conclusions and outlines future research activities.

2. Related Works

In more recent years, the use of exoskeletons for the hand and lower/upper-limb support applications has drastically increased [1–3]. In [4], an upper-limb power-assist exoskeleton by pneumatic muscle actuation with two metal joints was developed. An integrated cable-driven, low-cost, and light-weight wearable upper body orthotics system that can be worn over the upper body to generate effective torques to move the arm through a set of assistive motions was introduced in [5]. In [6] a wearable upper body exoskeleton system incorporating a model-based compensation control framework to assist robot-aided shoulder-elbow rehabilitation and power assistance functions is presented. An anthropometric seven-degree-of-freedom powered exoskeleton for the upper limb based on a database defining the kinematics and dynamics of the upper limb during daily living activities is presented in [7]. In [8], a cable-driven arm exoskeleton (CAREX) was developed to achieve desired forces on the hand, that is, both pull and push, in any direction as required in neural training while a 6-DOF upper-limb exoskeleton robot [9] and a muscle suit with providing human physical support was developed in [10].

The TSA concept implements a light-weight, low-cost, and compact linear transmission system [11]. With an appropriate choice of string parameters (in particular the string radius and length) and a rotative electric motor, it is possible to fulfill the generally stringent requirements for the implementation of miniaturized and highly integrated mechatronic devices. Moreover, the slender structure of TSA makes it particularly suitable for wearable devices. The measurement of the actuation force plays a crucial role in the control of this actuation system, as reported in [12]. Previous studies on integrated actuation module for TSA are reported in [13–15].

Many different uniaxial force gauges are available on the market, with the majority of them based on strain gauges that provide exceptional linearity at the expense of intricate electronics for signal acquisition and sensitivity to electromagnetic noise. However, several advantages can be seen through alternative solutions, such as piezoelectric-based force sensors [16] or force sensors based on optoelectronic components [17, 18], mainly because of easier integration and simpler electronics. Optoelectronic sensors exploit the modulation and/or reflection of a light beam emitted by a source and received by suitable detectors to directly measure the deformation of a compliant structure or the relative displacement between elastically coupled elements caused by the external forces. An optical micrometric force sensor based on the differential measures of the light intensity is reported in [19]. Moreover, a sensor based on changing the coupling of optical power between a photodiode and a vertical-cavity surface-emitting laser is presented in [20]. In [21] the movements induced on a compliant structure by external forces and torques along 6 axes are detected by exploiting multiple Photodetectors and a single LED light source. Sensors based on the use of discrete optoelectronic components for the measurements of the tendon force have been implemented in [22, 23].

3. System Description

The designed system is conceived for rehabilitation and assistive applications, as schematically reported in Figure 1, where a twisted string actuator with an integrated force sensor is mounted on the back of the user and connected to a forearm through a shoulder path. The designed TSA module fits very well with this kind of application due to its light and compact structure and the ability of acting similarly to human muscles, as will be shown in the experiments described in Section 5.

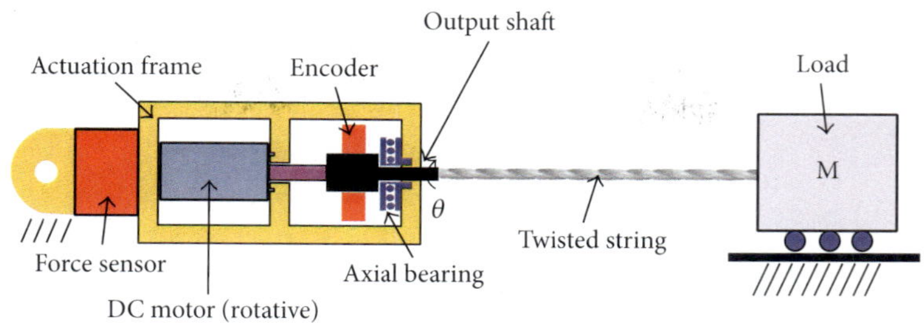

FIGURE 2: Schematic representation of the TSA structure.

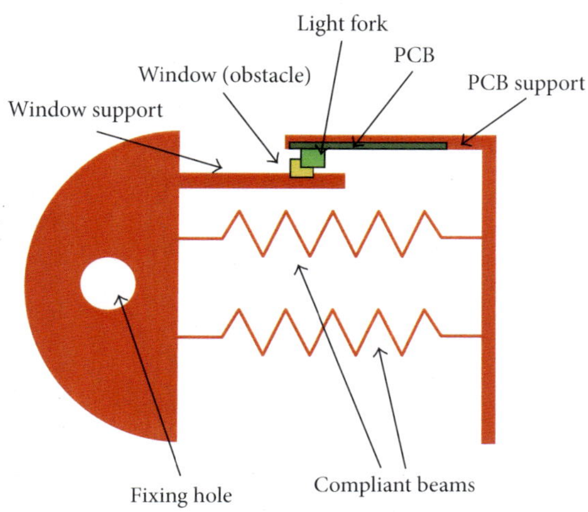

FIGURE 3: Scheme of the sensor structure and component arrangement.

3.1. Sensor Working Principle. The optoelectronic sensor is described in detail in [24], and therefore here only a brief summary of its main features is reported. The basic idea is to have LED illuminating a photodiode (PD), where the current flowing through the PD can be modulated by means of a mechanical component that partially intercepts the light emitted by the LED. The position of the mechanical component depends on the deformations occurring on the sensor's body as a consequence of the application of an external force, Figure 3.

The OMRON EE_SX1108, called a light fork because of its forked structure, is the sensitive element used for the implementation of the proposed sensor. This component is embedded with both LED and PD (a phototransistor) facing each other and is provided with a window restricting the light cone coming from the LED to the PD to a thin and well-focused region.

Several advantages accompany the use of this component. A compact sensor implementation is permitted because both the LED and the PD are embedded together in a compressed structure with precise relative position. The light cone is well shaped, providing a very steppy and linear transition region between the fully covered and the fully free light conditions and allowing good sensitivity and linearity to be obtained.

3.2. TSA Module Design. A schematic view of the designed TSA module is reported in Figure 2. With reference to this picture, the TSA module is composed, from left to right, by (i) a connection element to connect the module to the supporting frame; (ii) the force sensor to measure the actuation load; (iii) a frame hosting the DC motor, the output shaft where the twisted strings are connected, and all the electronics; (iv) the twisted string itself connecting the motor module with the load, and the load itself represented in Figure 2 as a translating mass. The basic TSA concept and modelling can be found in [12]. A couple of strings are attached to a rotative electrical motor and twisted on one end, whereas on the other end the strings are connected to a linear moving element, that is, the load. The overall string length is reduced by the rotation produced by the electrical motor. Therefore, the rotative motion of the electric motor is converted into a linear motion on the other side of the strings.

Figure 4 reports a detailed 3D view of the TSA module design. Its mechanical structure is manufactured by rapid prototyping in ABS plastic. A pair of axial-symmetric compliant beams has been integrated in the TSA module frame on the opposite side with respect to the twisted strings. These beams behave as a linear spring, granting a properly designed compliance to the structure required for converting the force exerted by the TSA module into a proper frame deformation and thus to the implementation of the force sensor. The integration of a force sensor into the TSA module is of paramount importance to successfully measure the force the actuator applies to the load. In the proposed TSA module, the force sensor is located in between the frame connection point on the robot structure and the frame hosting the DC motor, that is, on the opposite side of the twisted strings with respect to the rotative motor, as can be seen in Figure 4. A picture of the TSA module prototype developed in this work is reported in Figure 5. In particular, Figure 5(a) shows also the TSA module embedded controller based on an Arduino NANO board. In the TSA module, an optoelectronic device is then used to detect the frame deformation and convert it back into the applied force causing the deformation. As shown in Figure 5, a DC motor is hosted in the module frame along with an optical encoder for motor angular position sensing, while the output shaft is supported by a combined axial-radial bearing at the point of the twisted string connection to both reduce the friction and prevent the transmission force to damage the motor. A silicon tube is used to connect the DC motor and the module output shafts in order to solve problems related to misalignment of the rotational axes of the motor and the module output shaft. The designed TSA module structure allows the transmission force

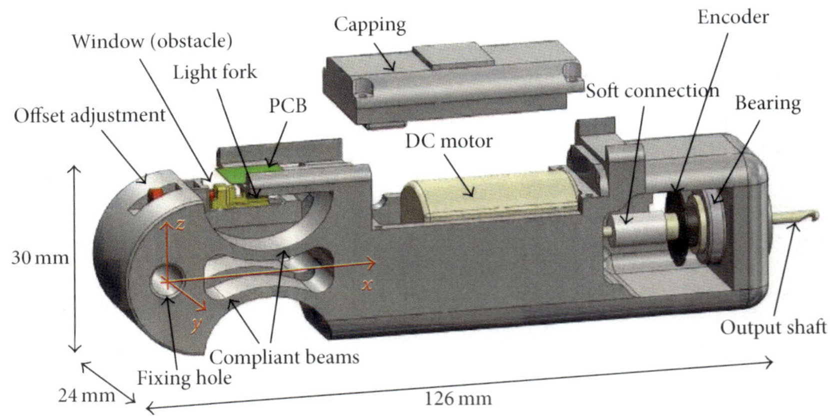

FIGURE 4: Detailed view of the twisted string actuation module.

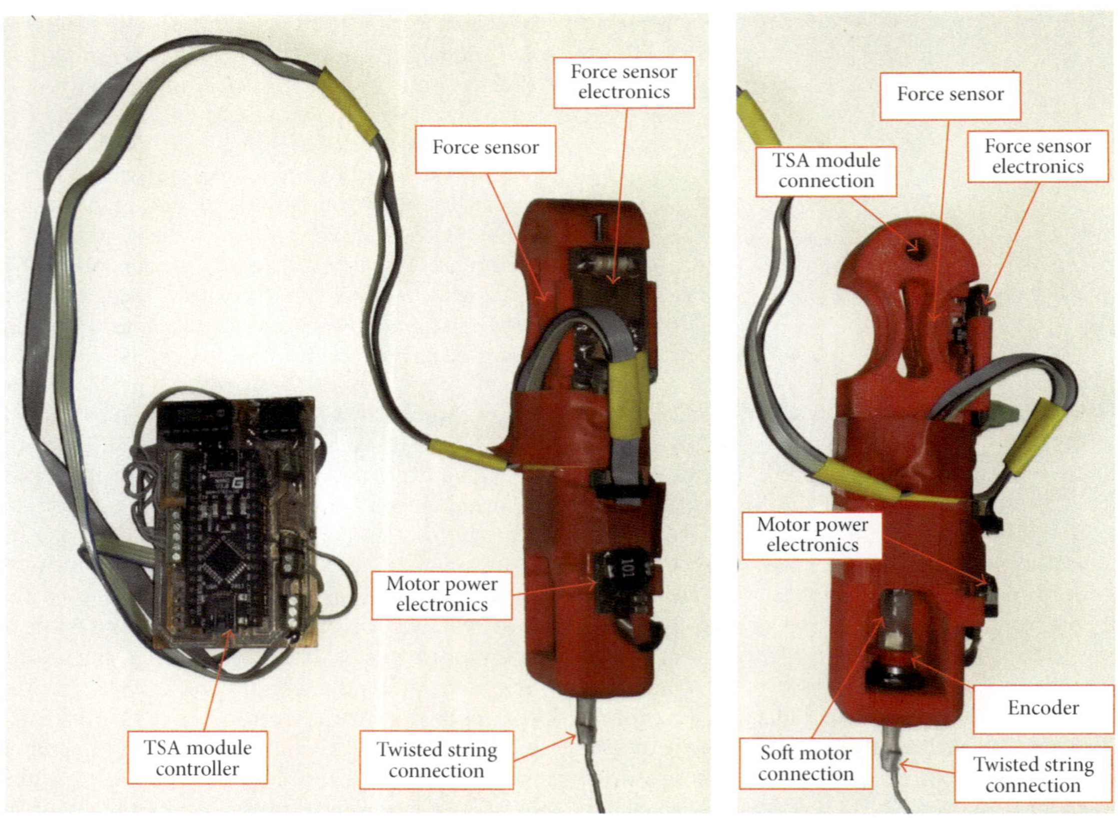

(a) Upper view of the TSA module prototype and control electronics (b) Lateral view of the TSA module prototype

FIGURE 5: Detailed view of the TSA module prototype and control electronics.

to be completely supported by the output shaft through the combined bearing, whereas the motor is only used to transmit the necessary torque for driving the twisted string actuation to the output shaft. In this setup, all of the components of the actuation system are integrated in a single element, thus optimizing the system encumbrance. A digital interface to input and output signals required for the actuation system control is provided by the actuator electronics, including both the motor power electronics, the conditioning system for both the optical encoder and the force sensor, and a microcontroller board for controlling the TSA module (see Figure 5). The communication of the TSA controller with external systems can be implemented either with UART, SPI, or I2C interfaces.

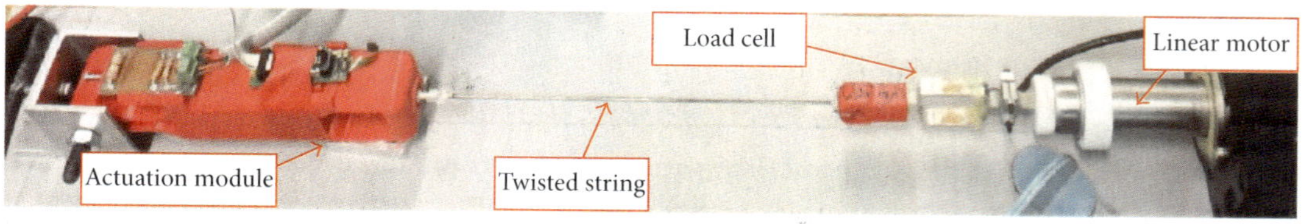

FIGURE 6: Overview of the experimental setup for validation of the actuation module.

FIGURE 7: Finite Element Analysis of the compliant frame.

4. Evaluation of the TSA Module

In this section, the experimental evaluation of the proposed actuation module is investigated.

4.1. Experimental Setup. Figure 6 depicts an overview of the experimental setup used for the validation of the proposed TSA module design. In this setup, the TSA module rotation axis is aligned with the translation axis of a linear motor (LinMot-37 × 160), and the twisted string is connected on one side to the TSA module and on the other side to the linear motor slider. The linear motor is able to apply a maximum continuum force up to 160 N and is equipped with a load cell able to measure a maximum force up to 100 N and an integrated encoder with resolution of 1 μm to measure the load position. The linear motor is therefore used to apply variable and controllable loads to the TSA module structure. A shielded cable for noise rejection connects the linear motor load cell to the amplifier. The linear motor is driven by a low-level control system able to generate different force profiles compensating for the friction acting on the motor slider [25]. Data collection and control of the system during the experiment are managed by a PC-104 running RTAI-Linux real-time operating system and a sensoray 526 data acquisition board. The same PC-104 is used to communicate commands and data with the TSA module embedded controller through a high-speed SPI serial connection.

4.2. TSA Design Validation. The TSA module structure is designed to exhibit a symmetric deformation along the direction of the measured force (i.e., the x-axis). Moreover, to fully exploit the characteristics of the optoelectronic component used for detecting the frame deformation, the mechanical structure has been designed to attain a proper linear displacement within the TSA module force range. Both experimental measurements and FEA have been used to validate the proposed actuation module design. In particular, Figure 7 reports the FEA of the TSA module at the maximum load condition, that is, with a load of 80 N. This limit is not given by the TSA capabilities, but it is limited by the control system to preserve a suitable lifetime of the twisted string. It is also important to say that even if the verification of the frame deformation is not significant in that case, the TSA module structure has been designed to sustain a maximum load 4 times higher than the previously mentioned force range to prevent damage to the mechanical structure in case of overload. Figure 8 portrays the comparison between the measured deformation and FEA. The result evidences that the measured deformation is slightly smaller than the FEA, which can be influenced by the manufacturing process. The maximum deformation with the maximum load of 80 N is $\Delta X_{max} \simeq 1.08 \cdot 10^{-4}$ m indicating the compliant frame deformation is within the goal working region $\Delta d < 0.12$ mm and can be considered linear.

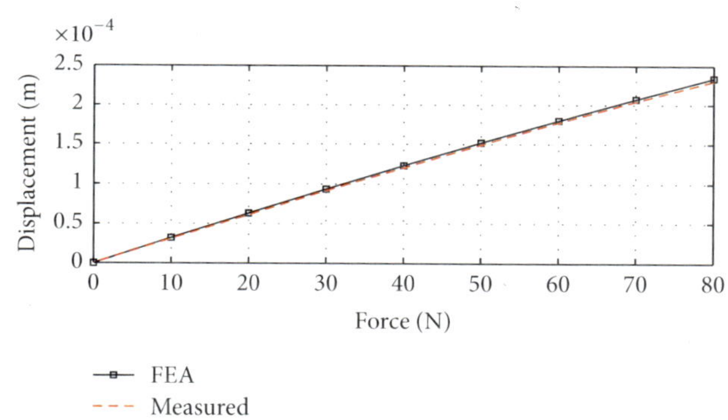

FIGURE 8: Deformation of the motor module in the direction of the transmission: comparison between experimental values and FEA.

4.3. Input-Output Characteristic. To evaluate the input-output characteristic of the system in dynamic conditions, suitable experiments have been performed. A force feedback loop based on the data gathered from the load cell mounted on the linear motor slider controls the force applied to the TSA module.

In the first experiment set, the linear motor is programmed to apply a constant load force of 5, 20, and 40 N, respectively, to the TSA module, and the rotative motor is then commanded to move with a trapezoidal trajectory from the initial zero position (untwisted string) to a maximum value of 540 rad, and the displacement caused on the load is then measured by means of the linear motor integrated encoder. Figure 9(a) presents the input-output position characteristic of the proposed TSA module with a constant load.

Different load conditions have been adopted during the second experiment set. In this case, the linear motor is programmed to act as a mass-spring-damper system with the load stiffness of 1000, 2500, and 5000 N/m, respectively. Figure 9(b) presents the input-output position characteristic of the proposed TSA module with spring-like load. It can be noted that the maximum position reached by the load visibly decreases as the input load increases.

For a better evaluation of the input-output characteristic of the TSA module over multiple working cycles, the relation between rotative motor position and load (i.e., linear motor) position has been reported in Figure 9(c) for all the performed experiments. These plots highlight that a certain variation of the input-output characteristic is present, due to settlements in the twisted string coil arrangement during the initial motion phases. As can be also seen in Figures 9(a) and 9(b), this phenomenon tends to vanish as the number of cycles increases.

5. Experimental Evaluation of the TSA Module as an Assistive Device

To show the effectiveness of the proposed system for rehabilitation and assistive applications, the TSA module has been tested on a user as a support to the biceps muscle activity during a load lifting task. In particular, in this test application, sEMG signal has been used to drive the assistive device. As a preliminary step toward the implementation of an elbow flexion/extension assistive system, in which we considered the installation of the device on the back of the user; see Figure 1, the TSA module has been fixed to a rigid structure (see Figure 10) to simplify the experimental test, in order to use the necessary string length without considering friction and curvature related phenomena introduced by the twisted string path across the shoulder, as can be seen in Figure 1. To solve this issue, dedicated experiments to study the effect of friction on the TSA transmission characteristics are under development [26]. In this preliminary experimental evaluation, the TSA module operates on the user's forearm and is regulated using a variable stiffness control approach. In this relation, the biceps sEMG signal is used within a specific PI (Proportional-Integral) control scheme in order to suitably adjust the device stiffness to obtain a load partial compensation, limiting the sEMG activity under a suitable threshold value.

5.1. Hardware and Setup Overview. The experimental setup used during this experiment is shown in Figure 10. With reference to this picture, a rigid structure has been implemented to allow the TSA module to be firmly fixed in a point above the user's shoulder, such that the string can be in tension with respect to its full length. In particular, the length of the string for this experiment has been chosen equal to 1 m, in order to exploit the mechanism contraction range related to the best module working condition, that is, the 20% of the full string length [12]. The string is therefore connected to an armband fixed on the forearm of the user.

The sEMG signal has been acquired by means of low-cost disposable surface electrodes connected to the acquisition board Cerebro [27]. The sEMG signals are acquired by a high-performance Analog Front End (AFE) [28] that is connected with an ARM Cortex M4 Microcontroller. In this application, the data are sampled at 1 kHz and streamed to a nearby PC using an onboard Bluetooth interface. Before using the sEMG signals in the TSA variable stiffness control scheme, a processing step is necessary. The following filtering procedure is applied to the signals [29]: (i) a 50 Hz notch filter for powerline interference cancellation, (ii) a 20 Hz high-pass filter, (iii) the rectification (absolute value of the signal), and finally (iv) the Root Mean Square (RMS) value is computed on a 200 ms window.

5.2. Experiment Configuration, Control Scheme, and Results. A healthy male subject is involved in the experiment, in particular, the user put on the TSA module, which is connected to the rigid structure and, on the other string end, to the armband fixed to his forearm. During the experiment, the subject is sat in a normal and comfortable position, holding his upper arm parallel to the trunk. A couple of differential sEMG electrodes have been placed in proximity of the biceps brachii muscle in the upper arm, referring to methods and best practices outlined in [30]. In this validation test, the goal is to assist the biceps during the flexion of the elbow for lifting tasks of a load of 2 Kg applied on the wrist. For this reason, only the biceps sEMG activity is acquired, considering the triceps activation negligible. Specifically, the subject is required to perform five consecutive forearm motions, each one consisting in a flexion followed by an extension, covering the elbow angle range between approximately 10° and 90°. The user is also requested to perform the movements smoothly and with a reasonably low velocity. Once the 2 Kg load is applied to the forearm by means of a wristband, in the first part of the experiment the user executes the flexion/extension motions freely, without the help of the TSA module. The behaviour of the biceps activity during this task can be observed in Figure 11 looking at the acquired sEMG signal and where the elbow angle is computed from the module's DC motor encoder signal and the setup geometry. In the second phase of the experiment, the subject connects his forearm to the TSA module in order to be assisted during the lifting task of the 2 Kg load. To achieve this goal, the TSA

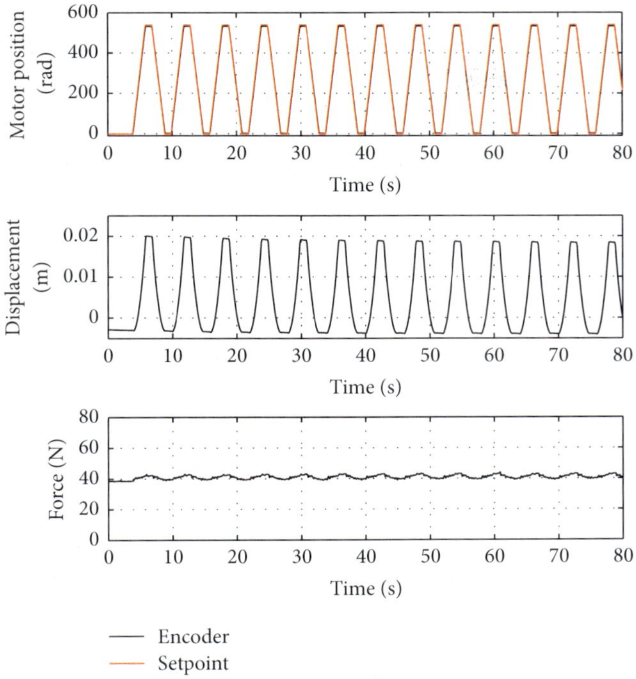

(a) Experimental evaluation of the constant load

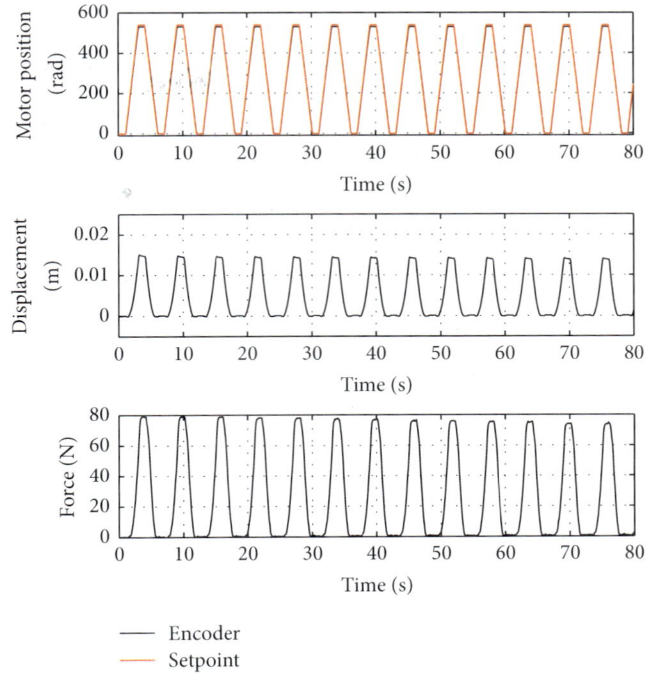

(b) Experimental evaluation of the spring-like load

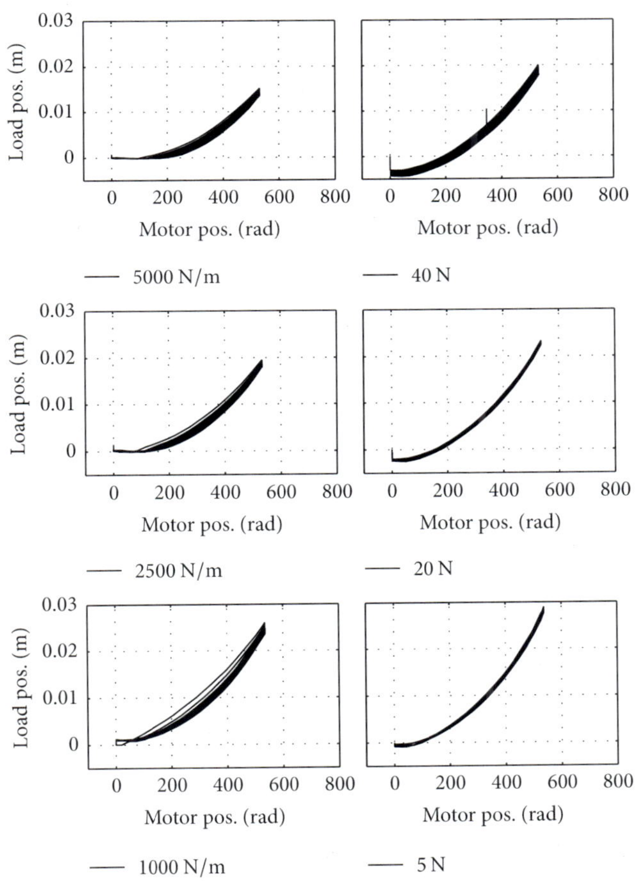

(c) Input-output position characteristic with constant load (right) and with spring-like load (left)

FIGURE 9: Dynamic response of the TSA actuation module.

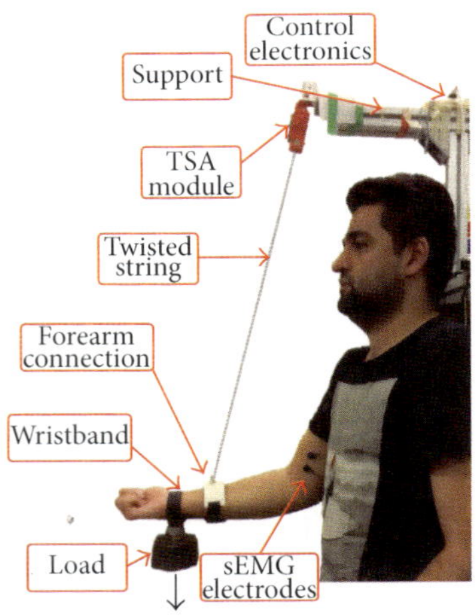

FIGURE 10: Experimental setup as an assistive device.

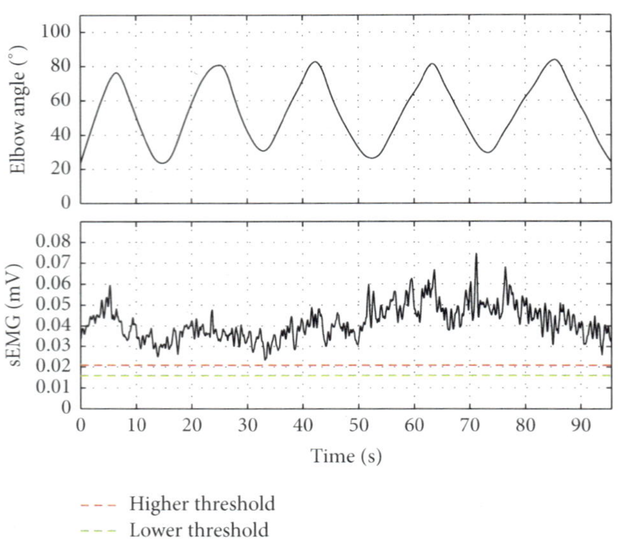

FIGURE 11: Plots for the lifting task without the assistive device support.

module is controlled through a specific sEMG-driven control scheme illustrated in Figure 12(a). With respect to this figure, a stiffness control is implemented for the TSA module, which means that the device is regulated in such a way to behave like a spring with stiffness K and contraction given by the difference between the actual string length (spring length) and an initial string length less than or equal to the length in the case of an elbow flexion of 90° (spring rest length). This is achievable because of the possibility of calculating the string length from the DC motor's encoder signal and thanks to the presence of a force control loop embedded on the TSA module. In particular, such lower-layer controller uses the force sensor measurement within a standard Proportional-Integral-Derivative (PID) control loop in order to track the requested reference force value. Therefore, in this way, the assistive device is equivalent to a spring attached on the user's forearm and the way it interacts with the human is determined by the value of K. Based on this observation, in our control scheme the value of K is determined by a sEMG control loop, which makes use of a PI controller modified according to a double threshold logic whose functioning is observable in Figure 12(b).

In detail, the controller is based on a sEMG error between the measured biceps sEMG signal and lower sEMG threshold, the latter used as reference value. The proportional action of the PI controller (identified by a properly chosen proportional gain K_P) continuously takes as input the sEMG error. On the other hand, the integral action (characterized by a properly chosen integral gain K_I) is activated only when the sEMG signal is continuously over a higher threshold for a period of 1 s, until the increasing of the stiffness K of the TSA module makes the sEMG signal touch the lower threshold value. Such double threshold logic has been implemented because of two considerable advantages: (i) a more reliable regulation of the sEMG signal (which presents a substantial inherent variability), with respect to a single reference value approach, and (ii) the capability of filtering sEMG peaks that surpass the higher threshold for less than 1 s, which are usually present in case of slightly faster motions or unwanted impulsive contractions, avoiding an increasing of the integral action not oriented to compensate the applied load. Furthermore, the threshold values are determined for the subject by means of a simple calibration phase. In particular, the sEMG signal of the user with the elbow flexed at 90° and with a load of 0.5 Kg is recorded. Then, denoting the thresholds as T_L and T_H, they are computed as

$$T_L = m_{\mathrm{EMG}} - \sigma_{\mathrm{EMG}},$$
$$T_H = m_{\mathrm{EMG}} + \sigma_{\mathrm{EMG}}, \tag{1}$$

where m_{EMG} is the mean value of the sEMG calibration recording and σ_{EMG} is the standard deviation computed over the same recording. The effects of this control strategy during the lifting task of the 2 Kg load by the user with the TSA module support are shown in the graphs of Figure 13. In particular it is possible to observe how the sEMG signal initially goes over the higher threshold (red), causing the activation of the sEMG controller integral action after one second. This makes the stiffness of the TSA module rapidly increase in such a way that the biceps sEMG activity decreases until it reaches the lower threshold (green). Then, for the subsequent flexions of the lifting task, the sEMG signals remain in the neighborhood of the lower threshold without surpassing the higher threshold except for peaks that are successfully filtered by the implemented double threshold logic, which means that the integral action is not activated is these cases. In other words, this means that the biceps exerts a muscular activity that never surpasses the sEMG value corresponding to the 0.5 Kg calibration recording, saving a quantity greater than or equal to 1.5 Kg-related muscle effort. On the light of this and, especially, comparing the sEMG values during the lifting of the 2 Kg load in the case with no support (Figure 11) and with support (Figure 13) the

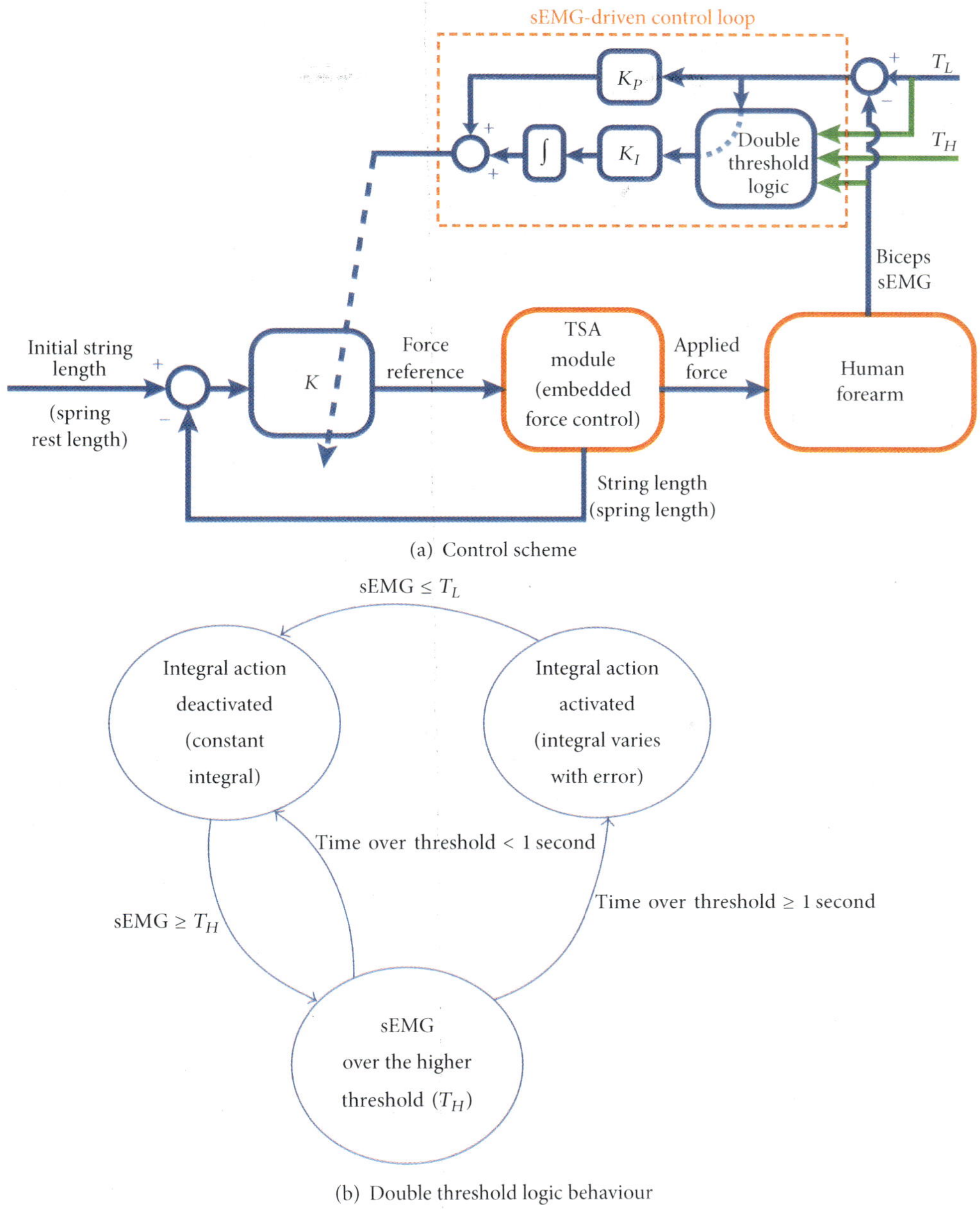

FIGURE 12: Variable stiffness control scheme of the TSA module for the assistive application.

effectiveness of the TSA module as an elbow assistive device is proved.

6. Conclusions

This paper presents an innovative wearable TSA module system with integrated force sensor which can be used for rehabilitation and assistive robotic applications. The proposed TSA module structure, the integrated force sensor, and the basic sensor working principle are discussed in this paper. To validate the TSA module compliant frame, the FEA has been compared with experimental measures. Different experiments have also been carried out to verify the properties of the proposed TSA module.

Furthermore, to show the capabilities of the proposed TSA module in real assistive applications, an experiment involving a male healthy subject has been carried out, where the TSA module has been used to provide load compensation to the user's elbow joint during lifting tasks of a 2 Kg load. The experimental results show that the device, together with a suitable control scheme, can be successfully used to help and limit the user's biceps muscle activity during dynamic motions when a load is applied in proximity of the wrist.

Future work will be devoted to the installation of the module on the back of the user, in order to obtain a wearable elbow assistive device and to the implementation of a sEMG-driven control scheme that makes use of both

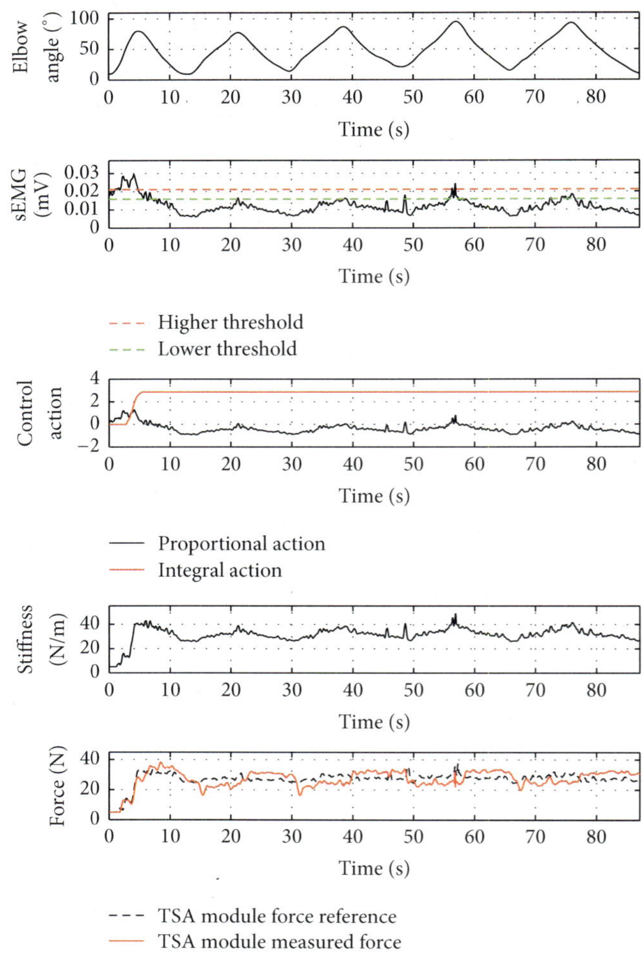

FIGURE 13: Plots for the lifting task with the assistive device support.

biceps and triceps, exploiting antagonistic muscle activation concepts.

Conflicts of Interest

The authors declare that there are no conflicts of interest regarding the publication of this paper.

References

[1] A. M. Dollar and H. Herr, "Lower extremity exoskeletons and active orthoses: challenges and state-of-the-art," *IEEE Transactions on Robotics*, vol. 24, no. 1, pp. 144–158, 2008.

[2] R. A. R. C. Gopura and K. Kiguchi, "Mechanical designs of active upper-limb exoskeleton robots: state-of-the-art and design difficulties," in *Proceedings of the IEEE International Conference on Rehabilitation Robotics (ICORR '09)*, pp. 178–187, IEEE, Kyoto, Japan, June 2009.

[3] J. C. Perry, J. Rosen, and S. Burns, "Upper-limb powered exoskeleton design," *IEEE/ASME Transactions on Mechatronics*, vol. 12, no. 4, pp. 408–417, 2007.

[4] Z. Tang, K. Zhang, S. Sun, Z. Gao, L. Zhang, and Z. Yang, "An upper-limb power-assist exoskeleton using proportional myoelectric control," *Sensors (Basel, Switzerland)*, vol. 14, no. 4, pp. 6677–6694, 2014.

[5] I. Galiana, F. L. Hammond, R. D. Howe, and M. B. Popovic, "Wearable soft robotic device for post-stroke shoulder rehabilitation: identifying misalignments," in *Proceedings of the 25th IEEE/RSJ International Conference on Robotics and Intelligent Systems (IROS '12)*, pp. 317–322, October 2012.

[6] B. Ugurlu, M. Nishimura, K. Hyodo, M. Kawanishi, and T. Narikiyo, "Proof of concept for robot-aided upper limb rehabilitation using disturbance observers," *IEEE Transactions on Human-Machine Systems*, vol. 45, no. 1, pp. 110–118, 2015.

[7] J. C. Perry and J. Rosen, "Design of a 7 degree-of-freedom upper-limb powered exoskeleton," in *Proceedings of the 1st IEEE/RAS-EMBS International Conference on Biomedical Robotics and Biomechatronics (BioRob '06)*, pp. 805–810, IEEE, February 2006.

[8] Y. Mao and S. K. Agrawal, "Design of a cable-driven arm exoskeleton (CAREX) for neural rehabilitation," *IEEE Transactions on Robotics*, vol. 28, no. 4, pp. 922–931, 2012.

[9] R. A. R. C. Gopura and K. Kiguchi, "Development of a 6d of exoskeleton robot for human upper-limb motion assist," in *Proceedings of the International Conference on Information and Automation for Sustainability*, pp. 13–18, December 2008.

[10] H. Kobayashi and H. Nozaki, "Development of muscle suit for supporting manual worker," in *Proceedings of the IEEE/RSJ International Conference on Intelligent Robots and Systems (IROS '07)*, pp. 1769–1774, San Diego, Calif, USA, November 2007.

[11] M. Shoham, "Twisting wire actuator," *Journal of Mechanical Design*, vol. 127, no. 3, pp. 441–445, 2004.

[12] G. Palli, C. Natale, C. May, C. Melchiorri, and T. Würtz, "Modeling and control of the twisted string actuation system," *IEEE/ASME Transactions on Mechatronics*, vol. 18, no. 2, pp. 664–673, 2013.

[13] G. Palli and S. Pirozzi, "Optical force sensor for the DEXMART hand twisted string actuation system," *Sensors and Transducers*, vol. 148, no. 1, pp. 28–32, 2013.

[14] G. Palli and S. Pirozzi, "Integration of an optical force sensor into the actuation module of the DEXMART hand," *International Journal of Robotics and Automation*, vol. 29, no. 2, pp. 193–201, 2014.

[15] J. J. Guzek, C. Petersen, S. Constantin, and H. Lipson, "Mini twist: a study of long-range linear drive by string twisting," *Journal of Mechanisms and Robotics*, vol. 4, no. 1, Article ID 014501, 2012.

[16] Y. Shen, E. Winder, N. Xi, C. A. Pomeroy, and U. C. Wejinya, "Closed-loop optimal control-enabled piezoelectric microforce sensors," *IEEE/ASME Transactions on Mechatronics*, vol. 11, no. 4, pp. 420–427, 2006.

[17] W. A. Lorenz, M. A. Peshkin, and J. E. Colgate, "New sensors for new applications: force sensors for human/robot interaction," in *Proceedings of the IEEE International Conference on Robotics and Automation (ICRA '99)*, pp. 2855–2860, May 1999.

[18] J. Peirs, J. Clijnen, D. Reynaerts et al., "A micro optical force sensor for force feedback during minimally invasive robotic surgery," *Sensors and Actuators A: Physical*, vol. 115, no. 2-3, pp. 447–455, 2004.

[19] M. Tada and T. Kanade, "Design of an mr-compatible three-axis force sensor," in *Proceedings of the IEEE/RSJ International Conference on Intelligent Robots and Systems (IROS '05)*, pp. 3505–3510, Alberta, Canada, 2005.

[20] J. Missinne, E. Bosman, B. Van Hoe et al., "Flexible shear sensor based on embedded optoelectronic components," *IEEE Photonics Technology Letters*, vol. 23, no. 12, pp. 771–773, 2011.

[21] G. Palli, L. Moriello, U. Scarcia, and C. Melchiorri, "Development of an optoelectronic 6-axis force/torque sensor for robotic applications," *Sensors and Actuators A: Physical*, vol. 220, pp. 333–346, 2014.

[22] G. Palli and S. Pirozzi, "Force sensor based on discrete optoelectronic components and compliant frames," *Sensors and Actuators A: Physical*, vol. 165, no. 2, pp. 239–249, 2011.

[23] G. Palli and S. Pirozzi, "A miniaturized optical force sensor for tendon-driven mechatronic systems: design and experimental evaluation," *Mechatronics*, vol. 22, no. 8, pp. 1097–1111, 2012.

[24] M. Hosseini, G. Palli, and C. Melchiorri, "Design and implementation of a simple and low-cost optoelectronic force sensor for robotic applications," in *Proceedings of the IEEE International Conference on Advanced Intelligent Mechatronics (AIM '16)*, pp. 1011–1016, Banff, Canada, July 2016.

[25] G. Palli and C. Melchiorri, "Velocity and disturbance observer for non-model based load and friction compensation," in *Proceedings of the 10th International Workshop on Advanced Motion Control (AMC '08)*, pp. 194–199, IEEE, Trento, Italy, March 2008.

[26] M. Hosseini, G. Palli, and C. Melchiorri, "Experimental evaluation of guided twisted actuation," in *Proceedings of the IFAC Symposium on Mechatronic Systems*, pp. 380–385, Loughborough University, Leicestershire, UK, September 2016.

[27] S. Benatti, B. Milosevic, F. Casamassima et al., "EMG-based hand gesture recognition with flexible analog front end," in *Proceedings of the 10th IEEE Biomedical Circuits and Systems Conference (BioCAS '14)*, pp. 57–60, Lausanne, Switzerland, October 2014.

[28] P. Schonle, F. Schulthess, S. Fateh et al., "A DC-connectable multi-channel biomedical data acquisition ASIC with mains frequency cancellation," in *Proceedings of the 39th European Solid-State Circuits Conference (ESSCIRC '13)*, pp. 149–152, September 2013.

[29] R. Meattini, S. Benatti, U. Scarcia, L. Benini, and C. Melchiorri, "Experimental evaluation of a sEMG-based human-robot interface for human-like grasping tasks," in *Proceedings of the IEEE International Conference on Robotics and Biomimetics (ROBIO '15)*, pp. 1030–1035, IEEE, Zhuhai, China, December 2015.

[30] A. Perotto and E. Delagi, *Anatomical Guide for the Electromyog-Rapher: The Limbs and Trunk*, Charles C Thomas, 2005.

16

Robust Feedback Control Design of Underactuated Robotic Hands with Selectively Lockable Switches for Amputees

Bilal M. Yousuf,[1] Asim Mehdi,[2] Abdul Saboor Khan,[2] Aqib Noor,[2] and Arslan Ali[2]

[1]*Electronics & Power Engineering, PN Engineering College, National University of Sciences and Technology, Karachi, Pakistan*
[2]*Electrical Department, Fast-National University of Computing and Emerging Sciences, Karachi, Pakistan*

Correspondence should be addressed to Bilal M. Yousuf; bilalmyousuf4@gmail.com

Academic Editor: L. Fortuna

In recent years, reproduction of human mechanical hand with upgraded abilities is one of the major concerns. This paper addresses the problems of underactuated robotic hand with low cost design as it avoids electromyogram (EMG) sensors. The main goal is to balance the hand in the way, like grabbing, speed, and power, and provide a more robust and cost effective solution. All fingers have some mechanical consistency for picking up objects in a better way. A Flex sensor is attached to all fingers and it is interfaced with a computer using Arduino UNO microcontroller. The sensor aids the arm in three different directions: at first it senses whether an object is grasped or not. In the second step, it determines the coefficient of friction between the objects. Finally it grasps the object and stops. One of the primary issues of prosthetic hand is to have the capacity to satisfy every detail of torque, speed, and latency. In this research, we have developed a model of robotic hand with some modifications. The adaptability of grasping is compared with the degree of freedom (DOF) along with the quantity of fingers. We are controlling our hands via sensors based signal controlling system. The idea is to design a robotic hand, which has low cost, is easy to use, and is light in weight, which helps the amputees to use it with ease in their daily lives. The efficacy of the proposed control is verified and validated using simulations.

1. Introduction

The field of robotics has always been an inspiration for researchers, due to its stability problem [1]. The robotic arm has many applications in the field of automation in industries, medical fields, etc.

The main problem which has influenced the advancement of this field is the stability and accuracy. Many researchers came across with different solutions for this problem. The main problems which are under consideration in this paper are the high cost, complex structure, control frameworks, and their synchronization. The quantity of the physical amputees has been expanded due to the accidents in traffic, workshop mishaps, and sickness. Our end goal is to enhance the quality of life for numerous amputees and tend to utilize their energy in their daily life [1, 2]. The hand is a unique part of the body that can perform different control undertakings with five fingers and with more than 20 DOF [3]. Despite the fact that a genuine human hand is more capable compared to some other manufactured control instruments, it still experiences a few challenges in certain condition. The outline of an automated prosthetic hand is an extremely dynamic field of research.

Over the years, it was simple two degrees of freedom with the limiting capacity of grasping and pulling mentioned in Figure 1. Nowadays, the researchers were working on improving the degree of freedom for the multifingered framework. The robotic hand is equipped with different sensors, which gives a feedback signal to the controller. The controller will control the speed and angle of the motor for better transient response [4, 5]. The hand needs a complicated controller, which stabilized the performance of the system. The proper usage of electrically powered prosthetic hand designs has been started from the twentieth century. The research found its ways for the electromyogram sensors to control the hand [5]. In recent days, the proposed design of robotic hand is becoming more simple, flexible, and generous without compromising the postures. Recently a researcher, Dollar et al. in [5], presented a new solution through elastomer material, which gained so much attention due to its low cost.

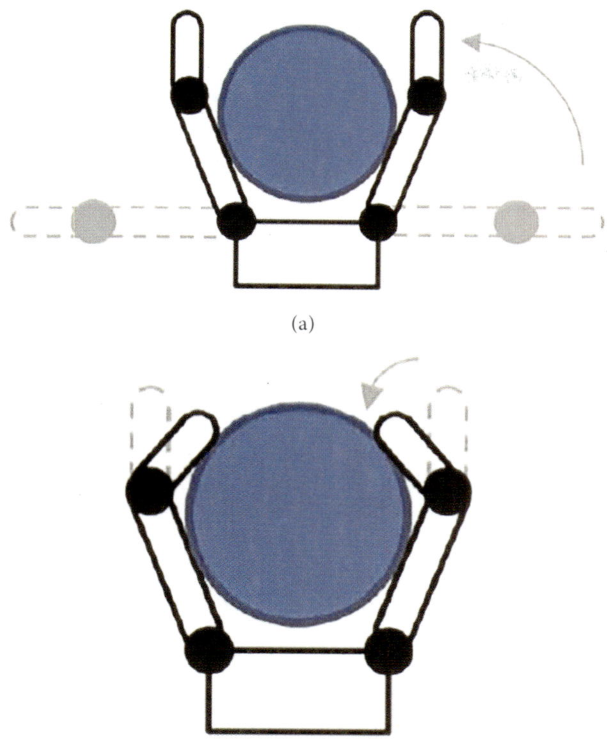

(a)

(b)

FIGURE 1: Grasping model of robot [2].

Robotic literature always discussed the major problems, which occur in the system over the time period. This may cause a certain fear for amputees. Therefore, they are avoiding using a prosthetic hand. As per survey the fact which comes in our mind is that they need a low cost, light weight, and easily repairable device. With these objectives in consideration, we focus on the development of the prototype in order to improve the following critical aspects of a biomedical device [6].

In this paper, the approach permits the most extreme design of control with an immense plan which required complex control calculations. It is an under demand system, which require less actuators. This approach improves the control calculations by reformulated hand capacity and quality according to the client. The level of flexibility of the human hand is routinely characterized. The primary objective of the prosthetic hand is to mimic the human hand in all viewpoints. As compared to this work, many researchers presented unique ideas about the robotic hand using EMG sensors to enhance their quality of living. But this paper is totally based on ex sensors, which can achieve seven different types of movements; the list of these movements is given below:

(1) **Hook grip**: secure and adaptable hook grip can hold everything from folder cases to purses for substantial shopping material.

(2) **Mouse grip**: someone is working in the office or browsing the Internet.

(3) **Open palm grip**: open palm grip gives a compelling method for conveying bowls or plates securely. Our projects palm is a bit larger than the natural human hand, which proves its uniqueness and allows a person to easily handle large things by carrying them on its palm [7].

(4) **Pinch grip**: with the pinch grip, the thumb and index finger meet up to give an adaptable, helpful approach to get and move an extensive variety of little objects, including auto keys, coins, tops, and pens.

(5) **Power grip**: it allows a person to shake someone's hand or to throw a ball while playing. It also helps while working with garden utensils or sometimes eating a piece of fruit [7].

(6) **Relax hand position**: when someone is not actively using the hand, the relaxed hand position helps to give our project a natural and gap appearance.

(7) **Key grip**: when someone desires to keep making a grip on small thing likes plate card, etc.

Simulations are provided to show the efficacy of the proposed idea of using the ex-sensor and validate the performance of the control algorithm

The rest of the paper is structured in this following manner. Section 2 discusses the literature review. Section 3 shows the problem formulation. Section 4 contains the hardware design of robotic hand. Section 5 contains the results that are shown based on simulation. Finally, Section 6 discusses the concluding remarks with future advancement.

2. Literature Review

Over the years, a number of researchers have proposed various modeling and control schemes for robotic hand starting from simple robotic platform to complex robotic hand. In the course of recent decades, a survey on automated hands has been led into the development of robotic hand, which roughly looks like the human hand [9].

A robotic hand can be viewed as a human hand, if all the definite properties and attributes are satisfied. These properties incorporate momentum, size, and the execution of operations, unwavering quality together with the common perspective. An anthropomorphic hand comprises fingers and a thumb. The arrangement of each of the four fingers is one next to the other opposite to the palm surface. In [10], the author presented a robotic hand with complete four fingers and a thumb. The palm is the medium which joins the fingers together with its corresponding positions. Another work of importance is the robotic hand motion which is shown in Figure 2 outlined by iRobot. Each finger controlled proximal and distal connections, which associated a finger to the base till the fingertip [2]. To prompt adaptability and solidness, overwhelming obligation and flexible joints were utilized which gave an enhanced grip around the items. The grip control is the utilizing link ligaments that provide solid grasping.

Chang in [8] presented the model, which manages to outline an automated hand that imitates a human hand. It

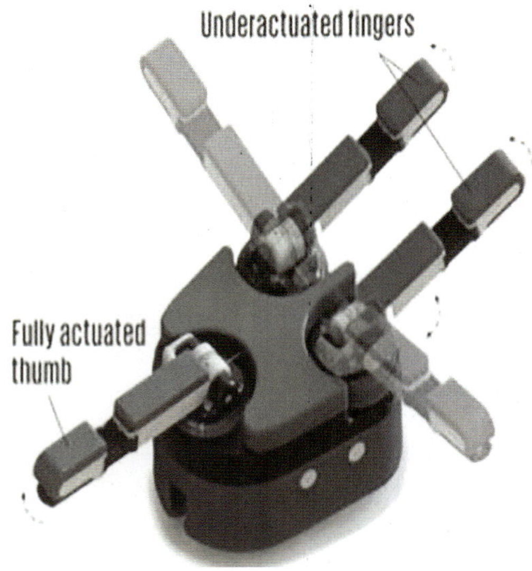

FIGURE 2: iHY hand design [4].

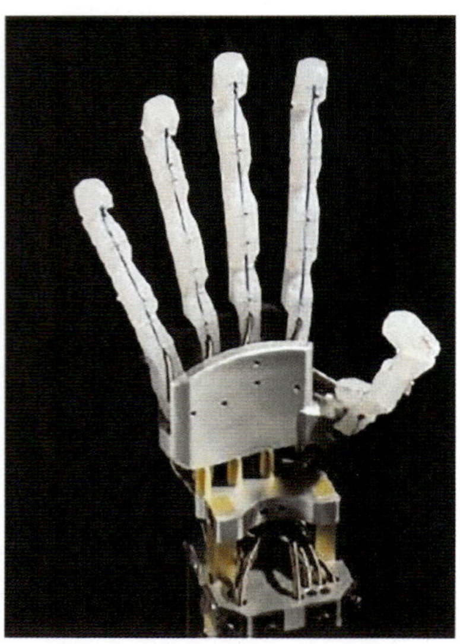

FIGURE 3: Hand gesture robot [4].

is made out of sensors and actuators. The automated glove controller is mainly based on the sensors, which tracked the development of the human hand. Five exion sensors are attached to the system; one for each finger which can be seen in Figure 3. Finally, in [11–13], the authors presented a unique design of robotic hand; one of them discussed the simple linear control based force feedback for the grasping. A force sensor is attached to one of the claws of the gripper, which sends a signal to the control for further processing. The other author presented a walking microrobot by piezoelectric materials. The last author presented the attitude control of a rigid body. This approach is based on two parallel PD controllers. The results are also compared with a classical PD controller and with an adaptive controller.

3. Problem Formulation

This section discusses the study of underactuated system and about prosthetic hand. The main problem is to design a low cost robotic hand. To come to a unique solution, we need to understand about what kind of sensors is required and their level of compatibility with the controllers.

3.1. Underactuated Design Hand Structure. The proposed design of the robotics hand is fully functional, same like human hands. The paper discusses the design of underactuated five-finger robots, with most complicated flexible joint incorporated by a single actuator and linking with a single sensor. Underactuated system is more reliable in terms of grasping items of different sizes and shapes robustly. The work can be described in two phases:

(i) Every finger is linked with an object and the other one is in free space where the distal link makes contact to fully encompass the object [7].

(ii) Every sensor sends an input signal to the control to operate the speed of the motor accordingly.

To achieve the desire purpose, it needs one single actuator which lies between the finger and the palm. The actuator can add force for grasping an object harder. The underactuated hand is found in different irregular shapes.

3.2. Prosthetic Hand. The prosthetic hand is an artificial device which is used to make an approximation to the appearance and function of a natural human hand [7]. Modernization in technology makes it possible to develop a prosthetic hand. The hand has the tendency of grasping objects, holding pens, and also moving like a natural hand. The process used in this research to make a prosthetic low cost robotic hand for amputees is present in Figure 4.

In this paper, the design of the robotic hand includes the ex-sensors to control the hand movements as shown in Figure 5. The involvement of ex-sensors makes it cost effective and easy to use. Flex sensors are attached to the analog pins of the Arduino UNO microcontroller, which processes and generates an output corresponding voltage. This signal is later used significantly to actuate the servo motors which respond to produce motion in the fingers.

The block structure of the entire system plays an important role in explaining the complete placement of components in system; it defines that every ex-sensor is working individually in every finger.

4. Problem

This section discusses the complete methodology based on computational geometry and graph theory. A method which is used to link multiagent or objects needs a set theory for quantifying the anthropomorphism of hand. The adopted scheme exploits a simple model based on graph theory for the simplification of the design and assumes a perfect communication link between the motors and the controller. The main problem is to maintain the stability of the system, which insures that it grasps the objects which are used in daily

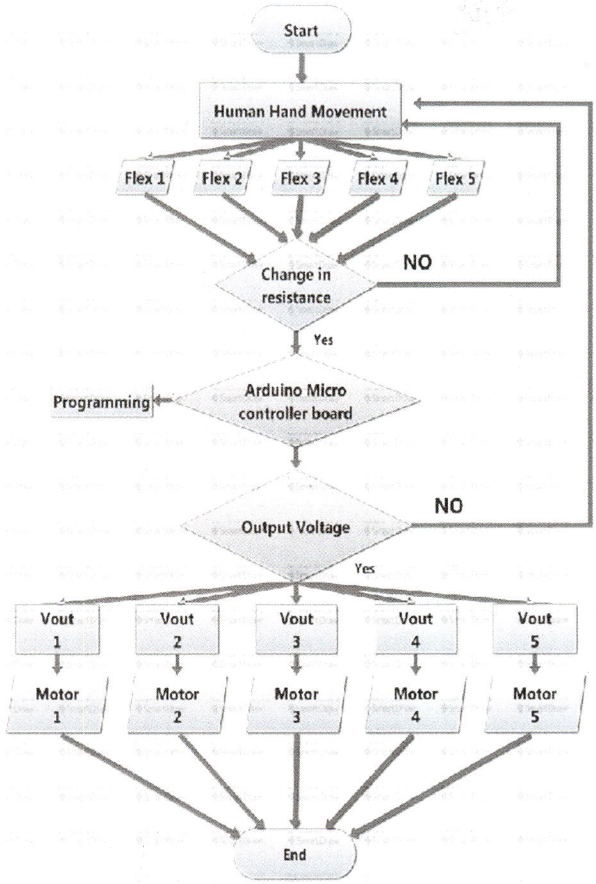

FIGURE 4: Process flowchart of prosthetic hand.

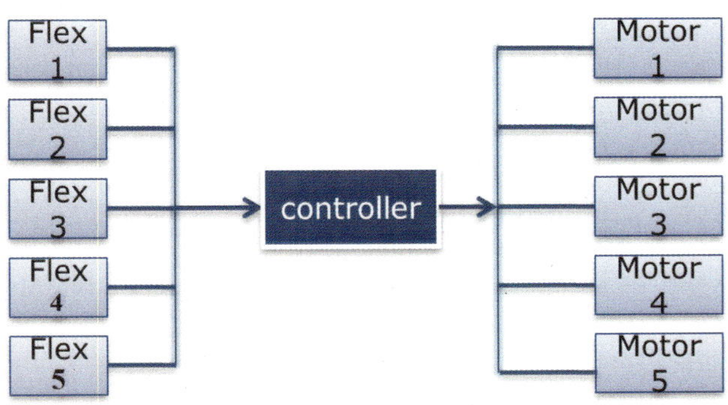

FIGURE 5: Block diagram of prosthetic hand.

routine. The main contribution of this paper mainly depends on the relationship of the ex-sensor and the motor.

4.1. Graph Theory. The basic idea of using a graph theory is described as the limitation in communication network, which does not allow each robot to communicate. For these limitations, the proposed scheme exploits a simple mathematical model based on graph theory to design such state feedback control. G (V, E), the digraph (directed graph) consists of node V= (1.....N) and edges E. In the diagraph, there are three protocols. The unidirectional symmetry protocol is utilized in this paper. As for this paper we are using directed graph. There will be a nonnegative adjacency matrix. The adjacency matrix basically describes the flow of the instruction of communication and maps the graph into the matrix, for the mathematical formulation. There is a Laplacian matrix Lij, which is sometime called admittance matrix, which gives us some special information like spanning tree. Spanning tree gives the details about undirected link. In this paper, only directed links are being used as the only controller sends the information to the motors, which means spanning tree should be eliminated. For this, Laplacian matrix is being utilized $A = [a_{ij}]\epsilon\mathfrak{R}^{N*N}$ and there will be a Laplacian matrix given as follows.

$L = [L_{ij}]\epsilon\mathfrak{R}^{N*N}$ of a diagraph corresponding to matrix contain elements such that

$$L_{ij} = \sum_{j=1}^{N} a_{ij} \tag{1}$$

if$i \neq j$ $L_{ij} = degout(A) - adj(A)$, where $degout(A)$ matrix shows each node connection with the fellow node and $adj(A)$ shows each node connection with the fellow node with the edges. After the subtraction we get the Laplacian matrix.

5. Design

The design of the robotic hand is presented in the following manner.

5.1. Hand Structure. The desired design incorporates four fingers and a thumb, joint to a settled palm. General joints are utilized in this approach, like Metacarpophalangeal (MCP) and the CMC (Carpometacarpal) joints. These joints make up the other 12 degrees of flexibility in this desired purpose plan. To accomplish the coveted movement with a lightweight, it needs quality pairs of thin gage stainless steel. Singular connections of each finger and the thumb are made up of twisting gage sheet metal that allows the finger, pulleys, and link channel to be tactfully covered up inside. The palm of the hand is built in a comparative form utilizing folded sheet metal. Back to back connections are associated with 3 mm pivot shafts that go through the metal balls.

5.2. Finger Designing. In this paper, we developed a multifingered robotic hand. The robotic hand has four fingers with a thumb [1]. All these flexible joints are implemented and the mechanical structure is developed. For the structure, we need the length and its particular angle. For this we used the aforementioned parametric model.

In this paper, a unique design that reproduces the flexibility and the extensile motion of the finger is presented. An extended mechanical structure was developed through the use of appropriate elastomer material. While flexibility is implemented through cable between the controller and the

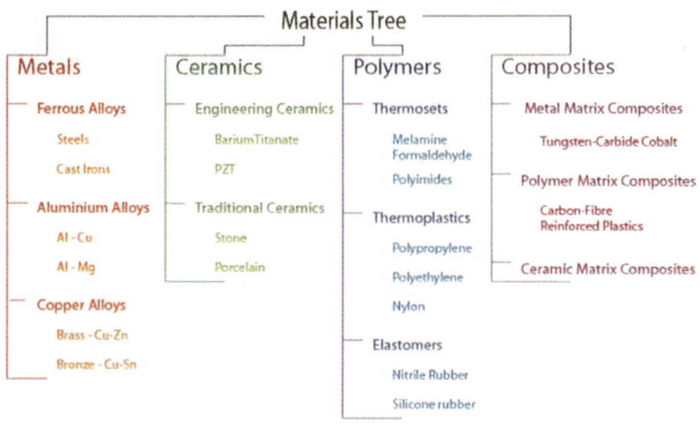

FIGURE 6: Material classification [5].

motor, the joints were made of the lightest material, but also stiff enough to produce a force range that corresponds to everyday life task.

The proposed design is completely stiff in contrast to friction based mechanisms, which is affected due to some external forces (these forces can create unstability). A completely different routing system is used for the design of thumb. The thumb is made with two different servo pulleys, which allows servo motors to make contact with the other fingers

After all the survey, we need to collect the desired parts of our project, which ensures the accuracy in making of robotic hand with low cost. The design should be easy to use for the amputees. Firstly, it needs a microcomputer which handles all the motor and sensor and makes them synchronize. The main working of the controller is mainly based on signals, which are coming from the sensor and it is making an output to the motors.

For the making of the prosthetic robotic hand, the material which was firstly implemented was Telon. The results are not as accurate as expected and also not fulfilling our requirements. For the better results, the design of the robotic hand is made up of acrylonitrile butadiene styrene (ABS) material. ABS is a polymer which consists of 3 monomers: acrylonitrile, butadiene, and styrene. ABS types can be manufactured offering a wide spectrum of different properties by means of branching or copolymerization [6]. It can often be quite straightforward to tell materials classes apart by look or feel. The class of material is shown in Figure 6. Metals are usually more reactive or metallic looking.

The motor selection was also really important for this project. Due to accuracy, a proper grip of the hand is entirely depending on the type of motor. For this research servo motor is selected, which was perfectly suitable. Typically, a servo motor is utilized to control a precise movement of somewhere around 0 and 180 degrees.

Flex sensors are the analog sensors which change the electrical resistance as they bend. The value of the resistance increases as the angle of curvature decreases. The design utilizes 5 ex-sensors. For this ex-sensor, it requires a microcontroller which has at least 5 analog pins. These pins work as analog voltage dividers.

6. Simulation Analysis

6.1. Momentum Analysis. This section discusses the graphical representation and the momentum analysis of the voltage of ex-sensor and the output of the controller. Let L1, L2, and L3 be the lengths of the phalanges and φ_1 to φ_4 be the joint edges of the 4 finger joints. The stance of fingertip (point Y) along with the values at point X can be communicated as a change, where the introduction of the fingertip is given by the quaternion:

$$T = \begin{bmatrix} \sin\left(\frac{\phi_1}{2}\right)\sin\left(\frac{\phi_2+\phi_3+\phi_4}{2}\right) \\ \sin\left(\frac{\phi_1}{2}\right)\cos\left(\frac{\phi_2+\phi_3+\phi_4}{2}\right) \\ \cos\left(\frac{\phi_1}{2}\right)\sin\left(\frac{\phi_2+\phi_3+\phi_4}{2}\right) \\ \cos\left(\frac{\phi_1}{2}\right)\cos\left(\frac{\phi_2+\phi_3+\phi_4}{2}\right) \end{bmatrix} \tag{2}$$

The area of point Y referenced from point X is given by the vector

$$Q_Y = \begin{bmatrix} \cos(\phi_1)\,L_1\cos(\phi_2) + L_2\cos(\phi_2+\phi_3) \\ -(2(L_1+L_2)+L_3)\cos(\phi_2+\phi_3+\phi_4) \\ L_1\sin(\phi_2) + L_2\sin(\phi_2+\phi_3) \\ -(2(L_1+L_2)+L_3)\sin(\phi_2+\phi_3+\phi_4) \\ \sin(\phi_1)(-L_1\cos(\phi_2)) - L_2\cos(\phi_2+\phi_3) \\ +(2(L_1+L_2)+L_3)\cos(\phi_2+\phi_3+\phi_4) \end{bmatrix} \tag{3}$$

ϕ_1, ϕ_2, and ϕ_3 joints are controlled by their own free actuator.

6.2. Simulation. As per proof of the theoretical concept, these simulations have been performed in order to validate the performance of the control scheme. The results are shown in terms of voltages between the controller and sensors, also with the controller and motors. These results show the behaviour of voltages according to the motion.

As shown in Figure 7, the curve is mainly based on the measured parameters, i.e., the curvature of a human hand on x-axis with the corresponding varying values of resistances of ex-sensors on the y-axis. This graph is proving that the theoretical concept between the changes occurs in resistance of the ex-sensor over the time. The mathematical equation gives us the following result:

$$y = 44x^2 + 32x + 15 \tag{4}$$

As shown in Figure 8, the graph is plotted between the resistances of ex-sensors on x-axis and input values of the voltages on the y-axis. This shows that the resistance is inversely proportional towards the input voltage. As the resistance of the ex-sensor increases, by the effect of resistance the input voltage at microcontroller terminal decreases. The mathematical model for this is defined as

$$y = 0.0071x^2 - 0.3x + 5.6 \tag{5}$$

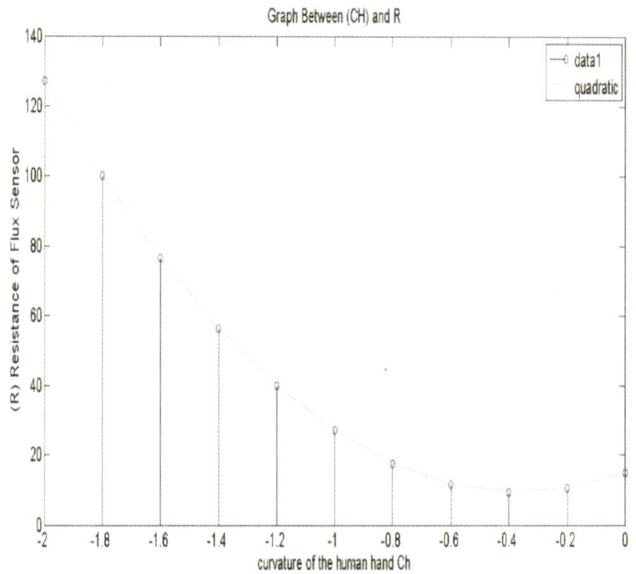

FIGURE 7: Curvature of human hand.

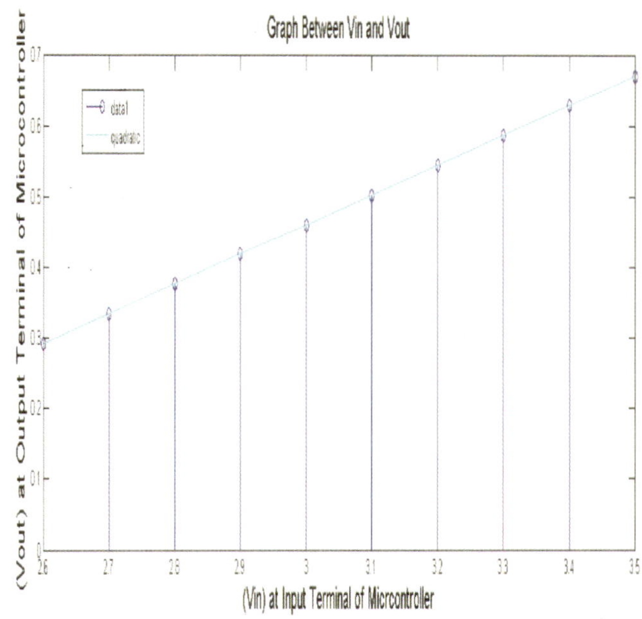

FIGURE 9: Input and output voltage of the controller.

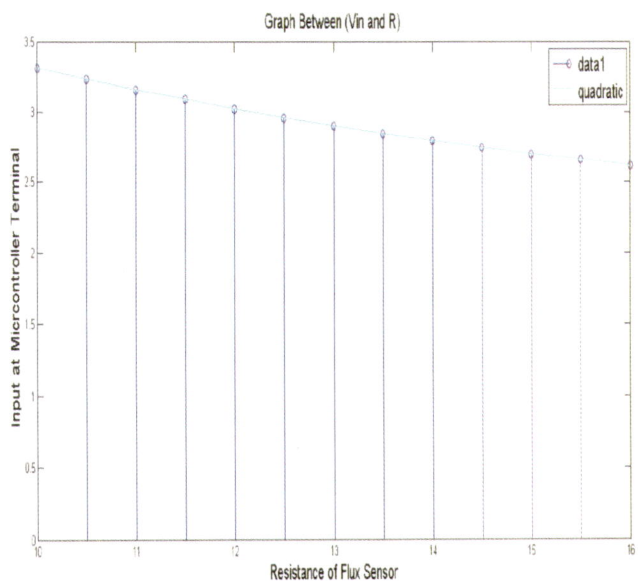

FIGURE 8: Resistance of flex sensor versus input voltage.

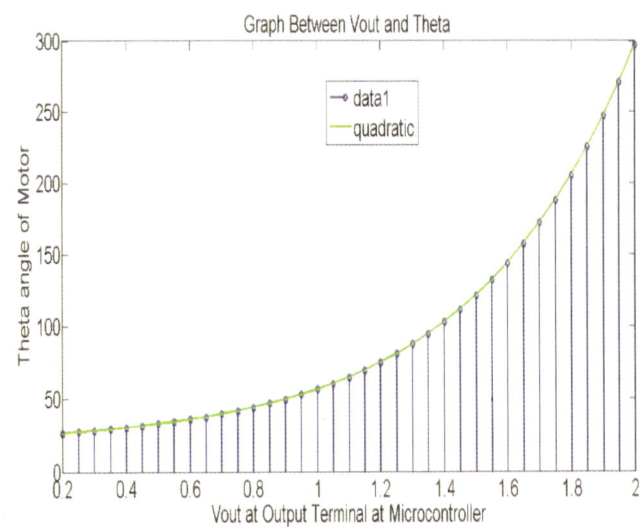

FIGURE 10: V(out) versus speed of the motor.

As shown in Figure 9, this curve is based on the following above equation, i.e., varying input values of the voltage on the x-axis and its corresponding output voltages of the micro-controller on the y-axis. This shows the linear relationship between the input and output voltages of the controller. The mathematical equation of this graph will be

$$y = 0.42x - 0.8 \tag{6}$$

Finally, Figure 10 shows us the curve, which is mainly based on the output voltages from the microcontroller on x-axis and the theta angle of the motor on the y-axis. This proves the concept that increase in voltage will also increase the rpm of the motor in nonlinear term which can be seen from this curve. This plot is defined by this mathematical equation:

$$y = 5.1 * e^{0.02x} - 1.8 * e^{0.02x} + 19 \tag{7}$$

In Figures 11-12, the relationship between the tendon and the finger motion has been shown, to evaluate how much the finger forces for different grasping positions. This shows the movement of one particular finger at a time.

7. Results

In this section, we validate the efficacy of the proposed design through extensive experimental paradigms that is included in the table. The table shows how the voltages have

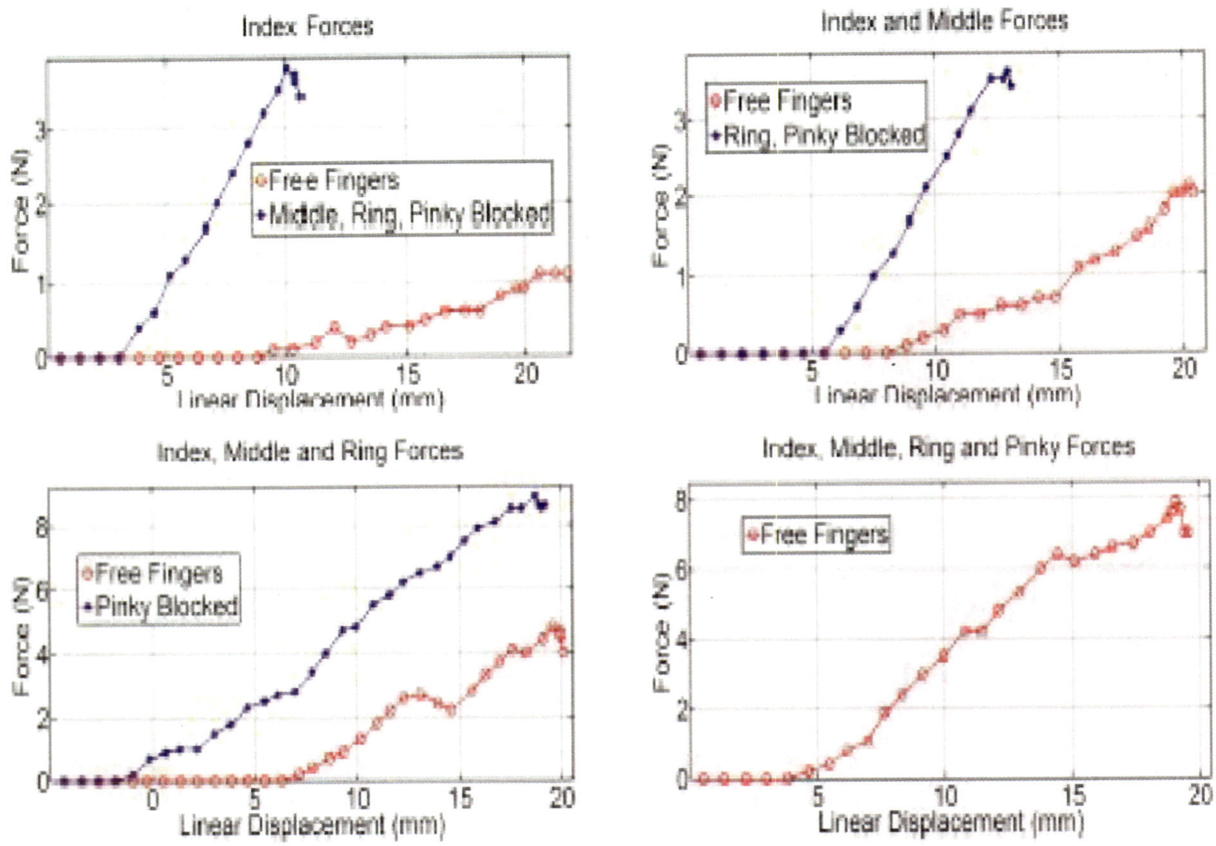

FIGURE 11: Linear displacement of fingers.

TABLE 1: Different voltage reading for different resistances of the Flux sensor.

S NO:	R(K ohm)	C(H)	Vin (V)	Vout (V)	Theta
1	15.2	0	2.65	0.295	0
2	14.88	-0.009612	2.676	0.3085	17.5
3	15.2	-0.019225	2.697	0.322	20
4	13.925	-0.03847	2.725	0.346	22.5
5	12.67	-0.0769	2.8	0.37	25
6	12.4	-0.0826	2.85	0.3955	27.5
7	12.21	-0.0884	2.913	0.42	30
8	12.055	-0.0826	3.006	0.4	32.5
9	11.9777	-01392	3.0532	0.469	42.2
10	11.8	-0.156	3.1	0.481	52
11	11.527	-0.1643	3.123	0.487	56.8
12	10.955	-0.1724	3.1468	0.493	61.7
13	10.588	-0.1805	3.1834	0.5315	71.3
14	10.01	-0.1866	3.22	0.57	81

been fluctuating which is measured in real time through the sensors. This signal is then fed to the controller by varying the control input parameter in real time, to produce such a control input which tackles the disturbances created due to the presence of uncertainties in the environment as shown in Table 1.

In order to verify our theoretical concept, different experiments are being conducted on the desired model. The main arrangement of tests concentrates on approving the viability of the proposed lockable differential component especially. The user can access different posture using a different switch which is placed on the fingers. Such functionality is not only

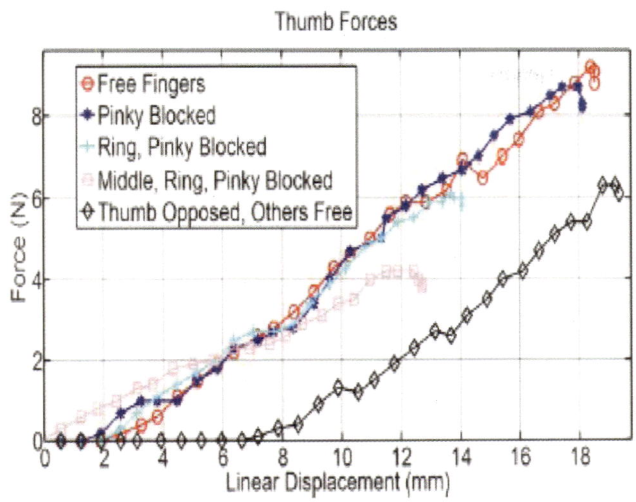

FIGURE 12: Thumbs displacement.

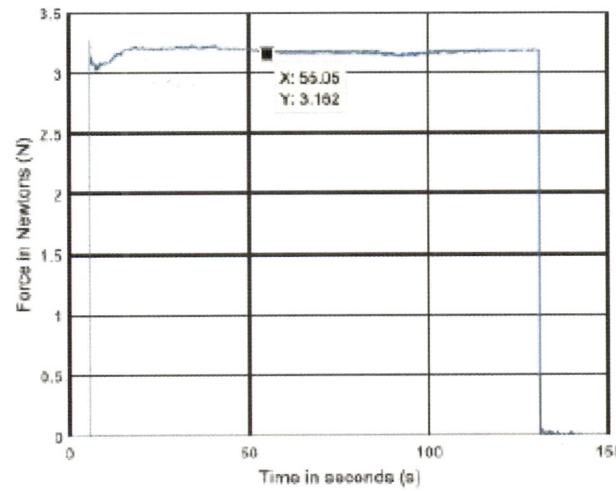

FIGURE 14: Object grasping at 200g.

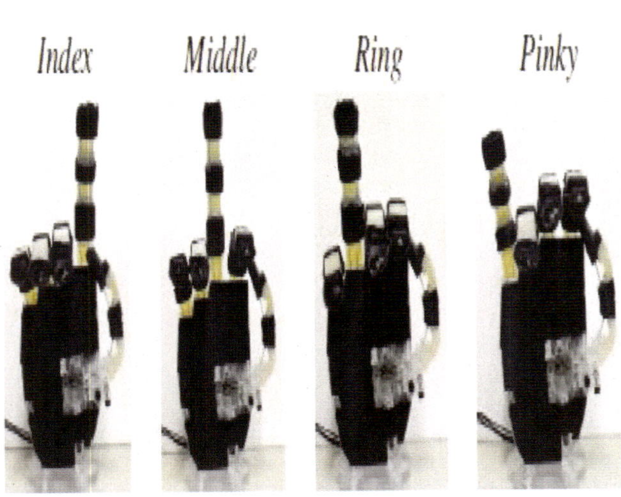

FIGURE 13: Movement of the fingers [8].

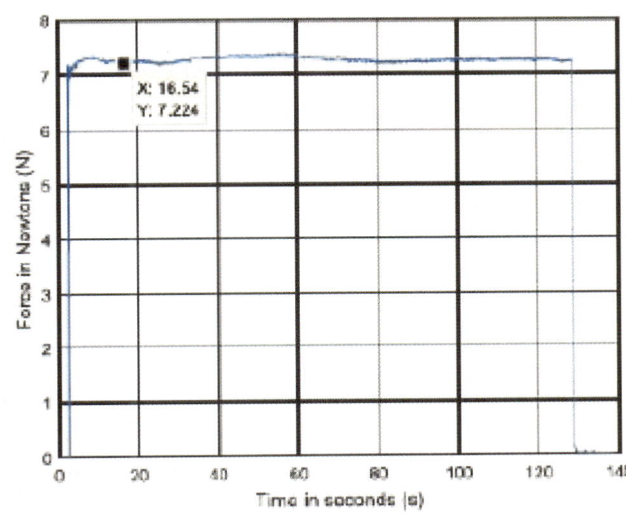

FIGURE 15: Object grasping at 400g.

important for grasping (where the user is able to choose the preferred grasping posture).

In Figure 13, different postures are depicted. The motion of each locked finger is constrained through the corresponding gesture which is changing according to user defined switches. The second set of examinations concentrates on getting a handle on the extensive variety of regular day to day objects, so as to execute everyday living exercises. Figures 14-15 show a real time graph, while grabbing weight of 200g and 400g, respectively. This figure shows the gripper successfully grasped objects. After a steady response of the system until 140 seconds, the force abruptly reaches toward zero as shown, indicating that the gripper released the object. In Figure 15, the weight is now being doubled as the mass, subsequent simulation result is shown. The result discusses that the increase in mass decreases the gripper power, as it converges to zero in 130 seconds.

8. Conclusion

In this paper, the design of an underactuated robotic hand is discussed using linear control and implemented as a prototype. According to the calculations and simulations readings during this research, it can be easily said that this mathematical model is now perfect. An ex-sensor is attached to every single finger for accurate measurement of the movement and grasping power of the robotic hand. Using the developed linear force, an object is handled by applying adequate force through the controller without damaging the object. The major advantage of this model can be seen as it easily allows all the fingers to move easily and independently using servo motors. A user can also change to different grasping postures by changing the programming of the controller. The thumb configuration can be easily adjusted by the user, using the lockable, stiff opposition mechanism. In conclusion, this robotic hand is a better device for helping amputees in their daily routine activities. This research shows how a prosthetic robotic hand can be developed with low cost. Flex sensor based hand gives proper hand control in grasping picking and holding things. The efficiency of the proposed robotic hands has been experimentally validated and the results were shown in the earlier sections. Hence, for future research investigation, it is recommended to use the output feedback controller for estimation of the states for better accuracy and low cost, as we eliminate the sensors from the system. Another recommendation is to include optimizing and building more intelligent control for such a robotic hand.

Conflicts of Interest

The authors declare that they have no conflicts of interest.

Acknowledgments

The authors are grateful for the help and support provided by the faculty and management of National University of Sciences and Technology-PNEC and FAST-National University of Computer and Emerging Sciences (NUCES), Karachi Campus.

References

[1] Y. Saito, T. Higashihara, K. Ohnishi, and A. Umemura, "Research on intelligent motorized prosthetic hand by functional analysis of human hand at near future," in *Proceedings of the 18th IEEE International Symposium on Robot and Human Interactive, RO-MAN 2009*, pp. 786–791, jpn, October 2009.

[2] S. S. Wankhede and R. Dharaskar, "Controlling Mouse Cursor Using Eye Movement," in *Proceedings of the International Journal of Application or Innovation in Engineering and Management (IJAIEM)*, pp. 2319–4847, 2009.

[3] M. H. M. Zaini, "Surgical and non-surgical prosthetic hands control: A review," in *Proceedings of the 2011 IEEE Symposium on Industrial Electronics and Applications, ISIEA 2011*, pp. 634–637, September 2011.

[4] H.-P. Huang and C.-Y. Chen, "Development of a myoelectric discrimination system for a multi-degree prosthetic hand," in *Proceedings of the 1999 IEEE International Conference on Robotics and Automation, ICRA99*, pp. 2392–2397, May 1999.

[5] R. R. Ma, L. U. Odhner, and A. M. Dollar, "3D Modular, Open-Source 3D Printed Underactuated Hand," in *Proceedings of the 2013 IEEE International Conference on Robotics and Automation (ICRA)*, pp. 2737–2743, Karlsruhe, Germany, May 2013.

[6] R. Kumar, S. D. Sharma, S. Despande et al., "Acrylonitrile Butadiene Styrene (ABS) plastic-based low cost tissue equivalent phantom for verification dosimetry in IMRT," *Journal of Applied Clinical Medical Physics*, vol. 11, no. 1, pp. 24–32, 2010.

[7] K. Scott and A. Perez-Gracia, "Design of a prosthetic hand with remote actuation," in *Proceedings of the 34th Annual International Conference of the IEEE Engineering in Medicine and Biology Society, EMBS 2012*, pp. 3056–3060, usa, September 2012.

[8] Y. Chnag, "A Survey of Robotic Hand- Arm Systems," *International Journal of Computer Applications*, vol. 109, no. 8, 2015.

[9] R. Wells and J. Schueller, "Forward and Feedback Control of a Flexible robotic Arm," in *Proceedings of the IEEE International Conference on Control System*, 1999.

[10] A. Crawford, J. Molitor, A. Perez-Gracia, and S. Chiu, "Design of a Robotic Hand and Simple EMG Input Controller with a Biologically-Inspired Parallel Actuation System for Prosthetic Applications," *Journal of the Franklin Institute, Elsevier*, vol. 344, pp. 36–57, 2007.

[11] G. Psihoyios and Z. Anastassi, "Robot Control Through Brain Computer Interface For Patterns Generation," in *Proceedings of the AIP Conference Proceeding Volume*, pp. 1031–1034, Halkidiki, (Greece), 2011.

[12] L. Fortuna, G. Muscato, and M. G. Xibilia, "A comparison between HMLP and HRBF for attitude control," *IEEE Transactions on Neural Networks and Learning Systems*, vol. 12, no. 2, pp. 318–328, 2001.

[13] R. Kumar, U. Mehta, and P. Chand, "A Low Cost Linear Force Feedback Control System for a Two-fingered Parallel Configuration Gripper," *Procedia Computer Science*, vol. 105, pp. 264–269, 2017.

17

Adaptive Foot in Lower-Limb Prostheses

Thilina H. Weerakkody, Thilina Dulantha Lalitharatne, and R. A. R. C. Gopura

Bionics Laboratory, Department of Mechanical Engineering, University of Moratuwa, Katubedda, Sri Lanka

Correspondence should be addressed to Thilina H. Weerakkody; thilinahweerakkody@gmail.com

Academic Editor: Gordon R. Pennock

The human foot consists of complex sets of joints. The adaptive nature of the human foot enables it to be stable on any uneven surface. It is important to have such adaptive capabilities in the artificial prosthesis to achieve most of the essential movements for lower-limb amputees. However, many existing lower-limb prostheses lack the adaptive nature. This paper reviews lower-limb adaptive foot prostheses. In order to understand the design concepts of adaptive foot prostheses, the biomechanics of human foot have been explained. Additionally, the requirements and design challenges are investigated and presented. In this review, adaptive foot prostheses are classified according to actuation method. Furthermore, merits and demerits of present-day adaptive foot prostheses are presented based on the hardware construction. The hardware configurations of recent adaptive foot prostheses are analyzed and compared. At the end, potential future developments are highlighted.

1. Introduction

Lower-limb assistive devices can be divided into two main categories as orthosis and prosthesis. The orthosis is an orthopaedic apparatus which is used as support for adjusting deformities to improve functionalities of moving body parts whereas prosthesis is an artificial replacement for a missing body part [1, 2]. According to the literature survey on amputation in 2005, the United States (USA) recorded about 1.6 million lower-limb amputees. It was predicted that the number of lower-limb amputees would be increased to 3.6 million over the span of the next 50 years [3]. Another Tanzania-based survey reported 86.4% of total amputees as lower-limb amputees [4]. A survey conducted in Brazil reports that 25% of total amputees require foot prostheses solutions [5]. Long-term passive flat foot prostheses users tend to suffer from physical injuries such as osteoarthritis, osteopenia, and subsequent osteoporosis due to musculoskeletal imbalances or pathologies [6, 7]. Foot prosthesis with flexible adaption capabilities is a precaution for the above-mentioned injuries [6, 7]. Statistical data and possible physical injuries reflect the necessity of suitable and reliable adaptive foot prostheses which could mimic the human foot functionalities in commercial level. Foot amputees' lives can be uplifted and made comfortable and more productive to the society by developing advanced, reliable prosthetic solutions. Currently, some passive [8–16], active [17–19], and hybrid [20–30] adaptive foot prostheses have been developed with a focus on different functional requirements and design mechanisms.

The human foot has the adaptive capability which enables the foot to withstand any uneven surface. Necessary kinematic and kinetic adjustments are done to the gait pattern during ambulation by pedestrians in order to maintain stability on slopped or uneven terrains [31]. Normally, the human walking decisions are taken upon on human vision sensors and neural sensors. Amputees lack certain neural sensors due to the loss of their body part. The inability of surface adaption of the foot has significantly increased the load on the residual limb. Additionally, pressure ulcers and deep tissue injuries can occur as a result of significant pressure on a residual limb [32]. Lack of stability causes prostheses users to fall when entering an uneven surface [33]. Lack of inversion-eversion in ankle prosthesis can cause instability due to partial contraction with the surface. Suitable solutions for these physical and practical problems have to be addressed when designing an adaptive foot prostheses. However, most of the existing lower-limb ankle prostheses have not focused on developing a proper adaptive foot prosthesis for their ankle prosthetic devices. Instead, the passive flat prosthetic foot has been commonly used as the end connector for commercial lower-limb prostheses such as Otto Bock. Since passive flat foot prostheses have limited functional capabilities and other physical side effects as mentioned above, adaptive foot prostheses are essential to be developed to regain natural foot motions in lower-limb prostheses [34–37].

In this paper, authors have reviewed designs and developments of adaptive foot prostheses that have been proposed for lower-limb prostheses since 1997. It is essential to study design features, merits, and demerits of existing designs in order to enhance the field of adaptive foot prostheses. Some of the available reviews are focused on lower-limb prostheses [38, 39], control methods of lower-limb prostheses [40, 41], and prosthetic feet devices [42]. Versluys et al. [42] classified conventional feet, energy storing feet, and bionic feet upon control, comfort, and cosmetics. They have reviewed only a limited number of existing bionic foot devices and also adaptive mechanisms have not been considered for the review article. Since 2009, a lot of active foot prostheses have been introduced with novel mechanisms. In-depth review on adaptive foot prostheses is rarely found with those novel mechanisms. A prompt review paper on adaptive foot prostheses is very useful, not only to identify the current status of research but also to provide information to anyone in the field of developing adaptive foot prostheses. This paper is prepared based on existing adaptive foot prostheses. Some passive prostheses are available with notable design functionalities and mechanisms. They can be transferred into active designs with suitable design changes which lead to adding those devices into this paper. The focus of this paper remains in existing designs, their favourable and adverse design issues, and common solutions available in adaptive foot prostheses.

The systematic review on recent developments in foot prostheses has been done based on sets of design criteria. The papers were chosen based on preselected search keywords. Out of many scientific databases, the following were selected due to the availability of a higher number of the relevant manuscripts: IEEE Xplore, Elsevier, SAGE, InTech, PLOS ONE, ASME (American Society of Mechanical Engineer's Journal), and Journal of Rehabilitation Research & Development (JRRD). The paper selection was compiled upon PRISMA criteria [43]. The selected papers were initially screened, then the duplications were removed, and the papers were further refined due to irrelevance. Later search keywords were readjusted in order to obtain a higher number of relevant results. Finally, the search keywords "adaptive foot prostheses" were selected. The detailed review methodology is explained in Section 5 below.

The paper is structured as follows. In Section 2, the anatomy of the ankle and foot has been explained briefly in order to clarify the adaptive foot prostheses functional requirements. Section 3 presents requirements and design difficulties encountered in adaptive foot prostheses development. Classification of adaptive foot prostheses is presented in Section 4. The extended details, a method of literature selection for the analysis, and comparison and review of existing prostheses are included in Section 5. Finally, discussion and future directions are included in Section 6.

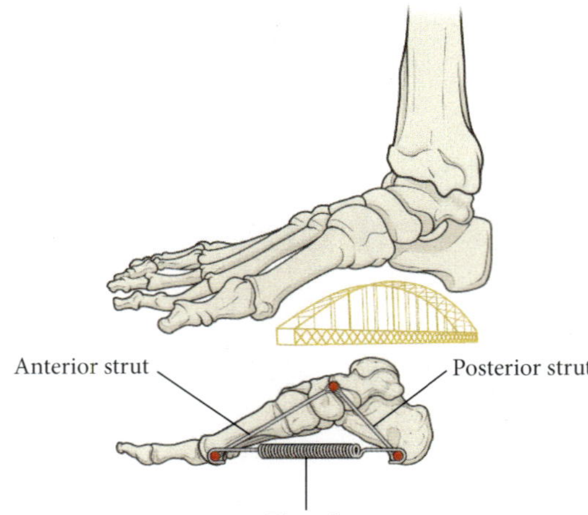

FIGURE 1: The arch of foot mechanism [45].

2. Anatomy of Ankle and Foot

The main function of the foot is performing gait cycle. Sufficient amount of mobility and stability is necessary for the foot to perform its tasks. Absorbing the ground reaction force is critically important for mobility. Stability is essential for well-balanced body posture [44]. The foot consists of 6 joints which can move along sagittal and transverse planes. Due to the complexity in foot joints, developing a foot prosthesis to mimic the human foot adaption capability is a challenging task. The anatomy of the human foot consists of 26 bones, 33 joints, 20 muscles, and over 100 ligaments [45, 46]. It can carry the human body weight due to its complex structure. The foot is capable of varying flexibility and elasticity of the complex structure to perform various challenging tasks such as running, climbing, balancing, jumping, hopping, and going up on the toes [45]. The foot bones are distributed along two main concurrent structures, known as the ache. There are three types which are medial longitudinal arch, lateral longitudinal arch, and transverse arch. The surface adaption (or flexibility and elasticity) of the foot occurs due to varying the arch angle of the foot (Figure 1). View along longitudinal (sagittal plane) arch is shown in Figure 2. The curvature of the bones of the foot provides a structure which is able to absorb high force repetitively similar to a bridge. Additionally, intrinsic and extrinsic muscles provide structural resilience by serving a tie rod as shown in Figures 1 and 2. As a result of contraction and relaxation of these muscles, the arch of the foot changes and increases the surface adaption capability of the foot. This geometric distribution combined with tendons and muscles creates foot windlass mechanism [47]. Windlass mechanism is used for moving heavy loads in engineering applications. Similarly, windlass mechanism provides the additional support for the foot arch to carry the load.

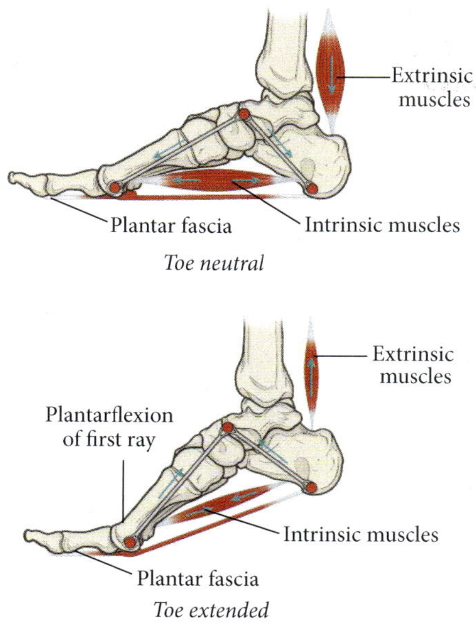

FIGURE 2: Windlass mechanism [45].

TABLE 1: Ranges of motions of human foot joints.

Motion	Human foot joint	Plane of movement	Range of motion
Dorsiflexion & plantarflexion	TC	Sagittal	N/A
	ST	Sagittal	−2.5° : 5°
Inversion & eversion	ST	Transverse	−10° : 20°
	MT - TN	Transverse	N/A
Abduction-adduction	ST	Frontal	−10° : 20°
Flexion & extension	MT-CC	Sagittal	N/A
	MTP (big toe)	Sagittal	(−) 80° : 40°
	MTP (toes 2–5)	Sagittal	(−) 60° : 40°
	Proximal IP (big toe)	Sagittal	0° : 90°
	Proximal IP (toes 2–5)	Sagittal	0° : 60°
	Distal IP	Sagittal	Hyper: 90°

The foot consists of three regions which are a hind foot (heel), midfoot, and forefoot (toe). The five major joints in the foot are ankle (or Talocrural (TC)) joint, Subtalar (ST) joint, Tarsometatarsal (TMT) joint, Metatarsophalangeal (MTP) joint, and Interphalangeal (IP) joint (Figure 3) [45, 46]. The hind foot consists of the calcaneus and talus. The midfoot consists of the navicular, cuboid, and the three cuneiforms. The forefoot consists of the metatarsals and phalanges. The ankle or TC joint is a hinge type joint which moves along the sagittal plane, providing the dorsiflexion and plantarflexion foot motions. The ST joint is a condyloid type joint which enables the movement along a transverse plane, providing inversion and eversion foot motions. The Midtarsal (MT) joint is in between ST joint and TMT joint which consists of two joints, namely, Talonavicular (TN) joint and Calcaneocuboid (CC) joint. The TN joint is a ball and socket type joint which enables the movement along transverse plane providing inversion and eversion foot motions. The CC joint is a modified saddle type joint which enables the movement along the sagittal plane, which provides flexion and extension foot motions. The TMT joint is a plane and synovial type joint which connects MTPs to the foot. The MTP joint is a condyloid type joint which moves along sagittal plane providing flexion/extension motions for proximal phalanges. This motion is essential when changing the arch of the foot on various surfaces. The IP joint is a hinge type joint which moves along sagittal plane which provides flexion/extension for middle and distal phalanges (Figure 3) [46].

There are rotation axes for each joint in the foot according to the plane of movement. The three cardinal planes of the human body are shown in Figure 4 which are a sagittal plane, transverse plane, and frontal plane. Some of the main rotation axes of the human foot are shown in Figure 5. Cardinal longitudinal axis of the foot is along the sagittal plane. Both ST joint and TC joint are joined to one another by talus bone, yet these two axes are more like perpendicular to one another due to the hinge and condyloid type joints. As a result, the toe can glide and roll. Knowledge of these movement planes and axes of rotation is important to understand the moving axes of existing foot prostheses. A better understanding of human foot anatomy is essential to identify design requirements. Table 1 summarized the ranges of motion of the above-mentioned human foot joints. (Consider supination as + direction and pronation as − direction). Distal IP has a small amount of extension which is known as hyperextension indicated in Table 1 as "hyper."

3. Requirements and Design Difficulties

The human foot consists of over 100 ligaments to control the five main joints. There are several design difficulties which can be incurred when developing an adaptive foot prosthesis. Complex nature of human foot anatomy makes it much difficult to mimic the adaptive nature of human foot through foot prostheses. The human foot maintains its stability by supinating/pronating along a longitudinal axis and plantar flexion/dorsiflexion along mediolateral axis. The surface contact area of the phalanges can be increased by flexing and extending them along mediolateral axis. A multi-DoF system with all the above-mentioned functionalities is a challenging task as actuators have to be arranged closer to each other while carrying the body load.

The human foot has arches along longitudinal and transverse axes which enable adapting to any surface by rotating along both directions. Developing a multidegrees of freedom system is a challenging task. The ankle joint is complex. Most of the existing prostheses have used high torque actuators for the ankle joint. Therefore sufficient space needs to be provided for ankle joint. Various mechanisms are available for transmitting the power to a prosthesis. Out of them, the most appropriate method has to be selected based on the power source, type of application, and expected weight of the prosthesis. The prosthesis should have the sufficient

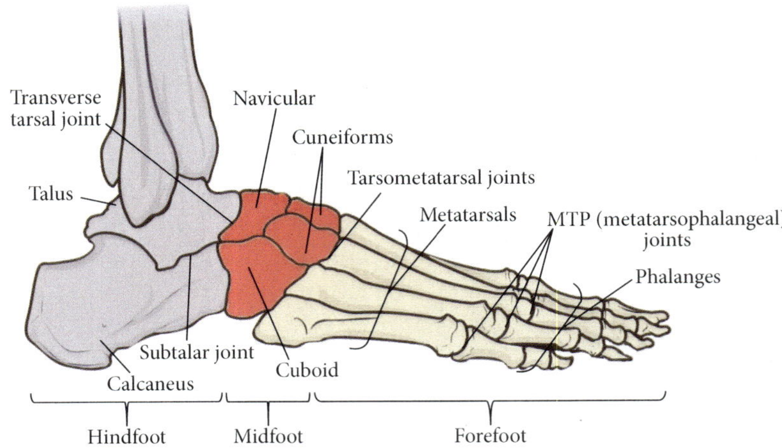

FIGURE 3: The human foot anatomy [45, 46].

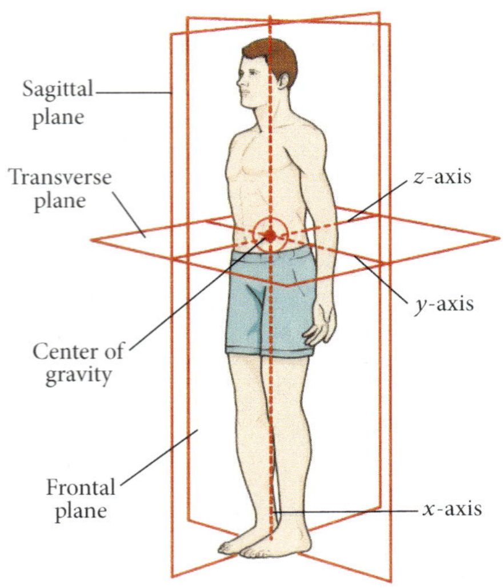

FIGURE 4: The three cardinal planes of the human body [45].

TABLE 2: Design requirements for a foot prosthesis development.

Requirement	Remarks
DOF	3DOF
Torque	Calculate by considering weight, type of mechanism, DOF, size, material (80–120 Nm)
Axis of rotation	Mediolateral axis, longitudinal axis, transverse axis
Type of mechanism	Coil spring, leaf spring, clutch, linkages, rolling joints, actuators, SEA, gears
Movable range	Refer to Table 1
Size	Approximately length 275 mm, width 100 mm, height 85 mm
Weight	Approximately 0.85–1.5 kg
Fabrication material	Carbon fiber or aluminum
Attachment method	Osseointegration, couplings, or pyramid adapters

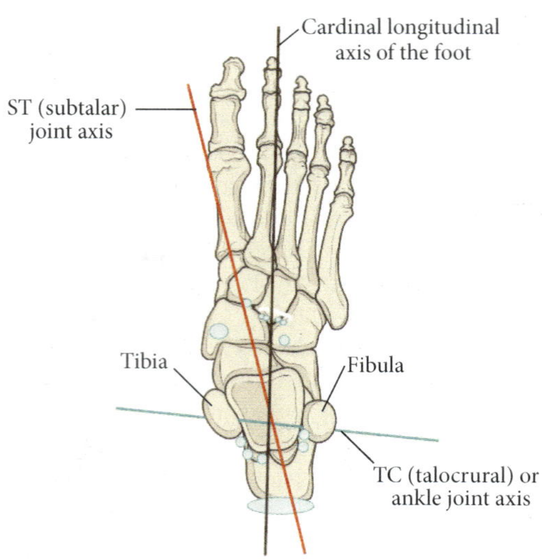

FIGURE 5: Human foot axes of rotation [45].

moving capability along each axis as given in Table 1. The dimensions of the device have to be within the limits of average human foot size. Adaptive foot prosthesis needs to be within average human foot weight. If it exceeds this, the amputee feels uncomfortable in long-term usage. High strength materials are needed for development as the foot needs to hold the total body load and large ground reaction forces for various activities of daily living (ADL) such as running, jumping, and hopping. Some developers have used lightweight, high strength polymer type materials instead of metals. The method of attaching the foot prosthesis to the remaining lower-limb or prosthesis device is another consideration that needs to be addressed. Table 2 provides a concise design requirements list.

4. Classification of Foot Prostheses

Prostheses can be classified according to the applications: upper limb prostheses, lower-limb prostheses, and other

TABLE 3: Methods of classification of adaptive foot prostheses.

Classification method	Parameters
Actuation method	Passive Active Hypird
DOF	Active DOF
Type of actuators	DC brushless motors DC servo motors AC servo motors
Power transmission method	Gear drives Chain drives Linkages Clutch drives
Energy regeneration methods	Series elastic actuators (SEA) Coil springs and clutch motors Linkages and camshafts Leaf springs
Attaching method	Coupling Pyramid adaptor

prostheses. Orthoses can be classified in two subcategories: exoskeletons and end effector connecting devices. There are different types of lower-limb prosthetic devices available based on an application which is for hip disarticulated amputees, above knee articulated amputees (transfemoral), knee articulated amputees, below the knee (transtibial), ankle disarticulated amputees, and partial foot amputees [1, 2]. Furthermore, adaptive foot prostheses can be classified into three categories which are a passive, active, and hybrid prosthesis. Passive prostheses are functionally lacking due to the mimicking of the human leg motions compared to the active prosthesis. Therefore the development of active prosthesis is essential. Yet they are still at the research level due to lack of design and control issues. Over the years a lot of transfemoral and transtibial prosthesis have been developed. However, there is a research gap in the field of development of an adaptive foot prosthesis. The hardware construction of adaptive foot prostheses can be classified into several categories which are classified upon actuation method, DoF, and types of actuators, based on power transmission method, energy regeneration method, and attaching method to the residual limb or transtibial prosthesis and so forth. Some of the classification methods are discussed below. Table 3 summarizes the classification of the hardware construction of adaptive foot prostheses devices.

(i) Actuation Method. Lower-limb adaptive foot prostheses are classified based on power source method. Passive prostheses are body-powered or use the power of user to actuate. Active prostheses are actuated using external power sources. Most of the present-day adaptive foot prostheses are combined with both passive and active joints. This method enhances the use of available energy during ambulation through passive joints and other required motions through active joints by external power sources.

(ii) DoF. Adaptive foot prostheses can be classified according to the number of active joints or externally power actuated joints like 1 DoF, 2 DoF, 3 DoF, and so forth.

(iii) Types of Actuators. There are various types of actuators that have been used in existing prostheses. They are DC motors, brushless DC (BLDC) motors, servo motors, and AC motors. Different types of DC motors are available such as brushless motors and servo motors.

(iv) Power Transmission Method. Prostheses transmit power using various methods such as gear drives, chain drives, and linkages mechanisms which are connected to actuators, clutch drives, and so forth. Additionally, belt drives, ball screw drives, and cable drive methods are possible.

(v) Energy Regeneration Method. Existing actuators have limited torque generation capability. Therefore some researchers have developed energy regenerative mechanisms to generate the required high torque. Series elastic actuator (SEA) is one of the most popular methods in modern days. Additionally, a combination of coil springs and clutch motors, linkages, camshafts with motors, and leaf springs with motors have been used in different existing devices.

(vi) Attaching Method. Attaching method of adaptive foot prostheses to the lower-limb prostheses is essential for the amputee's use. There can be two types of attaching ends which are connecting prosthesis to residual limb and attaching adaptive foot prosthesis to transtibial prosthesis of the transfemoral prosthesis. Ankle-foot couplings and pyramid adaptors are common among other attaching available methods. Additionally, socket attaching methods are possible.

5. Review on Adaptive Foot Prostheses

Foot prosthesis is used as the terminal device for the lower-limb prosthesis. There are passive and active adaptive foot prostheses. The passive prostheses are designed to be operated with user's body power and no actuator driven joints. Fully passive devices are relatively limited with motion capabilities. Active prostheses are designed to be controlled with externally powered actuator joints. It requires a well-designed control structure to control all the joints simultaneously to mimic the actual human foot movements. Active prostheses enhance the developers to focus more on the functionalities of the foot rather than mechanisms to power the device passively. The introduction of actuators to the active prostheses makes them much heavier compared to the passive prostheses. Due to the above-mentioned favourable and adverse drawbacks of passive and active prostheses, combined passive and active joints (or hybrid) prostheses have been developed by manufacturers lately. Hybrid prostheses have advantages over other prostheses which are improved workspace, higher functional capabilities, and larger range of motions.

The prosthesis gets bulky with the introduction of external power sources. As the prosthesis mass undergoes an increase, the user feels discomfort when using it over a

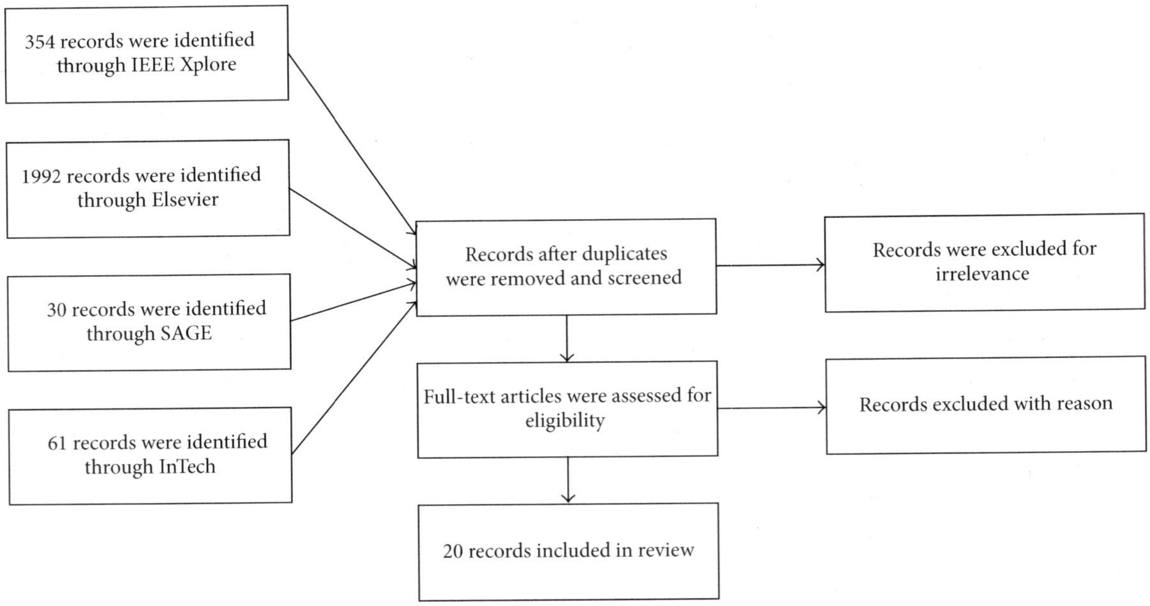

FIGURE 6: PRISMA flow diagram of the literature review process [43].

long period of time. Therefore energy saving mechanisms have been introduced by prosthesis developers to reduce the power requirement. The spring-based energy storing and regenerating methods like series elastic actuators (SEA) [48], parallel elastic actuators (PEA), clutchable series elastic actuators (CSEA) [49], continuously variable series elastic actuators [50], and so forth are some examples for such mechanisms. Attaching methods of adaptive foot prostheses to lower-limb prostheses are mainly coupling, pyramid type attachments. Aluminum is the commonly used material for the prototypes and expensive materials such as carbon fiber have been used in some of those developments. Tables 4, 5, and 6 provide a concise comparison of existing passive, active, and hybrid adaptive foot prostheses during 1997–2016. The weight of the adaptive foot prosthesis, actuation method and number of actuators, axis of rotation and equivalent human foot joint, working mechanism, moving range along each DoF, attaching method to remaining stump limb or transtibial prosthesis, and material used for the development are included.

In order to select databases for the paper, several generic keywords were searched such as "adaptive foot prosthesis, feet, ankle-foot prosthesis, lower-limb prosthesis, artificial limb, humanoid robots". IEEE Xplore, Elsevier, SAGE, InTech, PLUS ONE, ASME (American Society of Mechanical Engineer's Journal), and Journal of Rehabilitation Research & Development (JRRD) databases were chosen due to their high number of relevant search results. The search term of "adaptive foot prosthesis" was created in several iterations to obtain a larger number of relevant results. The search was limited to conference proceedings, journal papers, dissertations, and patents for the time period of 20 years from 1997 to 2017. Search results consisted of a significant number of control algorithms, medical researchers, and other robotic researchers. However, the basis was limited only to mechanical designs and developments. Most of those consisted of knee and ankle prostheses designs which had to be eliminated and only ankle-foot and foot were selected. With an in-depth study about available prostheses device designs and their focused area, most of them were refined and we retrieved the most appropriated few which were suitable for the topic of adaptive foot prostheses. Among the existing adaptive foot prostheses, flat foot designs were excluded. Only passive, active, and the combination of passive and active (i.e., hybrid prostheses) prostheses were adopted for the review. The number of search results obtained for each keyword in different academic databases is shown in Table 7.

The search retrieved a total of 2437 manuscripts from selected academic databases. The results were refined by manually screening for their relevance using the title and abstract. Selected remaining papers were studied further and we excluded the papers with no adaptive foot devices. A total of 20 papers were selected due to the high relevance to the topic of adaptive feet in prostheses. The PRISMA flow diagram in Figure 6 summarizes the review selection procedure. PRISMA (Preferred Reporting Items for Systematic Reviews and Meta-Analyses) is a method used in systematic reviews in contemplation of improving the reporting quality [43]. Since most of the novel designs have been developed based on available patents and some patents are beyond the scope of this review, only 4 patents have been included in the review.

5.1. The Heel Foot [8]. The Heel Foot (Figure 8(a)) was developed by the University of Twente, Enscheda, Netherlands, in 2003. This is a single DoF passive plantar flexion adaptive foot prosthesis rotating along a mediolateral axis. Plantar spring controls the arch angle of the Heel Foot for maintaining stability. The potential energy stored in compressed plantar and heel springs starting from heel-off phase is used to push forward at the toe-off point. Four-bar linkage mechanism has been used for compression springs and varying arch angle. The Heel Foot was validated for relative joint angle, joint

Table 4: Comparison of passive-adaptive foot prostheses/foot designs in lower-limb prostheses.

Country	Name/year/reference number	Weight	Axis of rotation	Type of mechanism	Movable ranges	Attaching method	Material
Netherlands	The Heel Foot (2003) [8]	0.5 kg	Mediolateral axis MTP joint axis	Spring based	$(-)\ 20^\circ : 20^\circ$	Knee ankle coupling	Toe-carbon fiber, forefoot, heel—aluminum
Netherlands	fully passive transfemoral prosthesis (2011) [9]	1.05 kg	Mediolateral axis at MTP joint	Spring based and linkages	$0 : 30^\circ$ (toe)	Prosthesis ankle joint	Carbon fiber
United States of America (USA)	Prosthetic ankle-foot system (2014) [10]	1.04 kg	Mediolateral axis	Linkages and camshafts	$87^\circ : 105^\circ$	Pyramid adapter	Nylon 6/6, polyurethane rubber, maraging steel
Japan	Bipedal walking robot with oblique midfoot joint in foot (2015) [11]	N/A	Oblique axis at MTP joint	Truss and windlass mechanism	N/A	Nut and bolt	N/A
Italy	SoftFoot (2016) [12]	N/A	Parallel to mediolateral axis	Series of rolling joints	vary with the surface	Coupling	Rapid prototyping material
United States of America (USA)	Hindfoot and forefoot stiff foot prostheses (2017) [13]	N/A	Mediolateral axis	Flexible composite forefoot keel and hindfoot of varying stiffness	sagittal declination angle 15°	Pyramid adapter	Aluminum 7075-T6
United States of America (USA)	One-piece mechanically differentiated prosthetic foot (1997) [14]	N/A	Mediolateral axis	Flex due to polymeric material	N/A	Flange type nut and bolt connector	Light weight polymeric material
United States of America (USA)	Instrumented prosthetic foot (2012) [15]	N/A	Mediolateral axis	Flex due to polymeric material	N/A	Pyramid adapter	Durometer polyurethane
United States of America (USA)	Ankle-foot prosthesis of automatic adaptation (2014) [16]	N/A	Mediolateral axis	Spring and link based mechanism	$(-)\ 45^\circ : 80^\circ$	Pyramid adapter	Elastomeric Materials

Table 5: Comparison of active adaptive foot prostheses/foot designs in lower-limb prostheses.

Country	Name/year/reference number	Weight	Actuator	Axis of rotation	Type of mechanism	Movable ranges	Attaching method	Material
China	PANTOE 1 (2010) [17, 18]	1.47 kg	2 DC motors	Mediolateral axis at MTP joint	SEA	$(-)\ 16^\circ : 27^\circ$	Socket adaptor	Aluminum alloy
USA	Universal prosthesis emulator (2014) [19]	0.96 kg	1 DC motor	Mediolateral axis at MTP joint	Spring base	$(-)\ 12^\circ : 12^\circ$	Universal adaptor	7075-T651 aluminum

Table 6: Comparison of hybrid adaptive foot prostheses/foot designs in lower-limb prostheses.

Country	Name/year/reference number	Weight	Passive Joint	Actuator	Axis of rotation	Type of mechanism	Movable ranges	Attaching method	Material
Japan	Parallel four-bar linkage humanoid robot (2007) [20]	0.76 kg	Toe mechanism	1 DC servo motor	Mediolateral axis at MTP joint	Actuator based	$0:44^{\circ}$	Humanoid ankle	Extra super duralumin
USA	Energy recycling foot (2010) [21, 22]	1.37 kg	Springs and clutches based mechanism	2 DC motors	Mediolateral axis at MTP joint	Coil Springs, clutch motor	N/A	Pyramid adapter	7075-T6 Aluminum, Stainless steel, Carbon/fiberglass
Japan	Adaptive bipedal deformable feet (2012) [23]	1.2 kg	Torsional Springs	2 Servo motors	Mediolateral axis at MTP joint	Torsional springs and servo motor combination	N/A	Nut and Bolt	Super soft urethane resin
Germany	An adaptive sensor foot for a bipedal and quadruped robot (2012) [24]	N/A	Bowden cables and dampers	2 BLDC motors	Mediolateral axis at MTP joint	Windlass mechanism	roll -20° to 10° pitch -30° to 20° yaw -10° to 10°	Coupling	Rapid prototyping material
Belgium	The AMP-foot 1.0 (2012) [25]	3 kg	Spring-based gear mechanism	N/A	Mediolateral axis at MTP joint	Spring, planter gear mechanism	$0:30^{\circ}$	Coupling	Aluminum
Belgium	The AMP-foot 2.0 (2013) [26, 27]	2.5 kg	Lever arm and spring combined mechanism	1 DC motor	Mediolateral axis at MTP joint	Springs and SEA	$0:45^{\circ}$	Prosthesis ankle joint	Aluminum
United Kingdom	Virtual prototyping of a semiactive transfemoral prosthetic leg (2015) [28]	2.3 kg	Springs	1 DC motor	Mediolateral axis at MTP joint	SEA and springs	N/A	Nut and bolt	N/A
Italy	Variable compliant humanoid foot (2016) [29]	0.52 kg	Leaf springs, cam followers based method	1 DC geared motor	Longitudinal axis of the foot	Leaf spring, motor actuated	N/A	Humanoid ankle	Aluminum, rubber
China	Bioinspired tunable stiffness robotic foot (2017) [30]	N/A	Spring	Stepper motor	Mediolateral axis and longitudinal axis	Spring and ball screw	N/A	Ball joint	N/A

TABLE 7: Results of keyword search in respective academic databases.

Keyword	IEEE Xplore	Elsevier	SAGE	InTech	PLOS ONE	ASME
Foot/feet	7,539	120,657	101,738	1,478	2,876	149
Lower limb prosthesis	267	12,434	2184	244	10,201	35
Humanoid robots	11,404	2,276	707	772	3,693	7
Ankle-foot prosthesis	48	3,864	865	244	880	25
Adaptive foot prostheses	354	1,992	148	61	237	399
Artificial limb	1404	26,272	3,496	691	37,951	50

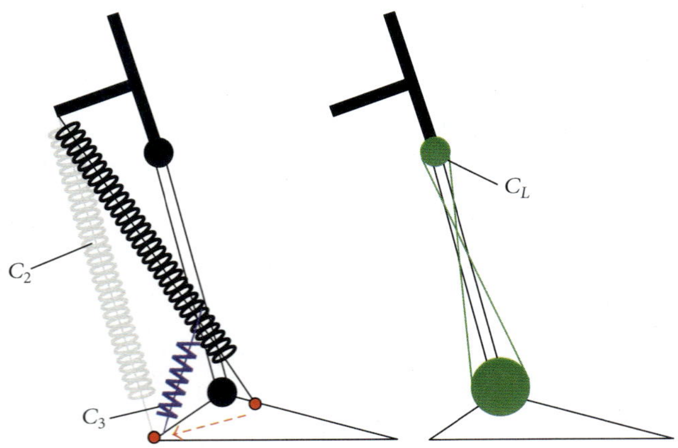

FIGURE 7: Spring arrangement of fully passive transfemoral prosthesis [9].

torque, joint power, and force variation for gait cycle for proving the prototype functionality.

5.2. Fully Passive Transfemoral Prosthesis Prototype [9]. Fully passive transfemoral prosthesis (Figure 8(b)) was developed by the University of Twente, Enscheda, Netherlands, in 2011. The adaptive foot prosthesis part is designed as spring-based linkage mechanism. The energy storing mechanism using springs is as in Heel Foot [8]. In this prosthesis adaptive foot manoeuvres with the aid of knee and ankle generating potential energy. During the stance phase, both knee and ankle absorb a certain amount of energy for carrying body weight. Then knee further absorbs energy for preswing and the ankle generates 80% of total energy for push-off. With the analysis of gait power requirement diagrams, this paper suggests that the knee is more like an energy absorber and ankle is more like an energy generator. This concept was the intuition for the conceptual design shown in Figure 7.

Two springs are crossed to each other and connected to ankle. During the preswing phase, knee absorbs the kinetic energy and stores it in C_L spring. Then when the swing phase arrives kinetic energy will be stored in C_2 spring. Stored energy during swing phase can be reused in stance phase while C_3 spring stored kinetic energy can be used in the next stage. Spring arrangement in the proposed mechanism is shown in Figure 7. Cable mechanism is used to govern the ankle and adaptive foot bends according to the knee flexion during the gait cycle. The conceptual design did not simulate. The prototype has been developed. However, prosthesis did not validate.

5.3. Passive Slope Adaption Prosthetic Ankle-Foot System [10]. This mediolateral direction rotating single DoF passive device was developed by a set of researchers from USA (Figure 8(c)). This prosthesis consists of link and cam on the passive ankle joint and the foot plate moves according to the slope of the surface. The moving range of the joint is only 18°. The prosthesis is validated with a set of experiments and the system has no energy regeneration method and comparatively the surface adaption mechanism is a basic method with a limited range of motions.

5.4. Bipedal Walking Robot with Oblique Midfoot Joint in Foot [11]. This foot was developed by a group of Japanese researchers in 2015. The bipedal walking robot in Figure 8(d) was developed to generate the adaptive nature of the foot with midfoot axis rotation nature. Oblique axis DoF foot prostheses are rarely used in foot prostheses due to the lack of strength. The bipedal walking robot is a humanoid robot that has been designed to replicate the human foot motions. The tendon wire mimics the arch of the foot. Yet the weight carrying capacity is limited in this design.

5.5. SoftFoot [12]. SoftFoot (Figure 9(a)) has been developed to improve the adaptive nature of the foot prosthesis. This is a complete passive foot prosthesis developed by studying the human foot arch and bone arrangement along the longitudinal direction. The prototype was developed by Research Center "Enrico Piaggio," University of Pisa, Italy, in 2016 using a rapid prototype method. SoftFoot was developed based on windlass mechanism [45]. Chain of connectors which can rotate parallel to mediolateral direction is used as foot links. The foot arch angle is fixed and no energy regeneration method is available with SoftFoot. The SoftFoot was validated with a compliant simulation for load distribution. Also the experiments were carried out to measure the performances on uneven terrains. The device was validated by comparing the surface adaption capability with a rigid flat foot.

5.6. Hindfoot and Forefoot Stiff Foot Prostheses [13]. This passive stiff foot prosthesis shown in Figure 9(b) was developed in the USA in 2017. The device consists of a rubber base which enables the prosthesis to function the push-off movement of the foot with varying arch and windlass mechanism. This

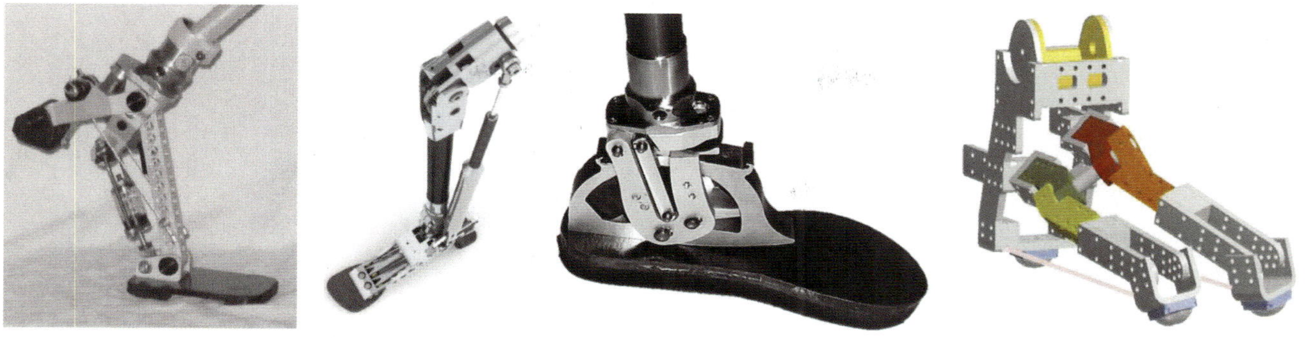

(a) (b) (c) (d)

FIGURE 8: Passive foot prostheses. (a) The Heel Foot [8], (b) fully passive transfemoral prosthesis prototype [9], (c) prosthetic ankle-foot system [10], and (d) bipedal walking robot with oblique midfoot joint in foot [11].

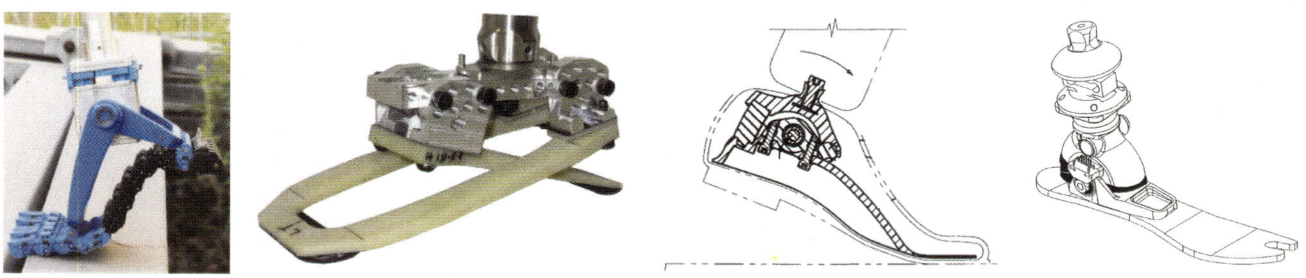

(a) (b) (c) (d)

FIGURE 9: Passive foot prostheses. (a) SoftFoot [12]. (b) Hindfoot and forefoot stiff foot prostheses [13]. (c) One-piece mechanically differentiated prosthetic foot [14]. (d) Instrumented prosthetic foot [15].

design has been validated and results have proved the compatibility of implementing it to an actual passive device.

5.7. One-Piece Mechanically Differentiated Prosthetic Foot [14]. This is one of the passive-adaptive foot prosthetic devices available in the patent database. These types of passive feet are very much similar to flat prosthetic feet. However, this foot prosthesis is fabricated by lightweight polymeric material which enables the foot flex on any surface. This device has been developed for ankle disarticulated amputees. Due to the material type and contact surface it has the limitation of walking along the rough uneven surface. Limitation of the bend along the longitudinal axis is another problem in this design. Load carrying capacity is limited due to the type of material used for this device. Figure 9(c) shows the design of this passive foot prosthesis which was patented in 1997.

5.8. Instrumented Prosthetic Foot [15]. The instrumented prosthetic device is a passive foot prosthesis which was developed by USA research team in 2012. This device is to be fitted to a lower-limb ankle controlled by sensors. Ankle prosthesis is connected to foot via a pyramid connector. This is made of a polymer material known as durometer polyurethane. Surface adaption of this prosthesis is obtained by the stiffness of the polymer material (refer to Figure 9(d)).

5.9. PANTOE 1 [17, 18]. PANTOE 1 is one of the advanced, energy regenerative active prostheses with mediolateral direction rotation. It has 1-DoF ankle and 1-DoF foot segment. It was developed by College of Engineering, Peking University, Beijing, China, in 2010 (refer to Figure 10(b)). PANTOE 1 consists of two series elastic actuators (SEA). SEA is one of the high torque generating actuation methods available in modern prosthesis world. Foot segment is actuated with one DC brush motor, ball screw, and SEA. PANTOE 1 was controlled by finite state control method [11] and the system was validated based on control method. The foot prosthesis segment lacks the adaption capability as PANTOE foot segment has 1 DoF. Adaption along the longitudinal axis is lacking in this design.

5.10. Universal Prosthesis Emulator [19]. This foot prosthesis was developed by Carnegie Mellon University, Pittsburgh, USA (Figure 10(c)). This prosthesis can bend through mediolateral joint the same as the human MTP joint. The significant difference compared to other prostheses is a user of chain mechanism to control foot arch angle and emulator based high-performance software environment use to control the prosthesis. This is an active foot prosthesis which has the ability to perform plantar flexion. 1.61 kW AC servo motor is used to control the arch angle of the prosthesis to maintain the stability. In this design while the prosthesis is at heel strike

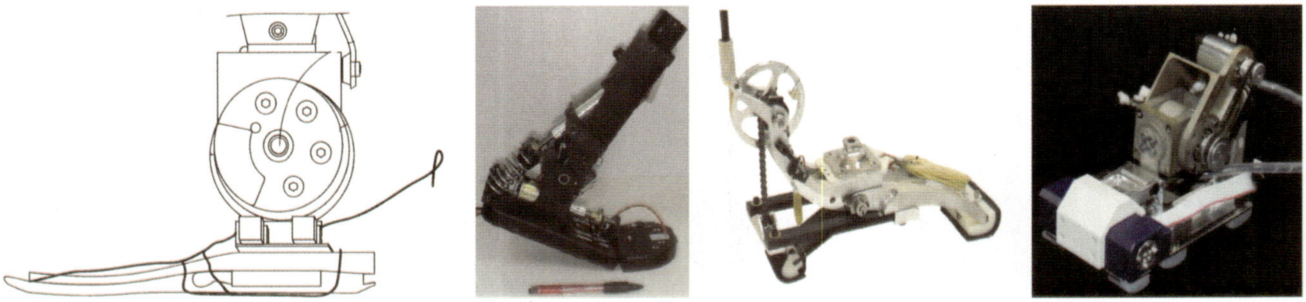

(a) (b) (c) (d)

FIGURE 10: Active and hybrid foot prostheses. (a) Ankle-foot prosthesis of automatic adaptation [16], (b) PANTOE 1 [17, 18], (c) universal prosthesis emulator [19], and (d) parallel four-bar linkage humanoid robot [20].

(a) (b) (c) (d)

FIGURE 11: Developed foot prostheses. (a) Energy recycling foot [21, 22]. (b) Adaptive bipedal deformable feet [23]. (c) An adaptive sensor foot for a bipedal and quadruped robot [24]. (d) The AMP-Foot 1.0 [25].

phase, passive heel spring bends and stores energy and pulley rotates to cause tension to the chain which is connected to passive heel to the other end.

As it is in Table 4, prostheses weighted around 1 kg range which is around average human foot weight [44, 45]. The majority of prostheses rotate along the transverse axis with no foot with a degree of freedom along both axes. Spring-based mechanisms are popular as an energy regenerative method. The motion ranges are closely following actual human joint ranges (Table 1). Lack of proper attachment methods to amputees can be seen in the majority of these designs. Aluminum and carbon fiber materials are more common due to the high strength and lightweight in the majority of these designs. Some foot prostheses have validated joint torques, forces, and angle for gait cycle [8, 9, 20] yet few simulated design performance [12].

5.11. Parallel Four-Bar Linkage Humanoid Robot [20]. Humanoid robots are too generating human foot motion. This 1-DoF humanoid (Figure 10(d)) was developed by the University of Tokyo, Japan, which attempted to mimic the toe joint motion through MTP joint of the human foot. Two parallel links have been used to connect to the toe link and foot link and developed four-bar parallel linkage mechanism. A DC servo motor (Maxon RE-max 17, 2.5 W) is used to control the toe mechanism. Toe-off can undergo maximum torque of 590 mNm. Three-axis force sensor has been attached to the base of the forefoot to detect ground reaction force and prevent the maximum torque. According to the validation results, toe-off motion can be performed with this mechanism and the toe can bend up to 44° while human MPT joint can bend about 40°.

5.12. Energy Recycling Foot [21, 22]. The University of Michigan, USA, developed this energy harvesting active prosthetic foot (Figure 11(a)) in order to introduce the control energy storage and return concept. This is a single active DoF prosthesis that stores the energy into springs and locks it during gait phase and releases it under clutch motor control based on sensory inputs. There are two DC electric motors to rotate the toe and heel. Force sensors connected to forefoot work as sensors which capture energy during heel contact phase and release it at toe-off phase. According to the validation results, this prosthesis has reduced net metabolic energy expenditure by 23% compared to normal walking.

5.13. The AMP-Foot 1.0 [25]. The AMP-Foot 1.0 was designed by the Department of Mechanical Engineering, Vrije Universiteit Brussel, Brussels, Belgium, in 2012 (Figure 11(d)). This was an initial design with a flat foot, yet with spring, locking mechanism, and planetary/epicyclical gear system to control the movement of joint. The locking mechanism was the

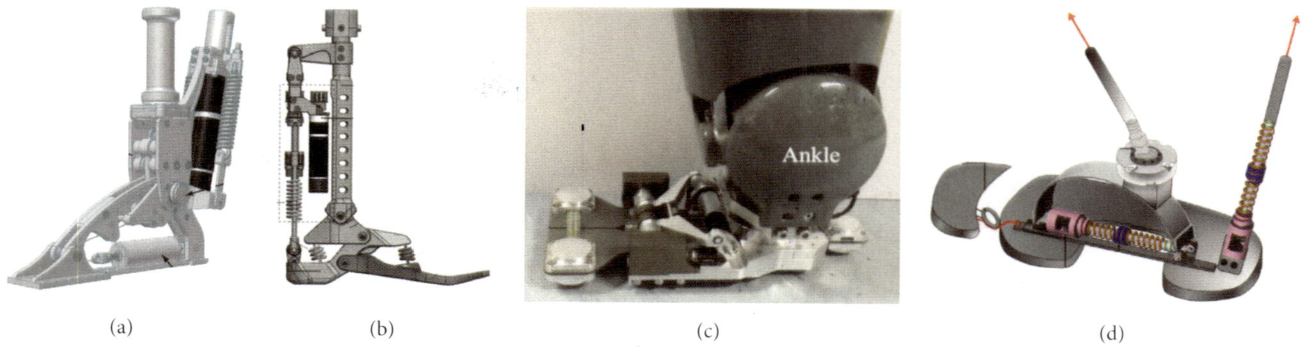

FIGURE 12: Developed foot prostheses. (a) The AMP-Foot 2.0 [26, 27]. (b) Virtual prototyping of a semiactive transfemoral prosthetic leg [28]. (c) Variable compliant humanoid foot [29]. (d) Bioinspired tunable stiffness robotic foot [30].

intuition for developing the AMP-Foot 2.0 [26, 27] later with foot mechanism. This design was validated experimentally to prove the functional capability of ankle foot.

5.14. The AMP-Foot 2.0 [26, 27]. This is a further development of AMP-Foot 1.0 [25] with energy regeneration adaptive foot (Figure 12(a)). Plantarflexion spring stores energy and regenerates the same as in other active foot prostheses. Two force sensing resistors are used as input sensors to detect the surface contact. The mechanism consists of lever mechanism to control the energy storing. AMP-Foot 2.0 is at the development stage and only the design was available. Simulated results were available based on the design. Lever and locking mechanism is novel in this design compared to other existing foot prostheses.

5.15. Variable Compliant Humanoid Foot [29]. This is another human foot in humanoid robots which was developed by the Department of Advanced Robotics, Istituto Italiano di Tecnologia, Italy, in 2016 (Figure 12(c)). The significance in this development is that it can adapt along the longitudinal axis of the foot. The variable compliant humanoid foot is an active robot that consists of small geared motor (Maxon), 6-axis force/torque sensor, leaf springs, and rubber ball with pressure sensors. Cam with leaf springs connected to transverse axis store energy when the toe leaf spring bends along the longitudinal axis. The humanoid robot was validated for motion experiments as well as spring stiffness experiments to prove the design functionality. Longitudinal adaption is the significance in this design. Some existing foot prostheses [25–27] and humanoid robots [20] use electronic sensor inputs to control motions.

6. Discussion and Future Directions

The anatomical structure of the human foot has been studied from a biomechanical perspective prior to the review of design and development. Several existing adaptive foot prostheses were reviewed in this paper upon different design criteria. Subsequently, the requirements and design difficulties were identified. In this paper, adaptive foot prostheses were classified as passive, active, and hybrid based on actuation method. The key parameters of existing adaptive foot prostheses were compared in Tables 4, 5, and 6 by indicating their country of origin, references, weight, actuation method, the axis of rotation, type of mechanisms used, movable ranges, attaching method to the remaining prosthesis or residual limb, and used materials.

The human foot consists of complex sets of joints. It undergoes significant impulsive force throughout the gait cycle due to the body weight and ground reaction force. It is essential to develop a device with the strong and lightweight material. Novel mechanisms and high torque lightweight actuators are necessary for adaptive foot prostheses to reduce the weight. Total weight of the device needs to be approximately closer to average human foot weight to avoid the baring of unnecessary weight. Most of the existing adaptive foot prostheses are 1 DoF or 2 DoF and can only be rotated along MTP joint. Only a few prostheses have the rotation capability along the longitudinal axis. Thus designing and developing an adaptive foot prosthesis which can be movable along both axes are a challenging task. Yet such development will improve the stability of lower-limb prostheses on any uneven terrain.

High torque-to-weight ratio actuators are essential for high-performance adaptive foot prostheses. The joint sizes are smaller and total number of joints is larger in the toe region of the human foot. Therefore miniature actuators are needed to actuate multiple DoF in the toe region. Currently available shelf actuators do not fulfil this requirement. Few developers have overcome this issue to a certain extent by using customized actuators. Yet, it is a costly method for small scale researches.

In order to reduce the external power usage and regenerate the power, mechanisms such as SEA, coil springs clutch motors, and springs can be used as actuators. These mechanisms can store the energy and release energy repetitively throughout the gait cycle. Additionally, spring effect enables the adaptive nature up to a certain extent. Furthermore, research needs to be carried out to develop energy regeneration. Authors foretell that future adaptive prostheses will consist of energy regenerative methods and will be more convenient for users.

As for not to feel discomfort by the amputees in long-term prostheses usage, attaching method of adaptive foot prosthesis to the lower-limb or to residual limb is crucial. Therefore, further research needs to be carried out to develop ergonomically friendly attaching sockets. Ultimately these robotic devices are to be used by humans as an artificial body part. Hence mechanical stoppers and control based safety precautions and manual maneuvering methods are necessary to be included for prostheses.

Prosthesis designs should fulfil the anatomical demands as well as the physiological demands of users. Adaptive foot prostheses are necessary to have an attractive elegant appearance with the portable facility. Some of the existing adaptive foot prostheses have managed to fulfil several of the above-mentioned design requirements, although none of them has combined all the essential functionalities to a single device. Most existing adaptive foot prostheses have limited torque, power, and ranges of motion. Unnecessary noise and vibration reduce the quality of device further. These general issues have to be addressed in future designs.

7. Conclusion

This review summarized existing design criteria of adaptive foot prostheses in order to develop an adaptive foot prosthesis. In this paper, systematic literature search approach was adopted. The scope of this paper, which is the adaptive nature of foot prostheses, has not been discussed in available review papers. This paper presented design classification parameters for each classification method of existing adaptive foot prostheses. In modern days, active and hybrid prostheses are more popular due to their high functional capabilities. Yet, in this paper some of the existing passive-adaptive foot prostheses have also been reviewed due to their significance in mechanisms and the possibility of transferring such mechanisms to hybrid devices. The adaptive foot prostheses have been classified based on actuation method and compared considering design requirements and design criteria. It enables the reader to compare and contrast the existing devices and choose the most appropriate method for their design requirements.

Conflicts of Interest

The authors declare that there are no conflicts of interest regarding the publication of this paper.

Acknowledgments

The authors gratefully acknowledge the support of the National Research Council (NRC), Sri Lanka, for the research grant (Grant no. 15-068).

References

[1] J. Martin, A. Pollock, and J. Hettinger, "Microprocessor lower limb prosthetics: review of current state of the art," *Journal of Prosthetics and Orthotics*, vol. 22, no. 3, pp. 183–193, 2010.

[2] R. A. R. C. Gopura, D. S. V. Bandara, K. Kiguchi, and G. K. I. Mann, "Developments in hardware systems of active upper-limb exoskeleton robots: A review," *Robotics and Autonomous Systems*, vol. 75, pp. 203–220, 2016.

[3] K. Ziegler-Graham, E. J. MacKenzie, P. L. Ephraim, T. G. Travison, and R. Brookmeyer, "Estimating the prevalence of limb loss in the United States: 2005 to 2050," *Archives of Physical Medicine and Rehabilitation*, vol. 89, no. 3, pp. 422–429, 2008.

[4] F. R. D. A. Senefonte, G. R. D. P. S. Rosa, M. L. Comparin et al., "Primary amputation in trauma: a profile of hospital center-west region of Brazil," *Brazilian Journal of Vascular Surgery*, vol. 11, no. 4, pp. 269–276, 2012.

[5] P. L. Chalya, J. B. Mabula, R. M. Dass et al., "Major limb amputations: A tertiary hospital experience in northwestern Tanzania," *Journal of Orthopaedic Surgery and Research*, vol. 7, no. 1, article no. 18, 2012.

[6] R. Gailey, K. Allen, J. Castles, J. Kucharik, and M. Roeder, "Review of secondary physical conditions associated with lower-limb amputation and long-term prosthesis use," *Journal of Rehabilitation Research and Development* , vol. 45, no. 1, pp. 15–30, 2008.

[7] A. H. Shultz, B. E. Lawson, and M. Goldfarb, "Walking on uneven terrain with a powered ankle prosthesis: a preliminary assessment," in *Proceedings of the 37th Annual International Conference of the IEEE Engineering in Medicine and Biology Society*, pp. 5299–5302, Milan, Italy, August 2015.

[8] F. te Riele, *The heelfoot-Design of a plantarflexing prosthetic foot [Ph.D. dissertation]*, Universiteit Twente, Enschede, The Netherlands, 2003.

[9] S. M. Behrens, R. Unal, E. E. G. Hekman, R. Carloni, S. Stramigioli, and H. F. J. M. Koopman, "Design of a fully-passive transfemoral prosthesis prototype," in *Proceedings of the 33rd Annual International Conference of the IEEE Engineering in Medicine and Biology Society*, pp. 591–594, Boston, Mass, USA, August 2011.

[10] E. Nickel, J. Sensinger, and A. Hansen, "Passive prosthetic ankle-foot mechanism for automatic adaptation to sloped surfaces," *Journal of Rehabilitation Research and Development* , vol. 51, no. 5, pp. 803–814, 2014.

[11] T. Kawakami and K. Hosoda, "Bipedal walking with oblique mid-foot joint in foot," in *Proceedings of the IEEE International Conference on Robotics and Biomimetics*, pp. 535–540, Zhuhai, China, December 2015.

[12] C. Piazza, C. Della Santina, G. M. Gasparri et al., "Toward an adaptive foot for natural walking," in *Proceedings of the 16th IEEE-RAS International Conference on Humanoid Robots*, pp. 1204–1210, Cancun, Mexico, November 2016.

[13] P. G. Adamczyk, M. Roland, and M. E. Hahn, "Sensitivity of biomechanical outcomes to independent variations of hindfoot and forefoot stiffness in foot prostheses," *Human Movement Science*, vol. 54, pp. 154–171, 2017.

[14] M. T. Wilson, "One-piece mechanically differentiated prosthetic foot and associated ankle joint with syme modification," US 5695526 A, December 1997.

[15] S. Bedard and P. O. Roy, "Instrumented prosthetic foot," US7815689 B2, December 2012.

[16] A. H. Hansen and E. A. Nickel, "Further improvements to ankle-foot prosthesis and orthosis capable of automatic adaptation to sloped walking surfaces," *US8696764 B2*, 2014.

[17] J. Zhu, Q. Wang, and L. Wang, "PANTOE 1: biomechanical design of powered ankle-foot prosthesis with compliant joints and segmented foot," in *Proceedings of the IEEE/ASME International Conference on Advanced Intelligent Mechatronics*, pp. 31–36, Montreal, Canada, July 2010.

[18] K. Yuan, J. Zhu, Q. Wang, and L. Wang, "Finite-state control of powered below-knee prosthesis with ankle and toe," in

Proceedings of the 18th IFAC World Congress, vol. 44, pp. 2865–2870, September 2011.

[19] J. M. Caputo and S. H. Collins, "A universal ankle-foot prosthesis emulator for experiments during human locomotion," *Journal of Biomechanical Engineering*, vol. 136, no. 3, Article ID 035002, 10 pages, 2013.

[20] K. Yamamoto, T. Sugihara, and Y. Nakamura, "Toe joint mechanism using parallel four-bar linkage enabling humanlike multiple support at toe pad and toe tip," in *Proceedings of the 7th IEEE-RAS International Conference on Humanoid Robots*, pp. 410–415, Pittsburgh, Pa, USA, December 2007.

[21] S. H. Collins, *Dynamic walking principles applied to a human gait [Ph.D. dissertation]*, Department of Mechanical Engineering, University of Michigan, Ann Arbor, Mich, USA, 2008.

[22] S. H. Collins and A. D. Kuo, "Recycling energy to restore impaired ankle function during human walking," *PLoS ONE*, vol. 5, no. 2, Article ID e9307, 2010.

[23] D. Owaki, H. Fukuda, and A. Ishiguro, "Adaptive bipedal walking through sensory-motor coordination yielded from soft deformable feet," in *Proceedings of the 25th IEEE/RSJ International Conference on Robotics and Intelligent Systems*, pp. 4257–4263, Vilamoura, Portugal, October 2012.

[24] K. Fondahl, D. Kuehn, F. Beinersdorf et al., "An adaptive sensor foot for a bipedal and quadrupedal robot," in *Proceedings of the 4th IEEE RAS & EMBS International Conference on Biomedical Robotics and Biomechatronics*, pp. 270–275, Rome, Italy, June 2012.

[25] B. Brackx, M. van Damme, A. Matthys, B. Vanderborght, and D. Lefeber, "Passive ankle-foot prosthesis prototype with extended push-off," *International Journal of Advanced Robotic Systems*, vol. 10, article 101, pp. 1–9, 2013.

[26] P. Cherelle, A. Matthys, V. Grosu, B. Vanderborght, and D. Lefeber, "The AMP-Foot 2.0: mimicking intact ankle behavior with a powered transtibial prosthesis," in *Proceedings of the 4th IEEE RAS and EMBS International Conference on Biomedical Robotics and Biomechatronics*, pp. 544–549, Rome, Italy, June 2012.

[27] S. Grosu, P. Cherelle, C. Verheul, B. Vanderborght, and D. Lefeber, "Case study on human walking during wearing a powered prosthetic device: effectiveness of the system 'human-Robot'," *Advances in Mechanical Engineering*, vol. 2014, Article ID 365265, 9 pages, 2014.

[28] Z. W. Lui, M. I. Awad, A. Abouhossein, A. A. Dehghani-Sanij, and N. Messenger, "Virtual prototyping of a semi-active transfemoral prosthetic leg," *Proceedings of the Institution of Mechanical Engineers, Part H: Journal of Engineering in Medicine*, vol. 229, no. 5, pp. 350–361, 2015.

[29] W. Choi, G. A. Medrano-Cerda, D. G. Caldwell, and N. G. Tsagarakis, "Design of a variable compliant humanoid foot with a new toe mechanism," in *Proceedings of the IEEE International Conference on Robotics and Automation*, pp. 642–647, Stockholm, Sweden, May 2016.

[30] Z. Qaiser, L. Kang, and S. Johnson, "Design of a bioinspired tunable stiffness robotic foot," *Mechanism and Machine Theory*, vol. 110, pp. 1–15, 2017.

[31] A. H. Hansen, D. S. Childress, and S. C. Miff, "Roll-over characteristics of human walking on inclined surfaces," *Human Movement Science*, vol. 23, no. 6, pp. 807–821, 2004.

[32] S. Portnoy, J. van Haare, R. P. J. Geers et al., "Real-time subject-specific analyses of dynamic internal tissue loads in the residual limb of transtibial amputees," *Medical Engineering & Physics*, vol. 32, no. 4, pp. 312–323, 2010.

[33] S. D. Prentice, E. N. Hasler, J. J. Groves, and J. S. Frank, "Locomotor adaptations for changes in the slope of the walking surface," *Gait & Posture*, vol. 20, no. 3, pp. 255–265, 2004.

[34] J. Friesen, J. R. Smith, and O. Pianykh, "Prosthetic foot," US20170135828 A1, May 2017.

[35] A. H. Hansen, D. S. Childress, and E. H. Knox, "Prosthetic foot roll-over shapes with implications for alignment of trans-tibial prostheses," *Prosthetics and Orthotics International*, vol. 24, no. 3, pp. 205–215, 2000.

[36] M. F. Eilenberg, H. Geyer, and H. Herr, "Control of a powered ankle-foot prosthesis based on a neuromuscular model," *IEEE Transactions on Neural Systems and Rehabilitation Engineering*, vol. 18, no. 2, pp. 164–173, 2010.

[37] J. Wernick and R. G. Volpe, "Lower extremity function and normal mechanics," in *Clinical Biomechanics of the Lower Extremities*, pp. 1–57, 1996.

[38] I. Díaz, J. J. Gil, and E. Sánchez, "Lower-limb robotic rehabilitation: literature review and challenges," *Journal of Robotics*, vol. 2011, Article ID 759764, 11 pages, 2011.

[39] S. Viteckova, P. Kutilek, and M. Jirina, "Wearable lower limb robotics: A review," *Biocybernetics and Biomedical Engineering*, vol. 33, no. 2, pp. 96–105, 2013.

[40] W. Huo, S. Mohammed, J. C. Moreno, and Y. Amirat, "Lower limb wearable robots for assistance and rehabilitation: a state of the art," *IEEE Systems Journal*, vol. 10, no. 3, pp. 1068–1081, 2016.

[41] R. Jiménez-Fabián and O. Verlinden, "Review of control algorithms for robotic ankle systems in lower-limb orthoses, prostheses, and exoskeletons," *Medical Engineering & Physics*, vol. 34, no. 4, pp. 397–408, 2012.

[42] R. Versluys, P. Beyl, M. van Damme, A. Desomer, R. Van Ham, and D. Lefeber, "Prosthetic feet: state-of-the-art review and the importance of mimicking human anklefoot biomechanics," *Disability and Rehabilitation: Assistive Technology*, vol. 4, no. 2, pp. 65–75, 2009.

[43] D. Moher, A. Liberati, J. Tetzlaff, and D. G. Altman, "Preferred reporting items for systematic reviews and meta-analyses: the PRISMA statement," *PLoS Medicine*, vol. 6, no. 7, Article ID e1000097, 2009.

[44] J. L. Johansson, D. M. Sherrill, P. O. Riley, P. Bonato, and H. Herr, "A clinical comparison of variable-damping and mechanically passive prosthetic knee devices," *American Journal of Physical Medicine & Rehabilitation*, vol. 84, no. 8, pp. 563–575, 2005.

[45] P. Houglum, D. Bertoti, and S. Brunnstrom, *Brunnstrom's Clinical Kinesiology*, F.A. Davis, Philadelphia, Pa, USA, 1st edition, 2012.

[46] J. Muscolino, *Kinesiology*, Elsevier Mosby, St. Louis, Mo, USA, 2nd edition, 2010.

[47] J. H. Hicks, "The mechanics of the foot: II. The plantar aponeurosis and the arch," *Journal of Anatomy*, vol. 88, no. 1, pp. 25–30, 1954.

[48] D. Paluska and H. Herr, "The effect of series elasticity on actuator power and work output: implications for robotic and prosthetic joint design," *Robotics and Autonomous Systems*, vol. 54, no. 8, pp. 667–673, 2006.

[49] E. J. Rouse, L. M. Mooney, and H. M. Herr, "Clutchable series-elastic actuator: implications for prosthetic knee design," *International Journal of Robotics Research*, vol. 33, no. 13, pp. 1611–1625, 2014.

[50] L. Mooney and H. Herr, "Continuously-variable series-elastic actuator," in *Proceedings of the 13th IEEE International Conference on Rehabilitation Robotics*, pp. 1–6, Seattle, Wash, USA, June 2013.

Permissions

All chapters in this book were first published in JR, by Hindawi Publishing Corporation; hereby published with permission under the Creative Commons Attribution License or equivalent. Every chapter published in this book has been scrutinized by our experts. Their significance has been extensively debated. The topics covered herein carry significant findings which will fuel the growth of the discipline. They may even be implemented as practical applications or may be referred to as a beginning point for another development.

The contributors of this book come from diverse backgrounds, making this book a truly international effort. This book will bring forth new frontiers with its revolutionizing research information and detailed analysis of the nascent developments around the world.

We would like to thank all the contributing authors for lending their expertise to make the book truly unique. They have played a crucial role in the development of this book. Without their invaluable contributions this book wouldn't have been possible. They have made vital efforts to compile up to date information on the varied aspects of this subject to make this book a valuable addition to the collection of many professionals and students.

This book was conceptualized with the vision of imparting up-to-date information and advanced data in this field. To ensure the same, a matchless editorial board was set up. Every individual on the board went through rigorous rounds of assessment to prove their worth. After which they invested a large part of their time researching and compiling the most relevant data for our readers.

The editorial board has been involved in producing this book since its inception. They have spent rigorous hours researching and exploring the diverse topics which have resulted in the successful publishing of this book. They have passed on their knowledge of decades through this book. To expedite this challenging task, the publisher supported the team at every step. A small team of assistant editors was also appointed to further simplify the editing procedure and attain best results for the readers.

Apart from the editorial board, the designing team has also invested a significant amount of their time in understanding the subject and creating the most relevant covers. They scrutinized every image to scout for the most suitable representation of the subject and create an appropriate cover for the book.

The publishing team has been an ardent support to the editorial, designing and production team. Their endless efforts to recruit the best for this project, has resulted in the accomplishment of this book. They are a veteran in the field of academics and their pool of knowledge is as vast as their experience in printing. Their expertise and guidance has proved useful at every step. Their uncompromising quality standards have made this book an exceptional effort. Their encouragement from time to time has been an inspiration for everyone.

The publisher and the editorial board hope that this book will prove to be a valuable piece of knowledge for researchers, students, practitioners and scholars across the globe.

List of Contributors

Markus Bajones, David Fischinger, Astrid Weiss, Daniel Wolf and Markus Vincze
Automation and Control Institute (ACIN), TUWien, Vienna, Austria

Paloma de la Puente
Universidad Politécnica de Madrid, Madrid, Spain

Tobias Körtner and Markus Weninger
Akademie für Altersforschung am Haus der Barmherzigkeit, Vienna, Austria

Konstantinos Papoutsakis, Damien Michel, Ammar Qammaz, Paschalis Panteleris, Michalis Foukarakis, Ilia Adami, Danai Ioannidi, Asterios Leonidis, Margherita Antona and Antonis Argyros
Institute of Computer Science, FORTH, Heraklion, Greece

Peter Mayer and Paul Panek
Institute for Design and Assessment of Technology, TUWien, Vienna, Austria

Håkan Eftring and Susanne Frennert
Department of Design Sciences, Lund University, Lund, Sweden

Van Dong Hai Nguyen, Xuan-Dung Huynh and Minh-Tam Nguyen
Ho Chi Minh City University of Technology and Education (HCMUTE), 01 Vo Van Ngan Street, Thu Duc District, Ho Chi Minh City, Vietnam

Ionel Cristian Vladu and Mircea Ivanescu
University of Craiova (UCV), 13 Alexandru Ioan Cuza Street, 200585 Craiova, Romania

Ivan Meza, Caleb Rascon, Gibran Fuentes and Luis A. Pineda
Instituto de Investigaciones en Matematicas Aplicadas y en Systemas, Universidad Nacional Autonoma de Mexico, Ciudad Universitaria, Coyocan, 04510 Mexico City, MEX, Mexico

Miguel A. Gálvez-Zúñiga and Alejandro Aceves-López
Tecnologico de Monterrey, CEM, 52926 Atizapan, Mexico City, Mexico

Michael Oluwatosin Ajayi and Karim Djouani,
F'SATI, Department of Electrical Engineering, Tshwane University of Technology, Staatsartillerie Road,Pretoria West, Pretoria, South Africa
LISSI, University of Paris Est Creteil (UPEC), 94400 Vitry Sur Seine, France

Yskandar Hamam
F'SATI, Department of Electrical Engineering, Tshwane University of Technology, Staatsartillerie Road, Pretoria West, Pretoria, South Africa
LISV, Bâtiment Boucher, Pole Scientifique et Technologique de Vélizy, 10-12 avenue de l'Europe, 78140 Vélizy, France

Rachel E. Stuck
Georgia Institute of Technology, 654 Cherry St., Atlanta, GA 30332, USA

Wendy A. Rogers
University of Illinois Urbana-Champaign, 1206 South 4th St, Champaign, IL 61820, USA

Roberta Alò, Francesco Bottiglione and Giacomo Mantriota
Dipartimento di Meccanica, Matematica e Management, Politecnico di Bari, Viale Japigia 182, 70126 Bari, Italy

Dugan Um and Jangwoon Park
School of Engineering and Computing Sciences, Texas A&M University-Corpus Christi, USA

Jeongsik Shin and Woo Ho Lee
Phomatix LLC, Keller, TX, USA

Yongli Feng, Qingxuan Jia and Wei Wei
School of Automation, Beijing University of Posts and Telecommunications, No. 10 Xitucheng Road, Haidian District, Beijing 100876, China

Takehito Kikuchi, Kohei Sakai and Isao Abe
Oita University, Dannoharu 700, Oita 870-1192, Japan

Ke We, Jiabo Zhang and JiZhi Yang
Beijing Spacecrafts, China Academy of Space Technology, Beijing 100194, China

Fuzhou Du
School of Mechanical Engineering and Automation, Beihang University, Beijing 100191, China

Michael F. Eilenberg, Jiun-Yih Kuan and Hugh Herr
Biomechatronics Group, MIT Media Lab, Massachusetts Institute of Technology, Cambridge, MA, USA

Mohssen Hosseini, Roberto Meattini, Gianluca Palli and Claudio Melchiorri
Università di Bologna (DEI), Viale Risorgimento 2, Bologna, Italy

Bilal M. Yousuf
Electronics & Power Engineering, PN Engineering College, National University of Sciences and Technology, Karachi, Pakistan

Asim Mehdi, Abdul Saboor Khan, Aqib Noor and Arslan Ali
Electrical Department, Fast-National University of Computing and Emerging Sciences, Karachi, Pakistan

Thilina H. Weerakkody, Thilina Dulantha Lalitharatne and R. A. R. C. Gopura
Bionics Laboratory, Department of Mechanical Engineering, University of Moratuwa, Katubedda, Sri Lanka

Index